Principles of Separation Processes

Principles of Separation Processes

Braylon Vincenzo

Larsen & Keller
www.larsen-keller.com

Principles of Separation Processes
Braylon Vincenzo
ISBN: 978-1-64172-141-7 (Hardback)

Larsen & Keller
Published by Larsen and Keller Education,
5 Penn Plaza,
19th Floor,
New York, NY 10001, USA

Cataloging-in-Publication Data

Principles of separation processes / Braylon Vincenzo.
p. cm.
Includes bibliographical references and index.
ISBN 978-1-64172-141-7
1. Separation (Technology). 2. Chemistry, Technical. 3. Chemistry, Analytic. I. Vincenzo, Braylon.
TP156.S45 P75 2019
660.2842--dc23

Printed and bound in China.

For more information regarding Larsen and Keller Education and its products, please visit the publisher's website www.larsen-keller.com

Table of Contents

Preface

Industrial separation processes are processes used in industry for the separation of a product from its impurities and other unwanted products. Separation operations are applied for the purification of raw materials, recovery of by-products, recycling of solvents and removal of contaminants from effluents. The separation processes of mixtures include extraction, drying, crystallization, filtration and evaporation, among others. This book is a compilation of chapters that discuss the most vital concepts of separation processes. The topics included herein on the principles and techniques of separation processes are of utmost significance and bound to provide incredible insights to readers. Those in search of information to further their knowledge will be greatly assisted by this book.

A short introduction to every chapter is written below to provide an overview of the content of the book:

Chapter 1, A separation process is a mechanism that converts a chemical solution or mixture into two or more constituent mixtures. This chapter has been carefully written to provide an easy understanding of the varied aspects of separation processes, such as industrial separation processes, electrophoresis, chromatography, centrifugation, distillation, extraction, etc.; **Chapter 2**, Electrophoresis refers to the motion of dispersed particles with respect to a fluid when acted by a uniform electric field. The diverse types of electrophoresis, such as capillary electrophoresis, dielectrophoresis, free-flow electrophoresis, gel electrophoresis, moving-boundary electrophoresis, etc. and their applications have been closely examined in this chapter; **Chapter 3**, Chromatography is a technique that is used for the separation of a mixture into its constituents. It can either be analytical or preparative. Some of the diverse topics covered in this chapter include high-performance liquid chromatography, electrochromatography, centrifugal partition chromatography, supercritical fluid chromatography, etc. which will provide an overview of the different chromatographic techniques; **Chapter 4**, Extraction is a technique that is used for the separation of a substance from a matrix. This chapter discusses in detail the most significant extraction techniques, such as electromembrane extraction, accelerated solvent extraction, leaching, raffinate, magnetic separation, etc.; **Chapter 5**, Fractionation is a separation process concerning the segregation of a mixture of solid, liquid, gas or suspension, into a number of smaller factions during a phase transition, in such a way that the composition varies relative to a gradient. This chapter closely examines the key concepts of fractionation and discusses topics like field flow fractionation, equilibrium isotope fractionation and polymer fractionation for an extensive understanding of the subject; **Chapter 6**, Membrane technology is the technology, concerning the transport of substances between different fractions through a permeable membrane. It is commonly used for the mechanical separation of liquid or gaseous streams. The topics elaborated in this chapter on mass transfer, concentration polarization, pervaporation, membrane gas separation, ultrafiltration, membrane distillation, reverse osmosis, etc. will help in providing a better perspective about membrane technology.

I extend my sincere thanks to the publisher for considering me worthy of this task. Finally, I thank my family for being a source of support and help.

Braylon Vincenzo

1

Separation Processes

A separation process is a mechanism that converts a chemical solution or mixture into two or more constituent mixtures. This chapter has been carefully written to provide an easy understanding of the varied aspects of separation processes, such as industrial separation processes, electrophoresis, chromatography, centrifugation, distillation, extraction, etc.

Separations are crucial in chemical engineering. A typical chemical plant is a chemical reactor surrounded by separators, as diagramed in the schematic flow sheet of Figure. Raw materials are prepurified in separation devices and fed to the chemical reactor; unreacted feed is separated from the reaction products and recycled back to the reactor. Products must be further separated and purified before they can be sold. This type of arrangement is very common.

Chemical plants commonly have from 40% to 70% of both capital and operating costs in separations.

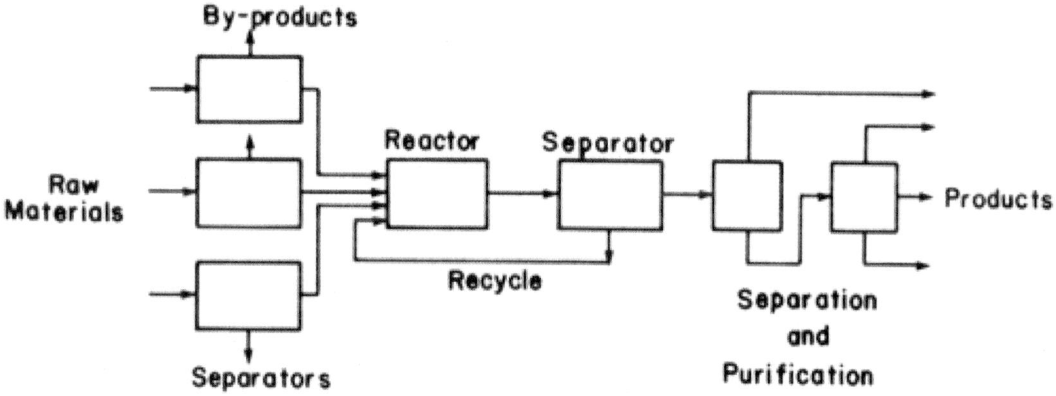

Figure: Typical chemical plant layout

Since separations are ubiquitous in chemical plants and petroleum refineries, chemical engineers must be familiar with a variety of separation methods. Some of the most common chemical engineering separation methods are flash distillation, continuous column distillation, batch distillation, absorption, stripping, and extraction. These separations all contact two phases and can be designed and analyzed as equilibrium stage processes.

The equilibrium stage concept is applicable when the process can be constructed as a series of discrete stages in which the two phases are contacted and then separated. The two separated phases are assumed to be in equilibrium with each other. For example, in distillation, a vapor and a liquid are commonly contacted on a metal plate with holes in it. Because of the intimate contact between the two phases, solute can transfer from one phase to another. Above the plate the vapor disengages from the liquid. Both liquid and vapor can be sent to additional stages for further separation. Assuming that the stages are equilibrium stages, the engineer can calculate concentrations and temperatures without detailed knowledge of flow patterns and heat and mass transfer rates.

Although this example shows the applicability of the equilibrium stage method for equipment built with a series of discrete stages, the staged design method can also be used for packed columns where there are no discrete stages.

A second useful concept is that of a unit operation. The idea here is that although the specific design may vary depending on what chemicals are being separated, the basic design principles for a given separation method are always the same. For example, the basic principles of distillation are always the same whether we are separating ethanol from water, separating several hydrocarbons, or separating liquid metals. Consequently, distillation is often called a unit operation, as are absorption, extraction, etc.

A more general idea is that design methods for related unit operations are similar. Since distillation and absorption are both liquid-vapor contacting systems, the design is much the same for both. This similarity is useful because it allows us to apply a very few design tools to a variety of separation methods.

Theoretical Plate

A chromatographic column does not consist of any physical plates which can be measured from time to time to rate a column's performance efficiency. It is only a theoretical concept drawing similarity to the huge fractional distillation columns used in refining of crude petroleum. Such distillation column actually comprises of separator plates which help separation and isolation of different petroleum fractions which are collected as separate streams.

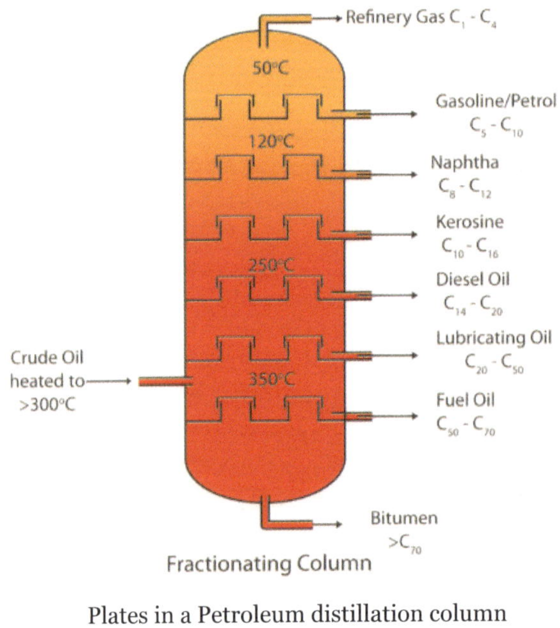

Plates in a Petroleum distillation column

However no such plates exist in the chromatographic columns.You have to imagine that your column is divided into a number of sections or plates. A sample component spends a finite time in each plate as it moves the column length and this is the time required to establish equilibrium over the length of the plate between its concentration in the stationary phase and the mobile phase. The movement of the analyte is assumed from plate to plate as a series of equilibrated mobile phase plugs. In other words the greater the number of such plates the more efficient is the separation power of the column.

The number of plates (N) over the column length (L) is dependent on the plate height (H)

$$N = L/H$$

Smaller plate height implies large number of plates in the column and higher is the column efficiency. H is also referred to as height equivalent to a theoretical plate (HETP) and smaller the value of HETP the greater is the column efficiency.

Calculation of the Number (N)

N is calculated from the peak of the injected standard under prescribed operational conditions

$$N = 16(\frac{t_r}{W_b})^2$$

where TR is the retention time of the peak and Wb is the peak width. In case peak width is measured at half height or W1/2 then the relation is expressed as:

$$N = 5.54(\frac{t_r}{W_{1/2}})^2$$

In above calculations retention time and width should be expressed in the same units.

Present day chromatographic softwares provide number N without the need for manual calculations but the basic assumption is that the peak shape should be Gaussian.

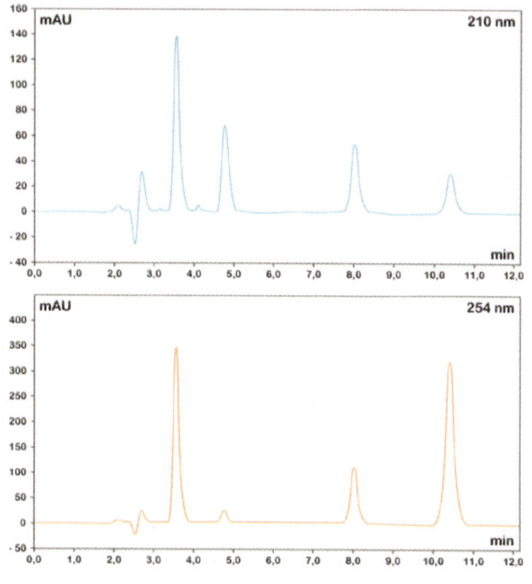

The number of theoretical plates is dependent on several operational parameters such as flow rate, analyte characteristics, temperature and sample size. Overloading a column rapidly deteriorates its performance.

A column manufacturer provides a certificate of column performance features along with the column and also specifies the minimum acceptable value of the number of theoretical plates. A deteriorating column should be reconditioned to improve the value and if this is not possible then it is suggested that the column be replaced.

Isopycnic Centrifugation

In isopycnic separation, also called buoyant or equilibrium separation, particles are separated solely on the basis of their density. Particle size only affects the rate at which particles move until their density is the same as the surrounding gradient medium. The density of the gradient medium must be greater than the density of the particles to be separated. By this method, the particles will never sediment to the bottom of the tube, no matter how long the centrifugation time.

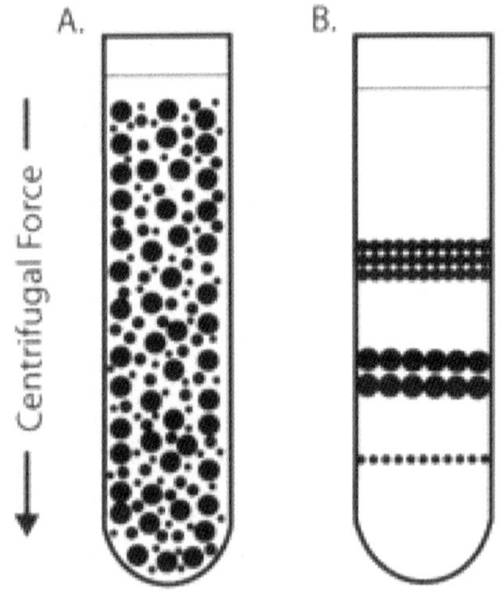

Isopycnic Centrifugation

Starting with a uniform mixture of sample and density gradient (A) under centrifugal force, particles move until their density is the same as the surrounding medium (B).

Upon centrifugation, particles of a specific density sediment until they reach the point where their density is the same as the gradient media (i.e., the equilibrium position). The gradient is then said to be isopycnic and the particles are separated according to their buoyancy. Since the density of biological particles is sensitive to the osmotic pressure of the gradient, isopycnic separation may vary significantly depending on the gradient medium used. Although a continuous gradient may be more suited for analytical purposes, preparative techniques commonly use a discontinuous gradient in which the particles band at the interface between the density gradient layers. This makes harvesting certain biological particles (e.g., lymphocytes) easier.

Aqueous Two-phase System

Aqueous two-phase extraction (ATPE) system is composed of either two different immiscible mixing polymers, or one polymer with salt, which are water-soluble in a certain concentration. It has been well known as a useful technique for separation and purification of biomolecules, such as proteins and antibodies. Partitioning of biomolecules in ATPE systems is affected by many factors, including molecular weight/size of polymer and concentration of polymer. Also, the ionic strength of salt and the addition of salts, such as NaCl, improve the hydrophobic resolution of the system. Furthermore, the partitioning of biomolecules can also be influenced by the degree of pH and affinity of the macromolecule for the phaseforming polymer.

The conventional liquid–liquid extraction using organic-aqueous phase systems was previously established. However, due to the poor solubility and possible denaturation of the protein in organic solvents, the technique leads to limit their application in partitioning of many bimolecular products. Now, the application of liquid–liquid extraction based on aqueous two-phase systems (ATPS) has been increased. The recoveries of high-value biomolecules were achieved from various plants using different applications, such as papain from Papaya fruit, α- and β -amylases from Zea mays malt, glutenin flour from special wheat and recombinant protein from alfalfa. In addition, the high-value bioproducts also are obtained from different fruits, such as bromelain (EC 3.4.22.33) from pineapple, serine protease from Mango, invertase from tomato and papain from Papaya fruit. In recent years, the high-quality biomolecules have been recovered from various sources, such as theanine from waste liquid of tea and flavonoids from pigeon pea roots. There are also other bioproducts, such as recombinant human serum albumin from Pichia pastoris broths, luciferase from fireflies and immunoglobulin G.

The ATPE has some advantages in the downstream processing of biomolecules, for instance, the system characterized with high-water content (70–80%, w/w) and low interfacial tension between conjugated phases. This, in fact, provides a secure separation and purification technique for biomolecules. In addition, it decreases energy consumption, reduces waste and minimizes costs due to a few steps of unit operation, which require low-energy input and easy to scale up. The ATPE also increases biomolecules recovered in a highly purified form. Moreover, it can be used in combination of other separation methods such as liquid chromatography, packed column, magnetic particle adsorption.

The ATPE in downstream processing of biomolecules in the bench-scale prototype has been successful with potentially commercial application. However, the scale-up of the ATPS of some biological products were not achieved. Most of those methods have some limitations such as low capacity, several steps and fewer chemicals and proteolytic stability, which may lead to the contamination of the final product. Moreover, there are maybe some difficulties during the large scale-up of the ATPS in the industry, especially polyethylene glycol (PEG)/salt system causing corrosion of equipment and precipitation of target product. Thus, the development of alternative methods is very important to achieve the desired biomolecules in high quality and purity.

Separation and Purification of Bioproducts by the ATPE System

The ATPS is considered a powerful and versatile technique, having low-cost and good efficiency downstream process, A high-water content (>70 % (w/w) water) generates low interfacial tension, nonflammable, slightly toxic and safe to the environment. This system has high selectivity and recovery yield of biomolecules. It has been widely used in the field of biotechnology for separation and purification of various biological products, such as proteins, amino acids, enzymes, cells, antibodies and other bioproducts. Also the ATPS method has received a special application in the area of non-biotechnology, including the recovery of glycosaminoglycans from tannery wastewater, crocins from saffron stigmas, papaverin from pericarpium papaveris.

Protein is one of the most important bio-molecules in the living organism, which is responsible for many reactions and functionalities, such as metabolism, bioprocess, signal transduction, cellular and extracellular structures. A protein in a purified form would be very useful in the applications (e.g. food, chemical and pharmaceuticals). The challenging problem is that the downstream processing mostly accounts for 50 – 80% of the total production costs of proteins. Conventional methods for separation and purification of proteins usually are expensive, time-consuming and difficult to scale up. The good news is that the ATPE, an achievable alternative method, has been recognized as an economical and effective technique for recovery and purification of proteins with a variety of advantages, such as simple, fast, low-cost and easily scaled-up. These make it possible strategy for purification of a desired protein in large-scale. Partitioning of proteins in the ATPS depends on many factors, namely, hydrophobicity, molecular size, weight and conformation, net electrical charge and environmental conditions. A protein purified by this method will be considered to be very important bimolecular product for the bioprocess as it can be used in commercial scale at low cost with relative reliability and accessibility.

Air Stripping

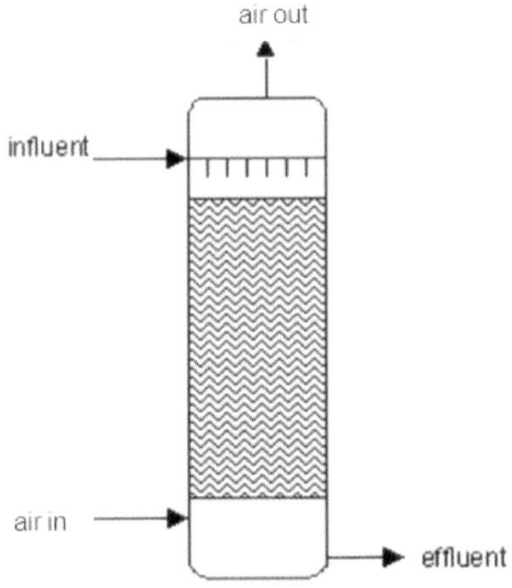

Method and Installation Description

Air stripping is a technique in which wastewater and air are intensively brought in contact with each other. This causes the volatile compounds present in wastewater to be transferred into the air.

The air containing VOC must be treated in an air treatment system (e.g. active carbon installation, bio-filter). A stripper only needs a small surface area; 5x5 m^2 is sufficient for a stripper with a capacity of 100 m^3/hour. The space taken by an air treatment unit varies greatly.

The main set-up types are the stripping tower or stripping column and the plate stripper. The stripping tower is based on the counter-flow principle, where a vertical column is filled with packing material. The plate stripper is based on the cross-flow principle, where the liquid flow is intensively aerated via a perforated plate.

The stripping process is cheap and reliable, and provides relatively good substance transfer. One of the disadvantages of this process is that it is susceptible to pollution.

Application

If sufficient attention is paid to pre-purification, stripping is a robust technique that can be easily implemented in a wide variety if sectors and in a wide volume range. Here are a few examples:

- In both inorganic and organic chemistry, stripping is used for the removal of volatile organic substances, sulphur compounds (H2S, phosphine) and NH_3. Stripping is normally carried out on the concentrated partial flow.
- Air stripping is used in the pharmaceutical sector for the removal of chlorinated solvents from wastewater.
- In viscose production, air stripping is the standard technique for the removal of CS_2 from wastewater.
- In glass engraving with ammonium-based solvents, pH supplementation and air stripping can be used for the removal of nitrogen from wastewater.
- In the graphics sector, stripping is used for the removal of toluene from condensate discharged by recuperation systems.
- In a variety of sectors, air stripping is implemented for the removal of chlorinated solvents from wastewater:
    - PER in the dry cleaning sector.
    - Methylene chloride in the removal of paint layers from wood.

Air stripping is currently being used for cleaning groundwater in soil remediation, possibly followed by an air-based active carbon filter or bio-filter. The following compounds are primarily removed when stripping groundwater: Aromatic compounds (BTEX) and/or volatile chlorinated hydrocarbons (VOCs, incl. trichloro ethene, perchlore ethene, tetrachloro methane, chloroform).

Boundary Conditions

In order to implement the technique, initially the main factor is the ratio between the partial vaporisation of pollution in the gas phase and the concentration of the pollution in the water phase. This ratio is a fixed substance property and is referred to as the Henry coefficient. In fact, this coefficient indicates the "stripability" of the compound. The Henry coefficient is determined by the temperature.

Further, a major role is played by the speed of substance transport from the water phase to the air phase. This substance transfer is determined by:

- The interacting surface between the air phase and water phase.
- The retention time in the system or the time that a substance is in contact with the surface (contact time).
- The concentration difference of the substance between the gas phase and the liquid phase (the driving force or the deviation from the thermo-dynamic balance).

The technique is used for volatile compounds that (generally) have a Henry coefficient greater than 0.001 atm.m^3/mol. The concentration range in which the technique can be implemented varies from around ten micrograms per litre to a few dozen milligrams per liter.

Iron-removal is needed prior to air stripping for iron contents above 2 mg/l in a stripping column and 5 mg/l in a plate stripper. If iron is not removed, iron deposits could result in blockages which, in time, would require the package to be replaced or cleaned with acid.

The pH must be corrected if the water leaves lime deposits.

The water must be free of suspended matter, in order to prevent blockages. The wastewater must be pre-treated so that all suspended matter is removed (for example, using a sand filter or by using coagulation/flocculation followed by sedimentation).

Effectiveness

Air stripping is primarily implemented for the removal of volatile organic matter (incl. chlorinated hydrocarbons, VOX and BTEX) from wastewater.

If the established condition for the Henry coefficient is met, air stripping can be used to realise high removal yield (e.g. >99% for VOX) per placed unit. If necessary, multiple units can be placed in a series to further increase total performance.

Support Aids

Ammoniac can be stripped in an air stripper. If ammoniac is interfering with other pollutants that need to be removed, this can be confirmed using neutralising substances.

Environmental Issues

Air containing VOC is released. Air treatment could possibly be needed (e.g. active carbon system, bio-filter). Depending on the influent, the location and the installation, extra measures may be needed to prevent odour and noise problems.

Costs

The costs for implementation of air stripping are determined by the desired purification yield, on the one hand, and the concentration desired in the effluent, on the other hand. These two aspects determine the number of modules (units) than must be placed in the series.

Stripping requires relatively large amounts of energy. Energy consumption with active carbon air purification amounts to circa 0.3 to 0.5 kW/m^3 water and to circa 0.1 to 0.2 kW/m^3 for a bio-filter.

Adduct Purification

Adduct purification is a technique for preparing extremely pure simple organometallic compounds, which are generally unstable and hard to handle, by purifying a stable adductwith a Lewis acid and then obtaining the desired product from the pure adduct by thermal decomposition.

Epichem Limited is the licensee of the major patents in this field, and uses the trademark EpiPure to refer to adduct-purified materials; Professor Anthony Jones at Liverpool University is the initiator of the field and author of many of the important papers.

The choice of Lewis acid and of reaction medium is important; the desired organometallics are almost always air- and water-sensitive. Initial work was done in ether, but this led to oxygen impurities, and so more recent work involves tertiary amines or nitrogen-substituted crown ethers.

Gravity Separation

Gravity separation uses differences in specific gravity (SG) between various minerals to achieve a separation, and is normally a wet process although examples of dry gravity separators exist. Gravity separation works best when there is a large difference in the SG in the minerals to be separated, and the particle size is similar and not excessively fine, and shape is rounded. The choice of separator is generally determined by these factors and includes the following:

- Conventional jigs
- Spirals
- Pinched sluices
- Centrifugal jigs
- Centrifugal concentrators
- Shaking tables (wet and dry)
- Mozeley gravity separators (MGS)

Conventional Jigs

Examples of conventional jigs include those manufactured by IHC, the Russell jig manufactured by Amtas, and the In Line Pressure Jig (ILPJ) manufactured by Gekkos.

Jigs are commonly used as a primary stage to recover coarse liberated minerals sized 2mm and above. Feed slurry is distributed into the hutch which consists of a moving slurry bed located above a screen and is subjected to a pulsating motion and upward hutch water flow which alternately causes dilation and compaction. The pulsation can be caused by a diaphragm (in the case of the Russell jig) or by mechanical movement of the screen (as in the ILPJ). Separation is also assisted by a ragging material of intermediate SG, with the result being that high SG (concentrate) particles pass through the ragging and screen and are removed via the hutch, while low SG particles (tailings) discharge over the lip of the hutch.

- Operating variables include:
- Hutch water flow
- Pulsation frequency
- Pulsation stroke length

- Ragging SG, size and shape
- Bed depth
- Screen aperture size
- Feed rate and density (20 tph / hutch for Russell jig at 40% solids)

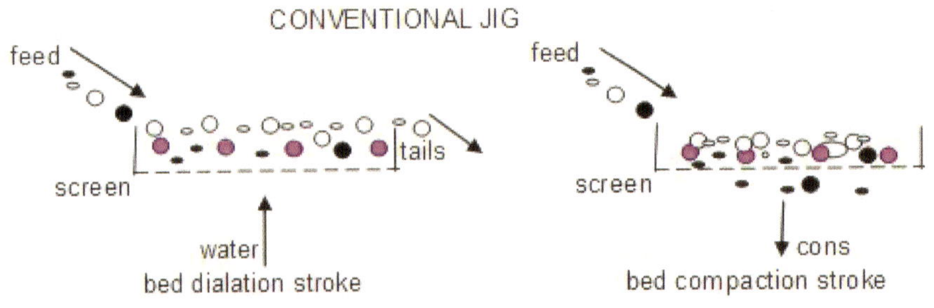

Spirals

Examples of spirals are manufactured by Mineral Technologies,and Multotec and are popular due to their low maintenance and simplicity, and higher separation efficiency than pinched sluice devices.

Spirals are commonly used to separate sand sized particles with moderate SG differential in the range -2mm +75um, although varieties exist that can separate reasonably efficiently down to 63um. Below this size range efficiency falls off rapidly, and enhanced fine gravity separators are generally required.

Spirals are made of fiberglass onto which smooth urethane surface is molded to form a trough in the shape of a spiral as the name suggests. Between individual spiral types, the profile of the trough and the pitch as well as the diameter and height and number of turns can vary according to duty. Feed slurry is introduced at the top and is subjected to a combination of gravitational and centrifugal forces imparted by its motion down the spiral. This causes high SG minerals to move towards the centre of the trough and water and low SG minerals towards the outside. The segregated slurry discharging from the spiral at the bottom can thus be separated by cutters into high SG (concentrate) and low SG (tailings) together with intermediate SG (middling).

The choice of spiral depends on the duty: rougher / scavenger or cleaner; the feed grade: low / medium / high; and the particle size: sand / fine. Typically rougher / scavenger spirals where the feed type favorable for easy separation and overall throughput is large are high capacity designs with triple or quadruple to reduce floor area. Cleaner and fine spirals where overall throughput is less and separation more difficult are typically operated at lower feed rates, and as twin starts to assist access for operational requirements. Cleaning spirals are also available with wash water.

Operating variables include:

- Feed rate (between 1 and 6 tph / start depending on type)
- Feed density (between 25% and 50% solids depending on duty)
- Splitter positions

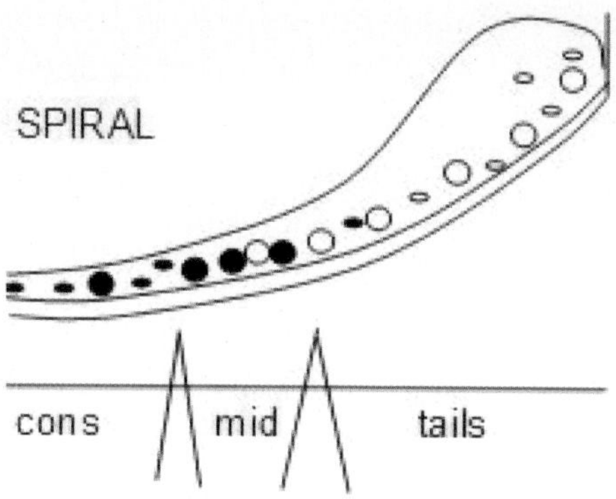

Pinched Sluices

An example of a pinched sluice is the Reichert cone, manufactured by Mineral Technologies, which is a derivative of the older tray pinched sluice separator.

Pinched sluices were commonly used to separate sand sized particles with moderate SG difference prior to the appearance of the modern day spiral in the 1980s. They operate on the principle of stratification of particles of differing SG on an inclined surface, area of which becomes smaller "pinched" as the slurry flows down the slope, with high SG grains on the bottom and low SG grains on the top. High SG (concentrate) grains are removed through a slot, while low SG (tailing) discharges off the end. Reichert cones normally have multi stage cons cleaning and water addition to maintain optimum density.

The operating variables include:

- Feed density (61% to 63% solids)
- Feed rate (between 75 and 90 tph for standard cone)
- Slot position

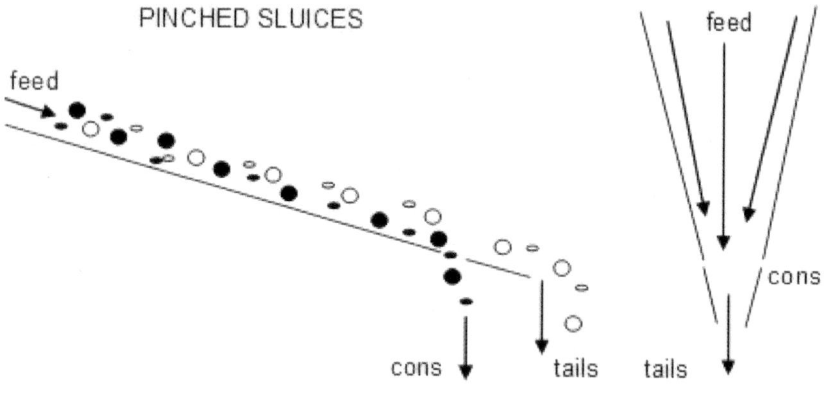

Centrifugal Jigs

An example of a centrifugal jig is the Kelsey jig, manufactured by Mineral Technologies. This incorporates the pulsing action, water injection and the use of a ragging medium and internal screen of a conventional jig, but in addition centrifugal spinning motion to enhance the gravity separation. This enables sand sized minerals of fairly similar SG as well as fine minerals to be separated more efficiently. Because of the additional processes used the Kelsey jig is significantly more expensive and complex to operate and maintain. The feed to a Kelsey jig must be screened at a size less than the internal screen to avoid pegging, and the tailings must be screened in order to recover ragging as oversize. In order to maintain operational and mechanical efficiency the Kelsey jig must have the internal screen cleaned daily and some of the moving parts checked and greased. This can now be done automatically, but the feed must still be stopped. The Kelsey jig is offered in two models: the J1300 and J1800, the latter having greater capacity.

Operating variables include:

- Hutch water flow
- Spin rate
- Pulsation frequency
- Pulsation stroke length
- Ragging SG, size and shape
- Bed depth
- Concentrate orifice diameter
- Internal screen aperture size
- Feed rate and density (typically 20 to 25 tph at 40% solids for a J1300 although throughput is dictated principally by concentrate rate)

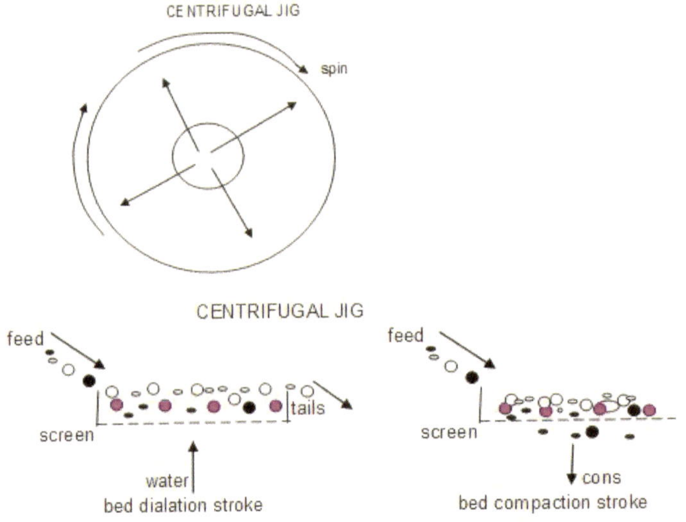

Centrifugal Concentrators

Examples of centrifugal concentrators include the Falcon C and Ultra Fine (U/F) types manufactured by Sepro and the Knelson CVD manufactured by Knelson.

The Falcon C and Knelson CVD are derivatives of the SB types (semi batch) used extensively in the gold industry, but provide continuous discharge of concentrate to enable higher grade feeds to be processed without having to shut down. The recently developed Falcon U/F is a batch machine and spins at extremely high (up to 600G) rates in order to target very fine particles. They all utilize the centrifugal force generated by spin to enhance gravity and enable SG separation on finer particles to be achieved. Feed slurry enters centrally and is distributed outwards at the base of the cone by centrifugal force and then flows up the inclined surface of the bowl, segregating in the process, with high SG particles on the outside closest to the bowl surface and low SG particles on the inside which discharge over the lip at the top of the bowl.

The Falcon C spins at high rates and can generate a G force of up to 200, and features a positioning valve for continuous concentrate discharge, while the Knelson CVD operates at lower G force (up to 150G), but has an injection water system which fluidizes the bed and assists separation. All three separators have different bowl profiles.

Operating variables include:

- Spin
- Concentrate valve pulsing frequency and duration (Knelson
- Injection water flow (Knelson)
- Concentrate valve position (Falcon C)

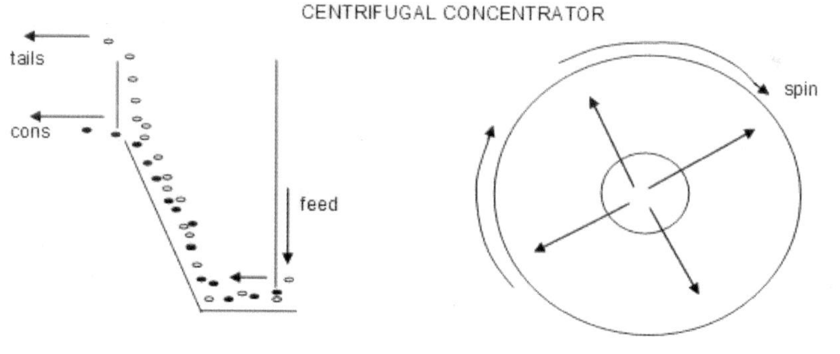

Shaking Tables (Wet and Dry)

Examples of wet shaking tables include the Wilfley made by Mineral Technologies, and the Deister made by Deisterconcentrators.

Shaking tables have been used for many years in gravity separation, and although limited in capacity, they produce a high upgrade, can handle a wide range of SGs, are very visual and operator dependent and are still used for cleaning stages to produce final concentrates.

Feed slurry is distributed at the head of the table via a launder, together with wash water, and spreads out across the inclined surface on the basis of particle SG, with high SG grains moving along the top of the flowing film to discharge off the far end as concentrate, while low SG grains move down the inclined slope of the table with the majority of the water to discharge at the bottom as tailings. The particle separation is assisted by the backwards and forwards motion (stroke) of the table, the tilt (both longitudinally and laterally), wash water applied along the length of the table, and riffles.

Operating variables include:

- Tilt angle
- Slope angle
- Splitter positions
- Wash water
- Stroke length
- Feed rate and density
- Riffle height

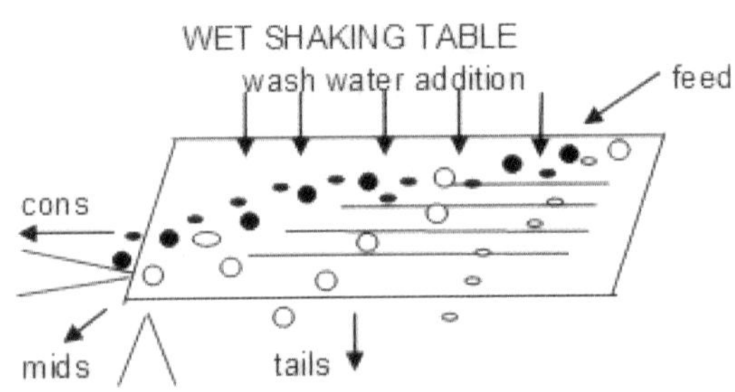

Dry shaking tables or air tables have specialized dry gravity separation applications such as fine quartz and zircon and consist of a table covered in cloth through which low pressure air is forced through by fans located underneath and controlled by dampers so that particles are partially fluidized and segregate according to SG and size, under the influence of stroke and tilt and discharge separately with the aid of splitters. The forces acting on the particles on air tables reverse the pattern of wet tables with the low SG grains discharging at the top of the table.

Operating variables include:

- Tilt angle
- Splitter positions
- Fan dampers (for air flow
- Stroke length
- Feed rate

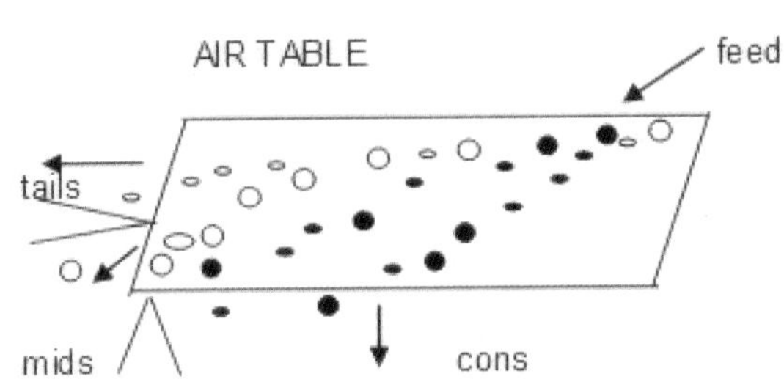

Mozeley Gravity Separators (MGS)

The Mozeley gravity separator, manufactured by Gravity Mining combines the centrifugal motion of a spinning drum to enhance fine gravity separation with the shaking motion of a table, making it suitable for upgrading of low tonnage rate fine particle slurries.

Feed slurry enters the MGS and is distributed onto a perforated feed ring mounted internally near the top end of an inclined spinning drum, together with wash water. The diluted slurry is thus subjected to centrifugal and shaking forces which cause the high SG particles to move up the inclined drum and low SG particles to move down the drum slope to discharge as tailings. Discharge of the high SG (concentrate) particles is assisted by internal scrapers which rotate at a speed slightly faster than the drum.

Operating variables include:

- Spin rate
- Feed rate
- Tilt angle
- Wash water
- Stroke frequency
- Stroke length

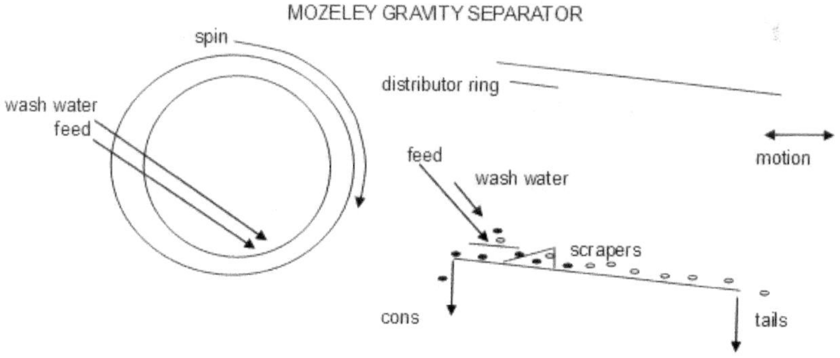

Sensor-based Sorting

Sensor-based sorting, is an umbrella term for all applications where particles are singularly detected by a sensor technique and rejected by an amplified mechanical, hydraulic or pneumatic process. The technique is generally applied in the three industries mining, recycling and food processing and used in the particle size range between 0.5 and 300 mm (0.020 and 11.811 in). Since sensor-based sorting is a single particle separation technology, the throughput is proportional to the average particle size and weight fed onto the machine.

Functional Principle of Sensor-based Sorting

The main subprocesses of sensor-based sorting are material conditioning, material presentation, detection, data processing and separation. Material conditioning includes all operations which prepare the particles for being detected by the sensor. All optical sensors need clean material to be able to detect optical characteristics. Conditioning includes screening and cleaning of the feed material. The aim of the material presentation is the isolation of the particles by creating a single particle layer with the densest surface cover possible without particles touching each other and enough distance to each other allowing for a selective detection and rejection of each single particle. There are two types of sensor-based sorters: the chute type and the belt type. For both types the first step in acceleration is spreading out the particles by a vibrating feeder followed by either a fast belt or a chute. On the belt type the sensor usually detects the particles horizontally while they pass it on the belt. For the chute type the material detection is usually done vertically while the material passes the sensor in a free fall. The data processing is done in real time by a computer. The computer transfers the result of the data processing to an ultra fast ejection unit which, depending on the sorting decision, ejects a particle or lets it pass.

Sensor-based Ore Sorting

Sensor-based ore sorting is the terminology used for sensor-based sorting in the mining industry. It is a coarse physical coarse particle separation technology usually applied in the size range for 25–100 mm (0.98–3.94 in). Aim is either to create a lumpy product in ferrous metals, coal or industrial minerals applications or to reject waste before it enters production bottlenecks and more expensive comminution and concentration steps in the process. In the majority of all mining processes, particles of sub-economic grade enter the traditional comminution, classification and concentration steps. If the amount of sub-economic material in the above-mentioned fraction is roughly 25% or more, there is good potential that sensor-based ore sorting is a technically and financially viable option. High added value can be achieved with relatively low capital expenditure, especially when increasing the productivity through downstream processing of higher grade feed and through increased overal recovery when rejecting deleterious waste.

Sensor-based sorting is a coarse particle separation technology applied in mining for the dry separation of bulk materials. The functional principle does not limit the technology to any kind of segment or mineral application but makes the technical viability mainly depend on the liberation characteristics at the size range 25–100 mm (0.98–3.94 in), which is usually sorted. If physical liberation is present there is a good potential that one of the sensors available on industrial scale sorting machines can differentiate between valuable and non-valuable particles. The separation is based on features measured with a detection technology that are used to derive a yes/no decision for actuation of usually pneumatic impulses. Sensor-based sorting is a disruptive technology in the mining industry which is universally applicable for all commodities. A comprehensive study examines both the technology's potential and its limitations, whilst providing a framework for application development and evaluation. All relevant aspects, from sampling to plant design and integration into mining and mineral processing systems, are covered. Other terminologeis used in the industry include *ore sorting, automated sorting, electronic sorting,* and *optical sorting.*

Sensor-based sorting has been introduced by Wotruba and Harbeck as an umbrella term for all applications where particles are singularly detected by a sensor technique and then rejected by an amplified mechanical, hydraulic or pneumatic process.

Liberation Characteristics

As for any other physical separation process, liberation is pre-requisite for possible separation. Liberation characteristics are well known and relatively easy to study for particulate lots in smaller size ranges, e.g. flotation feed and products. The analysis is essential for understanding the possible results of physical separation and relatively easy to conduct in laboratory on a couple of dozens of grams of sample which can be studied using optical methods or such as the QEMSCAN. For larger particles above 10 mm (0.39 in) it is widely known for applications that are treated using density separation methods, such as coal or iron ore. Here, the washability analysis can be conducted on sample masses up to 10 tonnes in equipped laboratories. For sensor-based sorting, where laboratory methods can only tell about the liberation characteristics where the describing feature is the density (e.g. iron ore, coal), hand counting, single-particle tests and bulk tests can reveal the liberation characteristics of a bulk material: Hereby, only single particle tests reveal the true liberation, while hand counting and bulk testing give a result which also incorporates the separation efficiency of the type of analysis.

Feed Preparation

A size range coefficient of approximately three is advisable. A minimum amount of undersized fine material must enter the machines to optimize availability. Moisture of the feed is not important, if the material is sufficiently dewatered and the undersize fraction is efficiently removed. For surface detection technologies sometimes spray water on the classifying screen is required to clean the surfaces. Surface detection technologies would otherwise measure the reflectance of the adhesions on the surface and a correlation to the particle's content is not given.

Main Machine Types

During the more than 80 years of technical development of sensor-based ore sorting equipment, various types of machines have been developed. This includes the channel-type, bucket-wheel type and cone type sorters. The main machine types being installed in the mining industry today are belt-type and chute-type machines. Harbeck made a good comparison of both disadvantages and advantages of the systems for different sorting applications. The selection of a machine-type for an application depends various case-dependent factors, including the detection system applied, particle size, moisture, yield amongst others.

Chute-type Machine

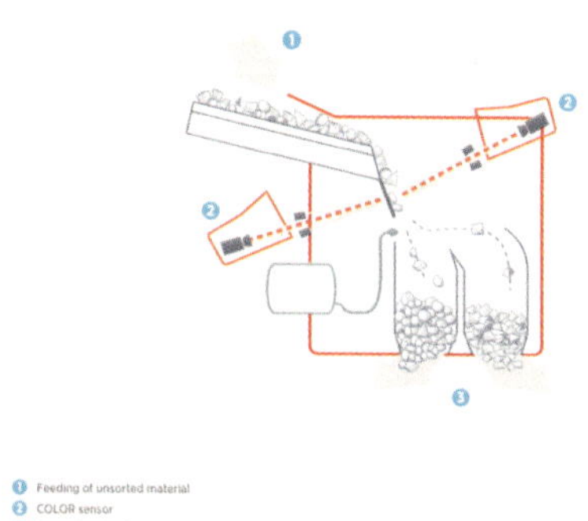

Side view and working principle of chute-type sensor-based ore sorting machine

The chute-type machine has a lower footprint and fewer moving parts which results in lower investment and operating costs. In general, it is more applicable to well liberated material and surface detection, because a double sided scanning is possible on a more reliable on the system. The applicable top size of the chute-type machine is bigger, as material handling of particles up to 300 mm (12 in) is only technically viable on this setup.

Belt-type machine

The belt-type machine is generally more applicable to smaller and to adhesive feed. In addition, the feed presentation is more stable which makes it more applicable for more difficult and heterogienous applications.

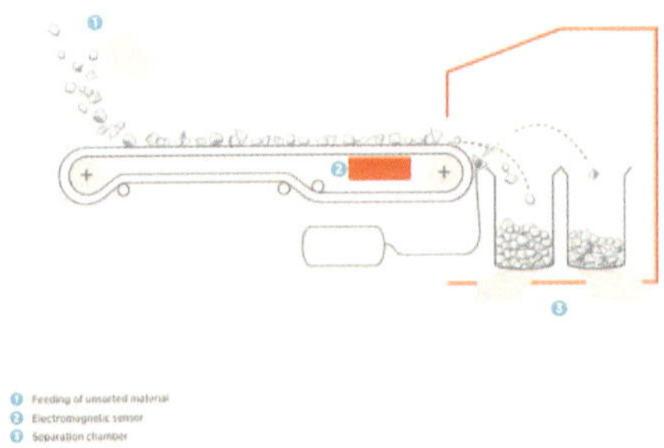

Side view and working principle of belt-type sensor-based ore sorting machine

Sub-processes

The separation in both machine types comprises the following sub-processes:

1. Feed distribution

A sized screen fraction with a size range coefficient (d95/d5) of 2-5 (optimal 2-3) is fed onto a vibratory feeder which has the function to create a mono-layer, by pre-accelerating the particles. A common misunderstanding in plant design is, that you can use the vibratory feeder to discharge from a buffer bunker but a separate units needs to be applied, since the feed distribution is very important to the efficiency of the sensor-based sorter and different loads on the feeder change its position and vibration characteristics.

2. Presentation

The feed is then transferred to the presentation mechanism which is the belt or the chute in the two main machine types respectively. This sub-process has the function to pass single particles of the material stream in a stable and predictable manner, thus in a unidirectional movement orthogonal to the detection line with uniform speed profile.

3. Detection

In the detection sub-process location and property vectors are recorded to allow particle localization for ejection and material classification for discrimination purposes. All detection technologies applied have in common to be cheap, contactless and fast. The technologies are subdivided in transmitting and reflecting groups, the first measuring the inner content of a particle while the later only uses the surface reflection for discrimination. Surface, or reflection technologies have the disadvantage that the surfaces need to be representing the content, thus need to be clean from clay and dust adhesions. But by default surface reflection technologies violate the Fundamental Sampling Principle because not all components of a particle have the same probability of being detected. The main transmitting technologies are EM (Electromagnetics) and XRT (X-ray-Transmission). EM detection is based on the conductivity of the material passing an alternating electromagnetic field. The principle of XRT is widely known through the application in medical diagnostics and airport luggage scanners. The main surface or reflection technologies are traditionally X-ray luminescence detectors capturing the fluorescence of diamonds under the excitation of X-ray radiation and color cameras detecting brightness and colour difference. Spectroscopic methods such as near-infrared spectroscopy known from remote sensing in exploration in mining for decades, have found their way into industrial scale sensor-based sorters. Advantage of the application of near-infrared spectroscopy is that the evidence can be measured on the presence of specific molecular bonds, thus minerals composition of the near-infrared active minerals.There is more detection technologies available on industrial scale sensor-based ore sorters. Readers that want to go into detail can find more in the literature.

4. Data Processing

Spectral and spatial is collected by the detection system. The spatial component catches the position of the particles distribution across the width of the sorting machine, which is then used in case the ejection mechanism is activated for a single particle. Spectral data comprises the features that are used for material discrimination. In a superseding processing step, spectral and spatial can be combined to include patterns into the separation criterion. Huge amount of data is collected in real time multiple processing and filtering steps are bringing the data down to the Yes/no decision – either for ejecting a particle or for keeping the ejection mechanism still for that one.

5. Ejection

The state-of-the-art mechanism of today's sensor-based ore sorters is a pneumatic ejection. Here, a combination of high speed air valves and an array of nozzles perpendicular to the acceleration belt or chute allows precise application of air pulses to change the direction of flight of single particles. The nozzle pitch and diameter is adapted to the particle size. The air impulse must be precise enough to change the direction of flight of a single particle by applying the drag force to this single particle and directing it over the mechanical splitter plate.

Positions in the Flow-sheet

Lumpy Ore Production

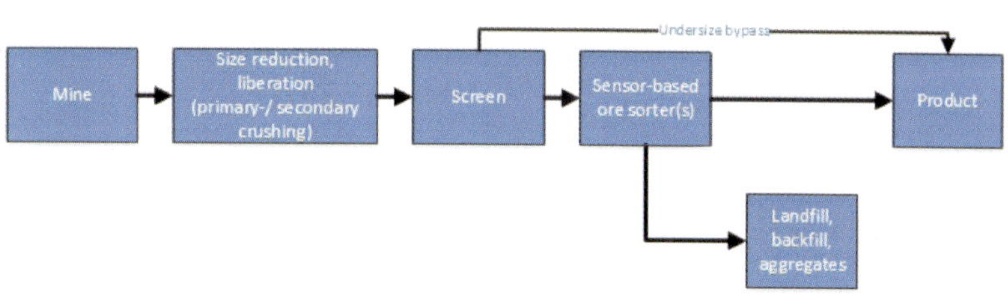

Simplified flow-sheet for lumpy ore production

For higher grade applications such as ferrous metals, coal and industrial minerals, sensor-based ore sorting can be applied to create a final product. Pre-condition is, that the liberation allows for the creation of a sellable product. Undersize material is usually bypassed as product, but can also be diverted to the waste fraction, if the composition does not meet the required specifications. This is case and application dependant.

Waste Rejection

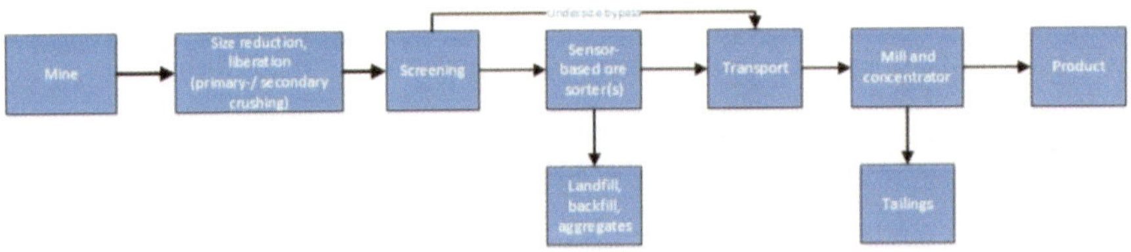

Simplified flow-sheet for waste rejection with sensor-based ore sorting

Most prominent example of the application of sensor-based ore sorting is the rejection of barren waste before transporting and comminution. Waste rejection is also known under the term pre-concentration. A discrimination has been introduced by Robben. Rule of thumb is that at least 25% of liberated barren waste must be present in the fraction to be treated by sensor-based ore sorting to make waste rejection financially feasible. Reduction of waste before it enters comminution and grinding processes does not only reduce the costs in those processes, but also releases the capacity that can be filled with higher grade material and thus implies higher productivity of the system. A prejudice against the application of a waste rejection process is, that the valuable content lost in this process is a penalty higher than the savings that can be achieved. But it is reported in the literature that the overall recovery even increases through bringing higher grade material as feed into the mill. In addition, the higher productivity is an additional source of income. If noxious waste such as acid consuming calcite is removed, the downstream recovery increases and the downstream costs decrease disproportionally as reported for example by Bergmann. The coarse waste rejected can be an additional source of income if there is a local market for aggregates.

Marginal Ore Treatment

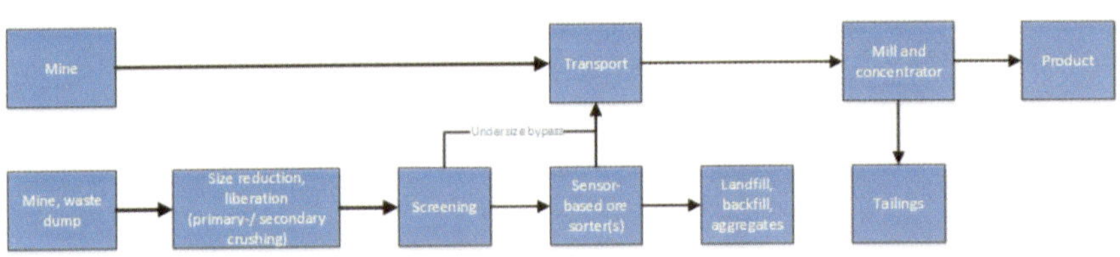

Simplified flow-sheet for marginal ore treatment with sensor-based ore sorting

Sensor-based ore sorting is financially especially attractive for low grade or marginal ore or waste dump material. This described scenario describes that waste dump material or marginal ore is sorted and added to the run-of-mine production. The needed capacity for the sensor-based ore sorting step is less in this case such as the costs involved. Requirement is that two crude material streams are fed in parallel, requiring two crushing stations. Alternatively, marginal and high grade ore can be buffered on an intermediate stockpile and dispatched in an alternating operation. The latter option has the disadvantage that the planned production time, the loading, of the sensor-based ore sorter is low, unless a significant intermediate stockpile or bunker is installed.

Treating the marginal ore separately has the advantage that less equipment is needed since the processed material stream is lower, but it has the disadvantage that the potential of the technology is not unfolded for the higher grade material, where sensor-based sorting would also add benefit.

Sensor-based Ore Sorting in Pebble-circuit of Autogenious and Semi-autogenious mills

Pebble circuits are a very advandageous location for the application of sensor-based ore sorters. Usually it is hard waste recirculating and limiting the total mill capacity. In addition, the tonnage is significantly lower in comparison to the total run-of-mine stream, the size range is applicable and usually uniform and the particles' surfaces are clean. High impact on total mill capacity is reported in the literature.

Spiral Separator

Spiral Separators are used on most of the free flowing products to sort round shapes from flat or oblong shapes.

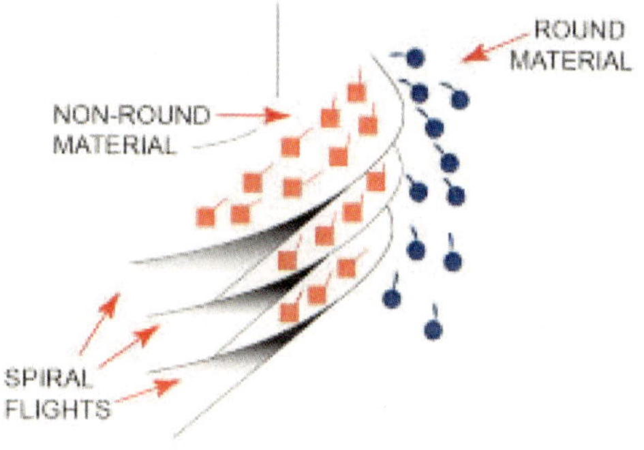

Spiral Separators are used in the seed industry for separating good soybeans from splits or weed seed.

Spiral Separators are often used in industrial applications such as separating metal abrasives, metal shot, ceramic beads, glass beads and other types of material.

Spiral separators operate on gravity only – product is loaded in the hopper, or direct feed tube on a single spiral.

The material flows over an adjustable cone and divides the material evenly onto the inner flights.

With increasing speed, the round material will separate and roll off the inner flights onto a larger outer flight and then exits a side discharge chute at the bottom.

The non-round material on the inner flights will exit a center discharge chute at the bottom.

Stripping (Chemistry)

Stripping is the removal of one or more components from a mixed system. Usually the components being removed are impurities or are considered less valuable than the material from which they are stripped. Examples of stripping operations are:

- The removal of overburden from a deposit of coal or ore.

- The removal of color from a dyed fabric using a reactive agent.
- The removal of an organic or a metallic coating from a solid surface.
- The removal of a dissolved component from a liquid by extraction using a second liquid which is insoluble in the first.
- The removal of a volatile component from a liquid by vaporization.

Strippers

Stripping columns are used to remove unwanted materials from process streams. The air stripping column shown below is used to remove ammonia from water and waste water feedstreams.

Air Stripping Columns

Air stripping columns are most widely used for the removal of volatile organic compounds (VOCs) from water. Air stripping columns are designed specifically for a given application.

The design of air stripping columns, such as the ones shown below, is governed by differences in volatility between the volatile organic compounds (VOCs) in the contaminated water.

The more volatile components transfer from the water to the air stream. The air exits the top of the column and contains the most volatile components, while the least volatile components exit with the water out the bottom of the column.

Equipment Design

After the contaminated water enters the column, it flows down the column and through the packing countercurrent to the air stream. The packing serves to increase the amount of contact between the two streams, and enhances the stripping as a result. After leaving the column, the air is either distributed into the atmosphere, or is collected for purification.

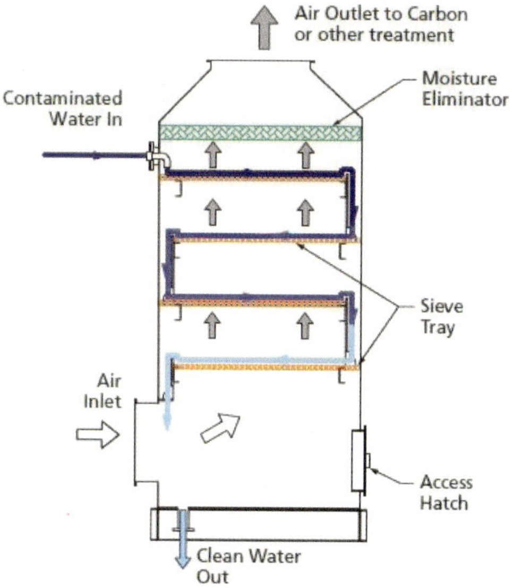

Sieve Tray Air Stripper

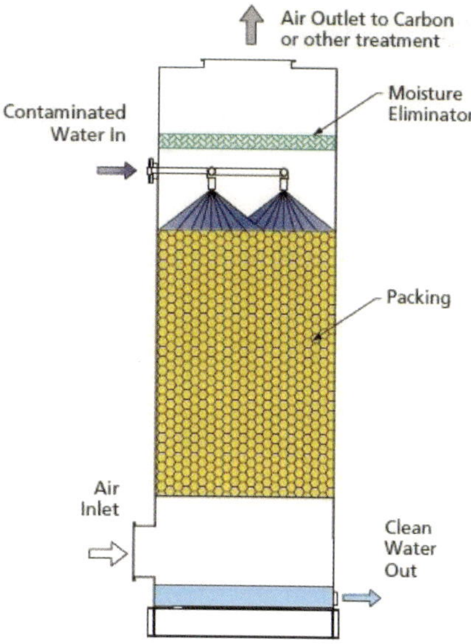

Packed Tower Air Stripper

Air stripping columns contain either trays or packing, depending on the application. These trays and packing are the same as those used in distillation and absorption columns.

Tray types include sieve, valve, and bubble cap. The geometry of the trays affects the extent and type of contact between the vapor and liquid streams. Downcomers channel the liquid flowing from one tray down to the tray below.

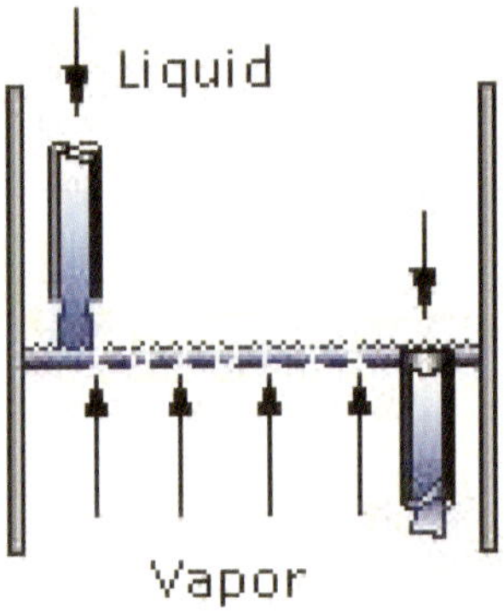

Packing types include dumped packing and stacked packing. The packing channels the liquid flowing from the top to the bottom of the column.

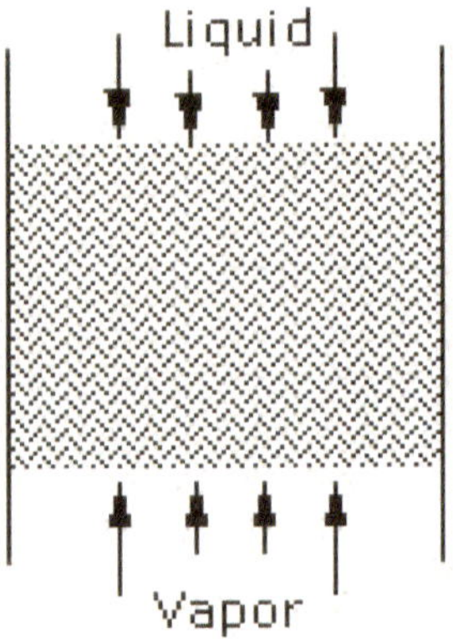

Usage Examples

In the system below on the left, air is used to remove benzene derivatives, chloroethylenes, and ammonia from contaminated groundwater. On the right is an air stripping column used to remove VOCs at a potable water treatment plant.

The air stripping columns pictured below are used to remove VOCs and dissolved gases from an incoming water feedstream.

The air strippers pictured below are used in a groundwater treatment plant. These units purify the incoming water through the use of trays rather than packing.

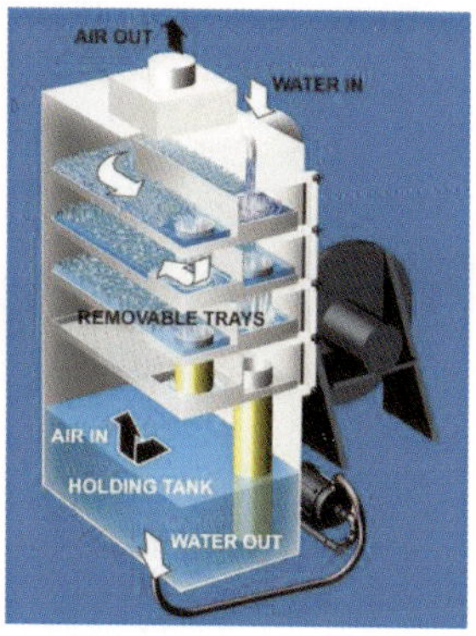

Advantages

- VOC removals in excess of 99.99% are possible with air stripping.
- Air strippers are more economical when low volatility VOCs are to be removed.
- Capable of achieving very high removal rates at very low pressure drops.
- Packed air strippers are the most economical alternative for flows above 50 gpm.

Disadvantages

- Treatment of the exiting VOC laden air needed to prevent air pollution.
- Fouling.

Steam Stripping Columns

Steam strippers are similar to air strippers with one exception: instead of air, they use steam to strip volatile components from a liquid stream.

Steam strippers allow for the removal of heavy soluble organic compounds that air strippers cannot remove. The temperature of these columns is usually very close to the boiling point of water.

In a typical steam stripping column, the liquid feed is introduced at the top of the column and the steam input is located at the bottom. When the liquid enters the column, it flows down the column and through the packing countercurrent to the steam. The packing enhances the stripping by increasing the amount of contact between the two streams. After leaving the column, the vapor is either distributed into the atmosphere or is collected for purification.

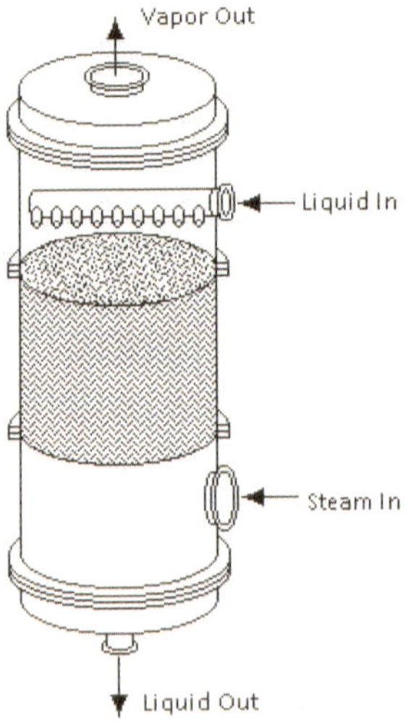

Equipment Design

Steam stripping columns contain either trays or packing, depending on the application. These trays and packing are the same as those used in distillation and absorption columns.

Tray types include sieve , valve , and bubble cap . The geometry of the trays affects the extent and type of contact between the vapor and liquid streams. Downcomers channel the liquid flowing from one tray down to the tray below.

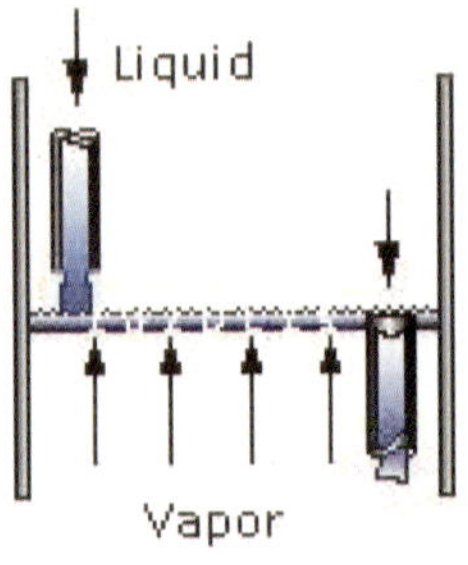

Packing types include dumped packing and stacked packing . The packing channels the liquid flowing from the top to the bottom of the column.

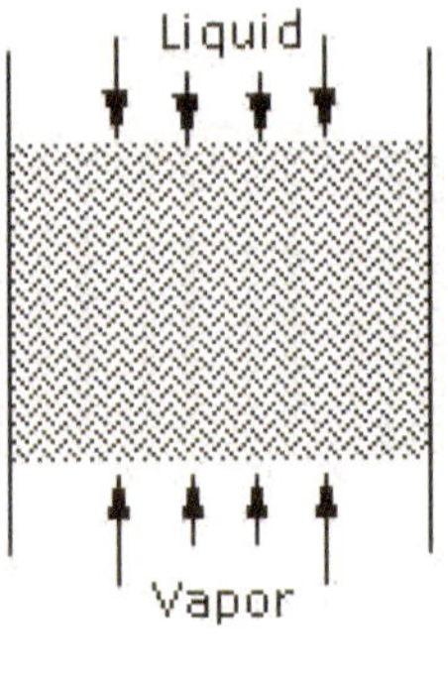

The operating pressure of a steam stripper can influence the efficiency and reliability of the column. At lower operating pressures, greater volatility and lower operating temperatures are achieved. Stream strippers operating at vacuum can be expensive but highly efficient; they allow the use of plastic internals to fight the effects of corrosive systems. Operating the column at or near atmospheric pressure results in higher volatilities and lower operating temperatures without the extra expense of a vacuum system. At higher operating pressures, the solubility of the solute increases and the separation becomes more difficult to perform. Also, at higher pressures, more steam input to the column is required for the same separation to occur.

Preheating the feed before it enters the column is important because it reduces the amount of heat added to the column to achieve the required separation. The feed is preheated by either a heat recovery exchanger that cross exchanges the feed without the effluent coming off of the bottoms, a steam heat exchanger that cross exchanges the feed with a steam stream, or a combination of both.

Usage Examples

Steam stripping columns are most widely used in the recovery of solvents and solutes, and also in the removal of Volatile Organic Compounds (VOCs) from water.

Steam stripping for wastewater purification is a distillation process where over 99% of VOCs are removed from water. Some typical VOCs are benzene, toluene, xylenes, ethyl benzene, styrene, and chlorinated hydrocarbons.

The wastewater feedstream is pre-heated and put in contact with steam in a packed or tray column. Trays or packing units facilitate contact between the contaminated water and the steam stream. The combined effects of the steam and heat cause the organic materials to transfer from the liquid into the vapor phase. As contacting proceeds down the column, the vapor becomes rich in organic materials while the wastewater gets cleaner. The organic vapor leaves the column at the top and is condensed and recovered. These hydrocarbons can be recycled back into the stripper or treated with an incinerator. The purified water that leaves the bottom of the column has very low organic contaminate concentrations and it can be recycled back into the plant's water system.

Advantages

- Capable of achieving very high removal rates and low effluent concentrations.
- Most economical removal technique at feed concentrations above 0.1% weight organics.
- Can be made very energy efficient with heat recovery.
- Minimizes air emissions and reduces loads to incineration.
- Easy solute recovery.

Disadvantages

- Not useful to strip phenol, glycols, ethylene, propylene, glycerine, ionic compounds such as sulfonated organics, inorganic compounds (except in free gaseous dissolved form), and low volatility, highly water soluble compounds such as acetic acid.

- Fouling.

Industrial Separation Processes

Voltammetry

Voltammetry is a method for determining the chemical makeup of a sample substance by measuring electrical activity, or the accumulation of chemicals, on electrodes placed in the substance.

Stripping Voltammetry is an analytical technique that is used to detect trace metals like arsenic, cadmium, lead, or mercury. These metals are usually toxic to humans and animals (and plants in some cases). It does not detect metals such as sodium, potassium, calcium and magnesium.

It works by electroplating these metals onto an electrode. This concentrates the metals. The metals on the electrode are then sequentially stripped off, which generates a current that can be measured.

- The current is proportional to the amount of metal being stripped off.
- The potential (voltage) at which the metal is stripped off is characteristic for each metal.

This means the metals can be identified, as well as quantified.

This diagram shows how different metals can be identified at different voltages.

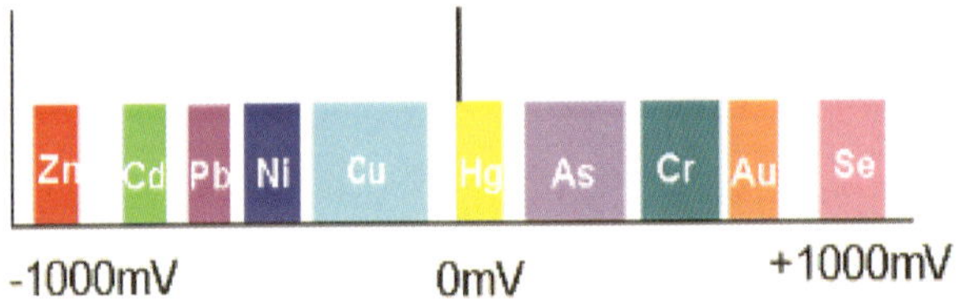

Electrowinning

Electrowinning is a process used to remove metal ions from liquid solutions such as rinse water, plating baths, and used process solutions. Most often used to recover precious metals, it relies on an electronic current to plate the metal ions onto a cathode. Recovered metals can then be sold for scrap or processed for use in plating. The solution can also be recycled with less stress on water treatment systems.

Electrowinning is an electrolytic process. An electrolytic process is based on the principles of electrolysis. Electrically conductive solution is created when an electrolyte is dissolved in water or other polar solvent. Electrolytes consist of positively and negatively charged ions (cations and anions). When a DC voltage is applied to the solution the positively charged ions (cations) move towards the cathode, while negatively charged ions (anions) move towards the anode. Processes like this are called electrolytic and electrowinning is based on this principle.

In electrowinning the electrolyte includes dissolved metals that have to be recovered. Another similar process is electrorefining which is strictly used in refining applications to improve the purity of the metals. Both processes use electroplating and are used to purify non-ferrous metals such as copper and silver.

A conventional electrowinning unit consists of a tank, rectifier and pump. Cathodes and anodes are aligned inside the tank. The pump fills the tank with electrolytic solution. Electrical current is provided by rectifier to cathodes and anodes in such way that difference in electrical potential creates a movement of cations towards the cathode. As time goes by the positively charged ions will plate on the cathodes. It is important to note that as metal builds-up on the cathode the deposition of metal in solution will decrease and the plating will slow down. Once the metal deposition decreases to a rate that is not sufficient for electroplating, the cathodes with pure metal deposited will be harvested. In the case of waste water treatment, the solution (waste water) will be cleaned or significantly purified of non-ferrous metals and can be further treated by chemical precipitation or reused within the industrial process.

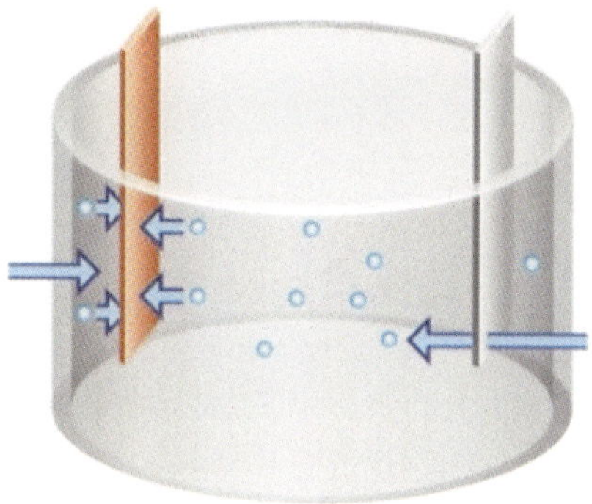

Electrowinning was mostly unchanged until 20 years ago with the advent of advanced electrowinning technologies that use cylindrical cells with high flow rate to enhance the rate of mass transfer, overcoming issues associated with depletion zones and enabling production of high purity metals even in presence of impurities. Electrometals was the very first to develop and commercialize the first cylindrical electrowinning cell known as emew, which stands for Electrometals electrowinning. The advent of cylindrical electrowinning cells has expanded the applicability of electrowinning beyond just metal refining to recycling, waste and wastewater treatment and even into hi-tech industries like semiconductors.

While electrowinning is mostly used to recover non-ferrous metals such as copper, nickel, tin, cadmium or precious metals like gold, silver and platinum, it also has a usage in industries that require waste water treatment. An electrowinning facility can operate 24 hours a day, 7 days a week and offers some great benefits:

- Reduction in waste and water use
- Enhanced process control
- Compliance with environmental permits
- Recovered metals sold for profit
- Stable recognized technology

Given all this electrowinning technology remains a core process in mining, refining and metal finishing industries providing great financial and environmental benefits.

Electrophoresis

Electrophoresis is an analytical method frequently used in molecular biology and medicine. It is applied for the separation and characterization of proteins, nucleic acids and subcellular-sized particles like viruses and small organelles. Its principle is that the charged particles of a sample migrate in an applied electrical field. If conducted in solution, samples are separated according to their surface net charge density. The most frequent applications, however, use gels (polyacrylamide, agarose) as a support medium. The presence of such a matrix adds a sieving effect so that particles can be characterized by both charge and size. Protein electrophoresis is often performed in the presence of a charged detergent like sodium dodecyl sulfate (SDS) which usually equalizes the surface charge and, therefore, allows for the determination of protein sizes on a single gel. Additives are not necessary for nucleic acids which have a similar surface charge irrespective of their size.

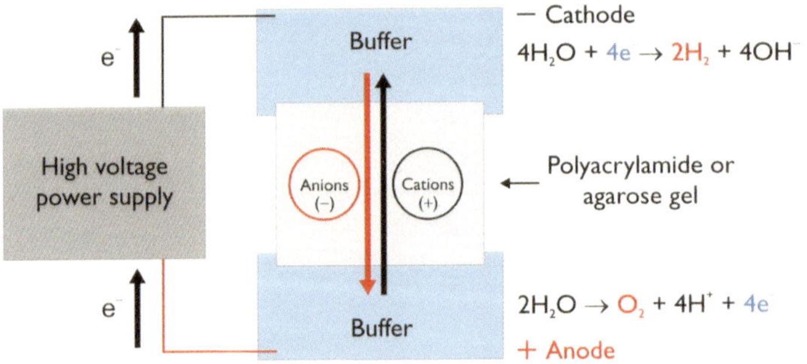

Principle of Electrophoresis

In the course of electrophoresis, two electrodes (typically made of an inert metal, e.g. platinum) are immersed in two separate buffer chambers. The two chambers are not fully isolated from each other. Charged particles can migrate from one chamber to the other in figure. By using an electric power supply, electric potential (E) is generated between the two electrodes. Due to the electric potential, electrons move by a wire between the two electrodes. More specifically, electrons move from the anode to the cathode. Hence, the anode will be positively charged, while the cathode will be negatively charged. As mentioned above, the two electrodes are immersed in two buffer chambers. Electrons driven to the cathode will leave the electrode and participate in a reduction reaction with water generating hydrogen gas and hydroxide ions. In the meantime, at the positive anode an oxidation reaction occurs. Electrons released from water molecules enter the electrode generating oxygen gas and free protons (which immediately form hydroxonium ions with water molecules). The amount of electrons leaving the cathode equals the amount of electrons entering the cathode. As mentioned, the two buffer chambers are interconnected such that charged particles can migrate between the two chambers. These particles are driven by the electric potential between the two electrodes. Negatively charged ions, called anions, move towards the positively charged anode, while positively charged ions, called cations, move towards the positively charged cathode.

The principle of electrophoresis. In the course of electrophoresis, two electrodes are immersed in two separate buffer chambers. The two chambers are connected such that charged particles can migrate from one chamber to the other. By using a power supply, electric potential difference is generated between the two electrodes. As a result, electrons flow from one of the electrodes, the anode, towards the other electrode, the cathode. Electrons from the cathode are taken up by water molecules of the buffer, resulting in a chemical reaction which generates hydrogen gas and hydroxide ions. In the other buffer chamber, water molecules transfer electrons to the anode an in another chemical reaction that generates oxygen gas and protons. (Protons are immediately taken up by water molecules to form hydroxonium ions.) As charged particles can migrate between the two chambers due to the electric potential difference, positive ions (cations) move towards the negatively charged cathode while negatively charged ions (anions) move towards the positively charged anode.

Different ions migrate at different speeds dictated by their sizes and by the number of charges they carry. As a result, different ions can be separated from each other by electrophoresis. It is very important to understand the basic physics describing the dependence of the speed of the ion as a function of the number of charges on the ion, the size of the ion, the magnitude of the applied electric field and the nature of the medium in which the ions migrate. By understanding these basic relationships, the principles of the many different specific electrophoresis methods become comprehensible. The fundamental principle of electrophoresis is illustrated in figure.

Centrifugation

Centrifugation is a mechanical process that utilizes an applied centrifugal force field to separate the components of a mixture according to density and/or particle size. The principles that govern particle behaviour during centrifugation are intuitively comprehensible. This may, in part, explain why centrifugation is seldom a part of post-secondary science curricula despite the broad range of scientific, medical and industrial applications in which this technique has been employed for well over 100 years. Applications that range from the mundane, industrial-scale dewatering of coal fines to the provision of an invaluable tool for biomedical research.

The first scientific studies conducted by Knight in 1806 reported the differences in orientation of roots and stems of seedlings when placed in a rotating wheel. However, it was not until some 60 years later that centrifuges were first used in industrial applications. The first continuous centrifuge, designed in 1878 by the Swedish inventor De Laval to separate cream from milk, opened the door to a broad range of industrial applications. About this same time, the first centrifuges containing small test tubes appeared. These were modest, hand-operated units that attained speeds up to 3000 rpm. The first electrically driven centrifuges were introduced in 1910, further accelerating centrifuge development. Svedberg's invention of the analytical ultracentrifuge in 1923, operating at 10 000.

Today, centrifuges are routinely used in a variety of disciplines ranging from large-scale commercial applications to laboratory-scale scientific research. The number of centrifuge designs and configurations used in the mineral, petrochemical, chemical, medical, pharmaceutical, municipal/industrial waste, dairy, food, polymer, energy and agricultural industries (to name a few) seem almost as numerous as the applications themselves.

Centrifugation separates on the basis of the particle size and density difference between the liquid and solid phases. Sedimentation of material in a centrifugal field may be described by:

$$\mathrm{v} = \frac{d^2(\rho_s - \rho_1)\omega^2 r F_S}{18\eta\theta}$$

where v is the rate of sedimentation, d is the particle diameter, ρ_s is the particle density, ρ_l is the solution density, ω is the angular velocity in radians s^{-1}, r is the radius of rotation, η is the dynamic viscosity, F_s is a correction factor for particle interaction during hindered settling and θ is a shape factor (=1 for spherical particles). F_s depends on the volume fraction of the solids present; approximately equalling 1, 0.5, 0.1 and 0.05 for 1%, 3%, 12% and 20% solids volume fraction respectively. Only material which reaches a surface during the flow through continuous centrifuges will be removed from the centrifuge feedstock, the efficiency depending on the residence time within the centrifuge and the distance necessary for sedimentation (D). This residence time will equal the volumetric throughput (φ) divided by the volume of the centrifuge (V). The maximum throughput of a centrifuge for efficient use is given by:

$$\theta = \frac{d^2(\rho_s - \rho_1)\omega^2 r V F_S}{18\eta\theta D}$$

The efficiency of the process is seen to depend on the solids volume fraction, the effective clarifying surface (V/D) and the acceleration factor (w^2r/g, where g is the gravitational constant, 981 cm s^{-2}; a rotor of radius 25 cm spinning at 1 rev s^{-1} has an acceleration factor of approximately 1 G). Low acceleration factors of about 1 500 g may be used for harvesting cells whereas much higher acceleration factors are needed to collect enzyme efficiently. The product of these factors (w^2rV/gD) is called the sigma factor (Σ) and is used to compare centrifuges and to assist scale-up.

Laboratory centrifuges using tubes in swing-out or angle head rotors have high angular velocity (ω) and radius of rotation (r) but small capacity (V) and substantial sedimentation distance (D). This type of design cannot be scaled-up safely, primarily because the mechanical stress on the centrifuge head increases with the square of the radius, which must increase with increasing capacity.

For large-scale use, continuous centrifuges of various types are employed in figure. These allow the continuous addition of feedstock, the continuous removal of supernatant and the discontinuous, semicontinuous or continuous removal of solids. Where discontinuous or semicontinuous removal of precipitate occurs, the precipitate is flushed out by automatic discharge systems which cause its dilution with water or medium and may be a problem if the precipitate is required for further treatment. Centrifugation is the generally preferred method for the collection of enzyme-containing solids as it does not present a great hazard to most enzymes so long as foam production, with consequent enzymic inactivation, is minimized.

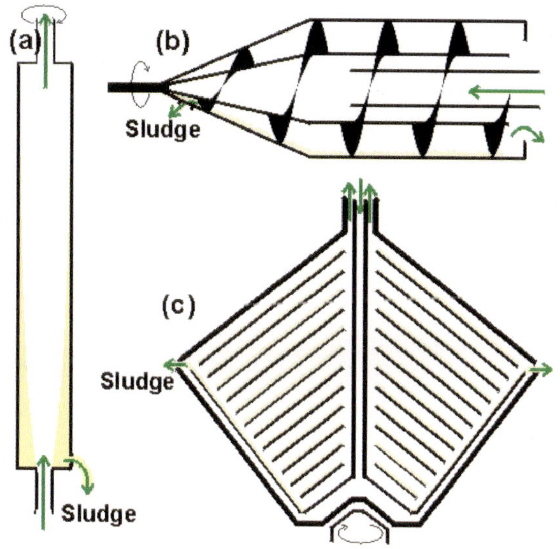

Basic designs of industrial centrifuges, showing the flow of material within the bowls. Motor drives, cooling jackets and sludge collection vessels are not shown. (a) Tubular bowl centrifuge. This is generally operated vertically, the tubular rotor providing a long flow path enabling clarification. The sludge collects and must be removed. (b) Continuous scroll centrifuge. This is operated horizontally. The helical screw scrolls the solids along the bowl surface and out of the liquid; the sludge being dewatered before discharge. The clarified liquor overflows over an adjustable weir at the other end of the bowl. The screw conveyer rotates at a slightly different speed to the bowl. (c) Continuous multichamber disc-stack centrifuge. The bowl contains a number of parallel discs providing a large clarifying surface with a small sedimentation distance. The sludge is removed through a valve.

Small particles of cell debris and precipitated protein may be sedimented using tubular bowl centrifuges, of which Sharples centrifuges (produced by Pennwalt Ltd.) are the best known. These semi-continuous centrifuges are long and thin enabling rapid acceleration and deceleration, minimising the down-time required for the removal of the sedimented solids. Here the radius and effective liquid thickness are both small allowing a high angular velocity and hence high centrifugal force; small models can be used at acceleration factors up to 50,000 g, accumulating 0.1 Kg of wet deposit whereas large models, designed to accumulate up to 5 Kg of deposit, are restricted to 16,000 g. The capacities of these centrifuges are only moderate.Multichamber disc-stack centrifuges, originally designed (by Westfalia and Alpha-Laval) for cream separation, contain multiple coned discs in a stack which are spun and on which the precipitate collects. They may be operated either semi-continuously or, by using a centripetal pressurising pump within the centrifuge bowl which forces the sludge out through a valve, continuously. The capacity and radius of such devices are large and the thickness of liquid is very small, due to the large effective surface area. The angular velocity, however, is restricted giving a maximum acceleration factor of about 8,000 g. A different design which is rather similar in principle is the solid bowl scroll centrifuge in which an Archimedes' screw collects the precipitate so that fluid and solids leave at opposite ends of the apparatus. These can only be used at low acceleration (about 3,000 g) so they are suitable only for the collection of comparatively large particles.

Although many types of centrifuge are available, the efficient precipitation of small particles of cell debris can be difficult, sometimes near-impossible. Clearly from Equation, the efficiency of centrifugation can be improved if the particle diameter (d) is increased. This can be done either by coagulating or flocculating particles. Coagulation is caused by the removal of electrostatic charges (e.g., by pH change) and allowing particles to adhere to each other. Flocculation is achieved by adding small amounts of high-molecular-weight charged materials which bridge oppositely-charged particles to produce a loose aggregate which may be readily removed by centrifugation or filtration. Flocculation and coagulation are cheap and effective aids to precipitating or otherwise harvesting whole cells, cell debris or soluble proteins but, of course, it is essential that the agents used must not inhibit the target enzymes. It is important to note that the choice of flocculant is determined by the pH and ionic strength of the solution and the nature of the particles. Most flocculants have very definite optimum concentrations above which further addition may be counter-effective. Some flocculants can be rapidly ruined by shear.

Chromatography

Chromatography is a technique for separating the components, or solutes, of a mixture on the basis of the relative amounts of each solute distributed between a moving fluid stream, called the mobile phase, and a contiguous stationary phase. The mobile phase may be either a liquid or a gas, while the stationary phase is either a solid or a liquid.

Kinetic molecular motion continuously exchanges solute molecules between the two phases. If, for a particular solute, the distribution favours the moving fluid, the molecules will spend most of their time migrating with the stream and will be transported away from other species whose molecules are retained longer by the stationary phase. For a given species, the ratio of the times spent in the moving and stationary regions is equal to the ratio of its concentrations in these regions, known as the partition coefficient. (The term adsorption isotherm is often used when a solid phase is involved.) A mixture of solutes is introduced into the system in a confined region or narrow zone (the origin), whereupon the different species are transported at different rates in the direction of fluid flow. The driving force for solute migration is the moving fluid, and the resistive force is the solute affinity for the stationary phase; the combination of these forces, as manipulated by the analyst, produces the separation.

Chromatography is one of several separation techniques defined as differential migration from a narrow initial zone. Electrophoresis is another member of this group. In this case, the driving force is an electric field, which exerts different forces on solutes of different ionic charge. The resistive force is the viscosity of the nonflowing solvent. The combination of these forces yields ion mobilities peculiar to each solute.

Chromatography has numerous applications in biological and chemical fields. It is widely used in biochemical research for the separation and identification of chemical compounds of biological origin. In the petroleum industry the technique is employed to analyze complex mixtures of hydrocarbons.

As a separation method, chromatography has a number of advantages over older techniques—crystallization, solvent extraction, and distillation, for example. It is capable of separating all the components of a multicomponent chemical mixture without requiring an extensive foreknowledge of the identity, number, or relative amounts of the substances present. It is versatile in that it can deal with molecular species ranging in size from viruses composed of millions of atoms to the smallest of all molecules—hydrogen—which contains only two; furthermore, it can be used with large or small amounts of material. Some forms of chromatography can detect substances present at the attogram (10−18 gram) level, thus making the method a superb trace analytical technique extensively used in the detection of chlorinated pesticides in biological materials and the environment, in forensic science, and in the detection of both therapeutic and abused drugs. Its resolving power is unequaled among separation methods.

Decantation

Decantation is a process to separate mixtures by removing a liquid layer that is free of a precipitate. The purpose may be to obtain a decant (liquid free from particulates) or to recover the precipitate. Decantation relies on gravity to pull precipitate out of solution, so there is always some loss of product, either from precipitate not fully falling out of solution or from leaving some liquid when separating it from the solid portion.

A piece of glassware called a decanter is used to perform decantation. There are several decanter designs. A simple version is a wine decanter, which has a wide body and a narrow neck. When wine is poured, solids stay in the base of the decanter. In the case of wine, the solid is usually potassium bitartrate crystals. For chemistry separations, a decanter may have a stopcock to drain away the precipitate or dense liquid or may have a partition to separate fractions.

Working of Decantation

There are two main decantation methods:

1. Separating Liquids and Solids

Decanting is done to separate particulates from a liquid by allowing the solids to settle to the bottom of the mixture and pouring off the particle-free part of the liquid.

For example, a mixture (possibly from a precipitation reaction) is allowed to stand so that gravity has time to pull the solid to the bottom of a container. The process is called sedimentation. Using gravity only works when the solid is less dense than the liquid. Clearwater may be obtained from mud by simply allowing time for the solids to separate from the water.

The separation may be enhanced using centrifugation. If a centrifuge is used, the solid may be compacted into a pellet, making it possible to pour off the decant with minimal loss of liquid or solid.

2. Separating Two or More Liquids

Another method is to allow two immiscible liquids to separate and the lighter liquid is poured or siphoned off. A common example is decantation of oil and vinegar. When a mixture of the two liquids is allowed to settle, the oil will float on top of the water so the two components may be separated. Kerosene and water can also be separating using decantation.

The two forms of decantation may be combined. This is especially useful if it's important to minimize loss of a solid precipitate. In this case, the original mixture may be allowed to settle or may be centrifuged to separate the decant and sediment. Rather than immediately drawing off the liquid, a second immiscible liquid may be added that is denser than the decant and that doesn't react with the sediment. When this mixture is allowed to settle, the decant will float on top of the other liquid and sediment. All of the decant can be removed with minimal loss of precipitate (except a tiny amount that remains floating in the mixture). In an ideal situation, the immiscible liquid that was added has a high enough vapor pressure that it evaporates, leaving all of the sediment.

Example: decantation of a mixture of water and soil

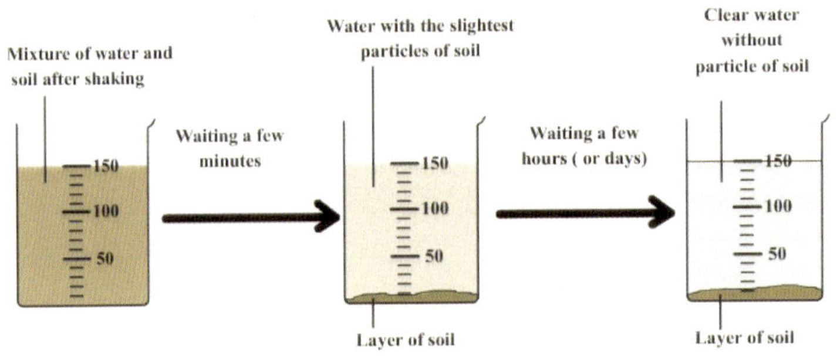

After shaking the mixture, the soil particles are dispersed in water.

We then observe:

- A layer of soil forms slowly to the bottom: it consists of particles that fall as a result of their weight.
- The liquid gradually brightens. Less dense particles fall slower.
- After a sufficiently long time, the liquid becomes clear since all particles have fallen to the bottom of the beaker.

Decantation is a simple separation to implement since it requires no special equipment but obtaining a complete separation and a totally clear liquid can takes a long time (sometimes several days) if the particles mixed to water are thin and of low density.

Distillation

Distillation is a process involving the conversion of a liquid into vapour that is subsequently condensed back to liquid form. It is exemplified at its simplest when steam from a kettle becomes deposited as drops of distilled water on a cold surface. Distillation is used to separate liquids from nonvolatile solids, as in the separation of alcoholic liquors from fermented materials, or in the separation of two or more liquids having different boiling points, as in the separation of gasoline, kerosene, and lubricating oil from crude oil. Other industrial applications include the processing of such chemical products as formaldehyde and phenol and the desalination of seawater. The distillation process appears to have been utilized by the earliest experimentalists. Aristotle (384–322 BC) mentioned that pure water is made by the evaporation of seawater. Pliny the Elder (AD 23–79) described a primitive method of condensation in which the oil obtained by heating rosin is collected on wool placed in the upper part of an apparatus known as a still.

Most methods of distillation used by industry and in laboratory research are variations of simple distillation. This basic operation requires the use of a still or retort in which a liquid is heated, a condenser to cool the vapour, and a receiver to collect the distillate. In heating a mixture of substances, the most volatile or the lowest boiling distills first, and the others subsequently or not at all. This simple apparatus is entirely satisfactory for the purification of a liquid containing nonvolatile material and is reasonably adequate for separating liquids of widely divergent boiling points. For laboratory use, the apparatus is commonly made of glass and connected with corks, rubber bungs, or ground-glass joints. For industrial applications, larger equipment of metal or ceramic is employed.

Uses of Distillation

Distillation is used for many commercial processes, such as the production of gasoline, distilled water, xylene, alcohol, paraffin, kerosene, and many other liquids. Gas may be liquefied and separate. For example: nitrogen, oxygen, and argon are distilled from air.

Types of Distillation

Types of distillation include simple distillation, fractional distillation (different volatile 'fractions' are collected as they are produced), and destructive distillation (usually, a material is heated so that it decomposes into compounds for collection).

Simple Distillation

Simple distillation may be used when the boiling points of two liquids are significantly different from each other or to separate liquids from solids or nonvolatile components. In simple distillation, a mixture is heated to change the most volatile component from a liquid into vapor. The vapor rises and passes into a condenser. Usually, the condenser is cooled (e.g., by running cold water around it) to promote condensation of the vapor, which is collected.

Steam Distillation

Steam distillation is used to separate heat-sensitive components. Steam is added to the mixture, causing some of it to vaporize. This vapor is cooled and condensed into two liquid fractions. Sometimes the fractions are collected separately, or they may have different density values, so they separate on their own. An example is steam distillation of flowers to yield essential oil and a water-based distillate.

Fractional Distillation

Fractional distillation is used when the boiling points of the components of a mixture are close to each other, as determined using Raoult's law. A fractionating column is used to separate the components used a series of distillations called rectification. In fractional distillation, a mixture is heated so vapor rises and enters the fractionating column. As the vapor cools, it condenses on the packing material of the column. The heat of rising vapor causes this liquid to vaporize again, moving it along the column and eventually yielding a higher purity sample of the more volatile component of the mixture.

Vacuum Distillation

Vacuum distillation is used to separate components that have high boiling points. Lowering the pressure of the apparatus also lowers boiling points. Otherwise, the process is similar to other forms of distillation. Vacuum distillation is particularly useful when the normal boiling point exceeds the decomposition temperature of a compound.

Distillation of Crude Oil

Crude oil is a mixture of many hundreds of liquid hydrocarbons. Dissolved in it are many other hydrocarbons some of which are solids and some gases (the lower members of the alkane family, predominantly methane and ethane but often with some propane and butane). There may also be some hydrocarbon gases trapped above the oil, as for example in some of the oil fields in the North Sea.

In the refineries the oil is distilled into liquid fractions with different boiling point ranges which are then further processed.

The crude oil is heated in a furnace (ca 650 K) and the resulting mixture fed as a vapour into a fractionating tower which can have a height of 25-100 m, handling volumes of over 40 000 m^3 a day. The column may contain 40-50 steel 'sieve trays' which fit horizontally across the column and are designed to ensure there is intimate mixing between the descending liquid, formed by condensation, and the rising vapour. To effect this close contact, the trays have holes in them ('the sieve') through which the vapour flows up into the liquid collecting on the trays.

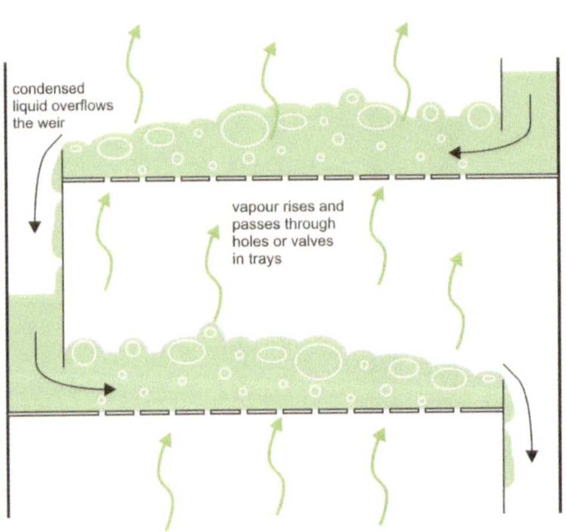

Figure: Trays in a fractionating column.

Alternatively, there may be valves fitting over the holes, which lift up when the pressure of the vapour below the tray is greater than the pressure on the tray (Figure). These are considered to be more efficient in fractionation than sieve trays without valves.

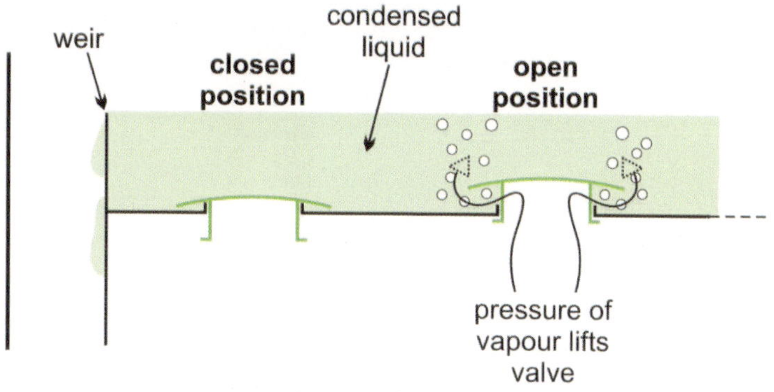

Figure: Valve trays.

A temperature gradient exists in the tower, the top being cooler than the bottom. When the rising vapour reaches a tray containing liquid whose temperature is below the boiling point (bp) of the vapour, it partially condenses. As some of the vapour condenses to a liquid, the dissipated latent heat then heats more liquid, and the more volatile components in the liquid evaporate joining the remaining vapour and passing up the tower. The less volatile liquid flows across the tray and down a pipe to the tray below.

Figure: This FLEXITRAYTM valve tray is a steel sheet on which liftable valves are mounted.

They are much more efficient than sieve trays.

This process occurs continuously in each tray, the least volatile vapour components condensing and the most volatile evaporating. This results in each tray containing products with a comparatively narrow boiling point range (called a 'close cut' of products). This leads to the low relative molecular mass products (low bp) accumulating near the top of the tower and high relative molecular mass constituents (high bp) collecting near the bottom.

The high boiling point residue from the crude oil is then transferred to another column and distilled under vacuum; lowering the pressure reduces the boiling point and ensures constituents distil at temperatures below their decomposition temperature. From this process, lubricating oils and waxes are obtained. The final residue from the process is bitumen. The constituents in crude oil vary from one oil field to another (for example, the proportion of naphtha from North Sea oil is considerably greater than that from Middle East oil) so the numbers given are approximate.

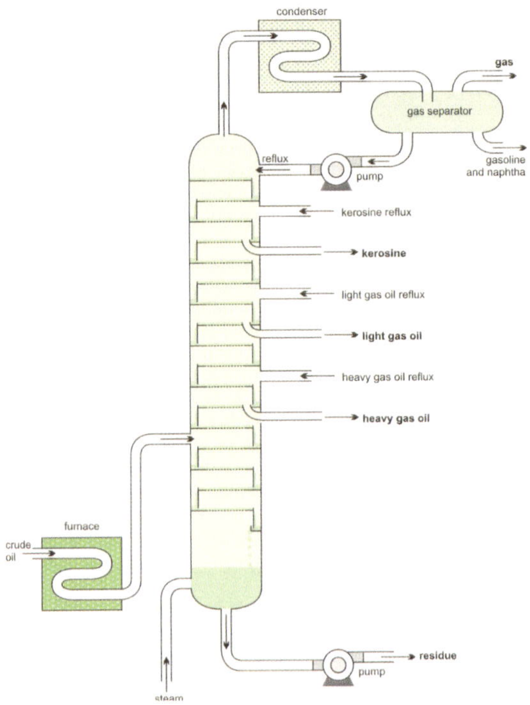

Figure: The fractional distillation of crude oil.

Elutriation

Elutriation is a separation method, which depends on the settling velocity of the particles in the fluid.

In this process of separation, the material is placed in a rising fluid having a fixed upward velocity, particles whose normal falling velocity is less than the velocity of the fluid will be carried upward and out of the vessel. If fractions obtained from a series of fluid velocities are collected and weighed, a complete size analysis may be obtained.

Extraction

Extractions are a way to separate a desired substance when it is mixed with others. The mixture is brought into contact with a solvent in which the substance of interest is soluble, but the other substances present are insoluble.

Extractions use two immiscible phases (these are phases that do not mix, like oil and water) to separate the substance from one phase into the other.

Tea-making is a very basic non-laboratory extraction. You boil tea leaves in water to extract the tannins, theobromine, polyphenols, and caffeine out of the solid tea leaves and into the liquid water. You are then free to drink and enjoy the water and the extracted substances it contains.

Laboratory Extractions

Typical lab extractions are to transport organic compounds out of an aqueous phase and into an organic phase. These are often carried out using a separating funnel, as illustrated below. The distribution of a solute between two phases is an equilibrium described by partition theory.

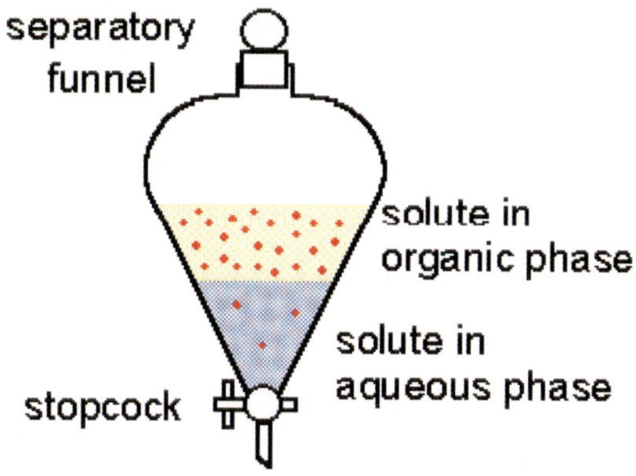

Analytical Extractions

Elemental analysis generally requires fairly simple (not necessarily easy) sample preparation.

Solids are usually dissolved or digested in caustic solution and liquids are sometimes extracted to separate the analyte from interferences.

Organic analysis is often much more complicated. Real-world samples can be very complicated matrices that require careful extraction procedures to obtain the analytes in a form that can be analyzed.

Flotation

Flotation, in mineral processing, is a method used to separate and concentrate ores by altering their surfaces to a hydrophobic or hydrophilic condition—that is, the surfaces are either repelled or attracted by water. The flotation process was developed on a commercial scale early in the 20th century to remove very fine mineral particles that formerly had gone to waste in gravity concentration plants. Flotation has now become the most widely used process for extracting many minerals from their ores.

Most kinds of minerals require coating with a water repellent to make them float. By coating the minerals with small amounts of chemicals or oils, finely ground particles of the minerals remain unwetted and will thus adhere to air bubbles. The mineral particles are coated by agitating a pulp of ore, water, and suitable chemicals; the latter bind to the surface of the mineral particles and make them hydrophobic. The unwetted particles adhere to air bubbles and are carried to the upper surface of the pulp, where they enter the froth; the froth containing these particles can then be removed. Unwanted minerals that naturally resist wetting may be treated so that their surfaces will be wetted and they will sink.

This ability to modify the floatability of minerals has made possible many otherwise difficult separations that are now common practice in modern mills. Flotation is widely used to concentrate copper, lead, and zinc minerals, which commonly accompany one another in their ores. Many complex ore mixtures formerly of little value have become major sources of certain metals by means of the flotation process.

Frothers

Frothers are liquids that produce the froth or foam on which the flotation process depends. The froth resembles soap suds and provides the physical separation between the mineral(s) floated and the pulp containing the waste. The froth must be strong enough to support the weight of the mineral floated and yet not be tenacious and non-flowing. It should have the tendency to break down when removed from the flotation cell. The frother should not enhance the flotation of unwanted material. Many other characteristics are required for a foaming agent to be a good flotation frother. Typical frothers include:

- Pine oil
- Certain alcohols
- Low molecular weight polypropylene glycols

Collectors

A collector is a chemical that attaches to the mineral surface and produces a hydrophobic (water-fearing) surface. While certain minerals are naturally hydrophobic and do not require a collector, recovery is often improved when a collector is used. This water-repellent film facilitates the attachment of the mineral particle to the air bubble. Many different chemicals are used as collectors, such as:

- Oils
- Xanthates
- Dithiophosphates
- Petroleum sulfonates
- Fatty amines

Depressants

Depressants are chemicals that inhibit the flotation of minerals. They are used to improve the selectivity of a flotation process. They often make it feasible to separate minerals that were initially floated together.

The response of many minerals to the flotation process is often dramatically affected by pH. Flotation circuits are often operated at a pH range of 7.5 to 11.5. The exact range at any given plant is optimized for the ore at that site. Lime is often used to raise the pH of the pulp and also reduce the flotation of iron pyrite.

Particle Size

The particle size to which the ore is ground depends on the nature of the ore. The grind must be fine enough to liberate the mineral grain from associated rock, but producing too small a particle size is both expensive and detrimental to recovery.

Froth flotation is generally limited to size fractions between roughly 65 and 100 mesh. Particles larger than 65 mesh Tyler (210 micron) are difficult for the air bubble to lift while particles smaller than 400 mesh (37 micron) often will not attach to the air bubble.

Reagent Dosage Rate

Many factors influence the amount of reagent required for a particular application. Variations in particular size, mass of particle, quantity of mineral, and the character of the host rock are some of the many factors that influence reagent usage.

Water quality, flotation equipment size and type, temperature, and ore body variation are just a few of the variables affecting the flotation process.

Copper/Molybdenum Ore

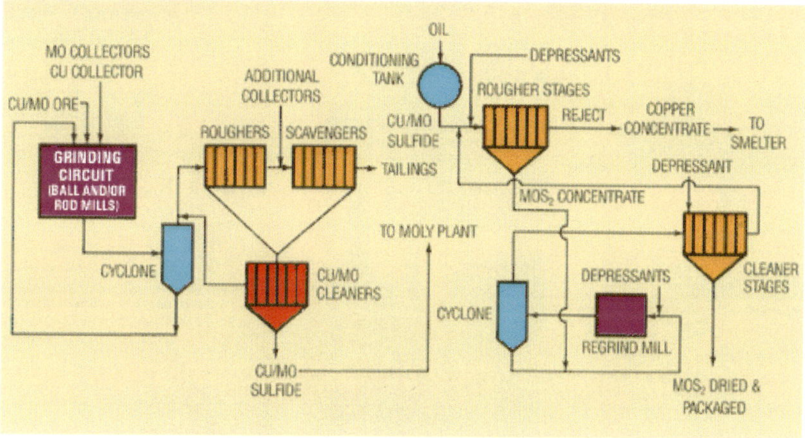

The figure above is a simplified flow sheet showing the basic steps involved in a copper mill producing the byproduct molybdenum. The crushed copper ore (containing, for example, 0.75% Cu and 0.015% Mo) is fed to the grinding circuit where water is added and the ore is ground to the proper size. Lime and an oil for collecting molybdenum are usually added in the grinding circuit. A copper collector may also be added at this point.

The pulp progresses to the flotation cells. The frother and copper collector are usually added to the pulp just before the roughers. Additional copper collector is often added to the scavenger flotation cells and may be added at several other points. The froth, enriched in copper and molybdenum, is removed from the flotation roughers and scavengers. This enriched material is now called the rougher concentrate or cleaner feed. The combined rougher-scavenger concentrate may be 5% Cu and 0.06% Mo (a concentration factor of 3 to 10 times is typical).

The scavenger or combined rougher/scavenger concentrate may go to regrind where it is ground to further liberate the minerals. The rougher and scavenger concentrates ultimately reach the copper molybdenum cleaners. The cleaners, which represent another cycle of froth flotation, provide additional cleaning and upgrading of the rougher scavenger concentrate.

The concentrate, which is now approximately 30% Cu and 0.35% Mo, is pumped to the molybdenum plant where the molybdenum disulfide is separated from the copper sulfide. Chemicals referred to as depressants are added to the pulp to reduce the flotation of the copper minerals while allowing the molybdenum minerals to float. The tailings from the molybdenum plant are the final copper concentrate (typically greater than 30% copper). Following molybdenum rougher flotation and several stages of cleaning, a final molybdenum concentrate is produced. The final molybdenum concentrate is typically greater than 55%.

Lead/Zinc Ore

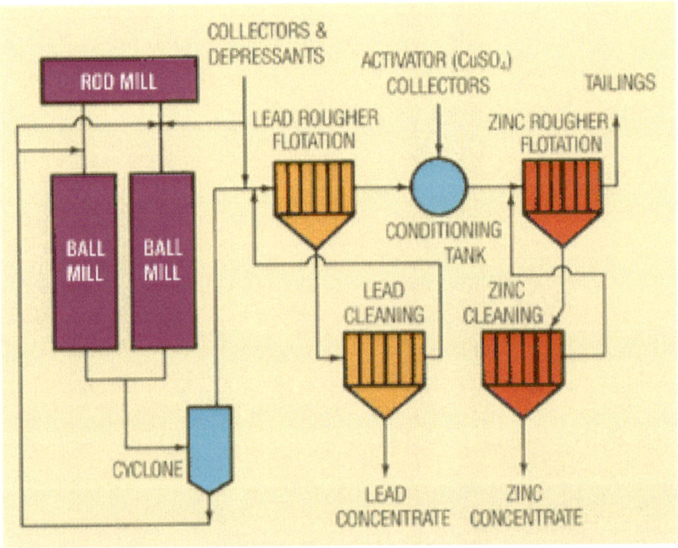

Lead and zinc often occur in the same ore. Copper is also present in some ores. Froth flotation is again employed to produce marketable concentrates of both. The process is identical in principle to the flotation of other minerals. The figure on the right is a simplified version of a Pb/Zn flotation circuit. The main difference in lead/zinc flotation is that zinc is usually depressed while lead (and copper) is floated. The tailing from the lead flotation circuit is the feed to the zinc circuit. The ability to selectively float the separate minerals depends on the ore, the chemicals, and other factors.

Solvent Extraction

Solvent extraction is a hydrometallurgical process. The process entails:

1. Dissolution of the metal in an aqueous, typically acid, solution.
2. Transfer of the dissolved metal to an organic solution.
3. Transfer of the dissolved metals to a second aqueous solution.
4. Production of a product.

The product is usually electroplated to produce the pure metal or precipitated from the solution to produce a metal salt.

Copper, uranium, vanadium and other metals are produced by solvent extraction.

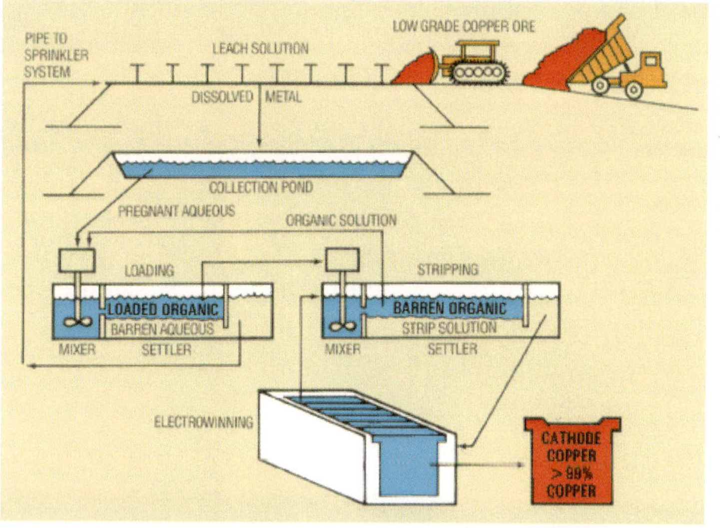

The figure is a simplified schematic of a typical copper solvent extraction process.

Oxide copper ore, low-grade copper sulfide overburden, and other copper-containing materials are stacked into a heap. Leach solution (pH approximately 2) is sprinkled or flooded on top of the heap and percolates through. The leach solution containing the dissolved metal eventually flows out of the heap into a collection pond. This material is referred to as pregnant leach solution. Oxide copper minerals are very soluble in the leach solution. Sulfide minerals are rendered soluble by certain bacteria that feed on the sulfide minerals and convert them to soluble sulfates and other soluble forms. This biological action tends to maintain the acidity of the leach solution. Typical operations will leach some areas of the heap, enlarge other sections, and let others rest.

The pregnant leach solution is pumped to the extraction mixer where it is intimately mixed with organic solution. The organic solution is a chelating agent diluted in a hydrocarbon solvent. The chelate reacts with the copper and removes it from the aqueous phase. The mixture flows to the settler where the organic phase and aqueous phase separate. The aqueous phase, now called raffinate, is pumped to the raffinate ponds and eventually back to the heap to leach more copper.

The loaded organic phase goes to the strip circuit where it is mixed with a strong acid solution. The chelate releases the copper to this aqueous phase, called the strip solution. Again the mixture goes to a settler where the organic and aqueous phases separate. The organic solution returns to the extraction circuit to load more copper, thus repeating the cycle.

The strip solution is pumped to the tankhouse where the copper is electrowon (electroplated). This solution then returns to the extraction circuit to repeat the cycle.

Direct electrical current is applied to the electrowinning solution in the tankhouse. A copper starter sheet functions as the cathode, and additional copper deposits on the surface of the sheet. When the copper cathode has reached a certain size it is removed from the electrowinning tank, washed, and prepared for sale. The cathode will be copper of greater than 99% purity.

Filtration

Filtration is the process of separating suspended solid matter from a liquid, by causing the latter to pass through the pores of some substance, called a filter. The liquid which has passed through the filter is called the filtrate. The filter may be paper, cloth, cotton-wool, asbestos, slag- or glass-wool, unglazed earthenware, sand, or other porous material.

Filtration is very frequently employed in chemical technology, and it often presents great difficulties. In most technical operations, cotton cloth is the filtering material, but occasionally woolen or hair cloth is necessary. The cloth may be fastened on a wooden frame in such a way that a shallow bag is formed, into which the turbid liquid is poured. The filtrate, in this ease, is cloudy at first, but soon becomes clear, and then the turbid portion is returned to the filter. Filtration is often retarded by the presence of fine, slimy precipitates, or by the formation of crystals in the interstices of the cloth, from the hot solution. Any attempt to hasten filtration, by scraping or stirring the precipitate on the cloth, will always cause the filtrate to run turbid.

A better form is the "bag-filter," which is a long, narrow hag of twilled cotton, supported by an outside cover of coarse, strong netting, capable of sustaining a considerable weight and hydrostatic pressure. These bags are often five or six feet long, and eight inches or more in diameter. The open end of the bag is tied tightly around a metallic ring or a nipple, by which the whole is suspended, and through which the liquor to be filtered is introduced. "When hot liquids are filtered, the bags are often hung in steam-heated rooms, the temperature being nearly that of the liquid.

In pressure filtration, the liquid is forced through the interstices of the filter by direct atmospheric pressure, the air being exhausted from the receiver; or by hydrostatic pressure, obtained either by means of a high column of the liquid, or by a force pump. By the first method, called suction filtration, the liquid may be forced downward through the filter into a receiver; the precipitate collects on the top of the filter and becomes a part of the filtering layer. This sometimes causes difficulty, for the particles of certain precipitates unite to form an impervious layer. Or the filtrate may be drawn upwards through the filter, which is suspended in the liquid to be filtered; thus clogging does not occur so easily, as a large part of the precipitate settles to the bottom of the vessel and does not come in contact with the filter until most of the liquid has been drawn off.

Fig. 8.

In technical work, pressure is usually obtained by the filter press (Figs). This is a strong iron frame, in which a number of east-iron or bronze filter cells are supported. Each cell is made up of two flat metal plates with raised edges, separated by a hollow "distance frame" of the same metal. There is a hole in the centre of each plate, and grooves on each surface leading to an opening at the lower edge of the plate. A filter is made of two pieces of cloth, slightly larger than the plates, sewed together along the margin of a small circular opening cut in the centre of each. One piece of the cloth is passed through the hole in the plate, and then both pieces are spread out smoothly, one on either side of the plate.

Another plate is prepared in the same way, and a distance frame having been placed between them, the cell thus formed is set vertically in the press frame, where it is supported by lugs on each plate and distance frame. 'When the desired number of cells are ready, they are tightly clamped together by means of a heavy screw, which passes through one end of the press frame. Thus a series of cells, lined with filter cloth and connected by a straight channel through the central holes, is formed. A powerful force pump drives the liquid to be filtered into the cells, where it passes from one to the other until they are all filled. The hydrostatic pressure forces the liquid through the filters into the grooves in the plates, along which it flows, and escapes through the openings at the lower side of the plate. The sediment retained by the cloth collects ill the cell and forms a solid cake, which finally fills each cell completely. The process is then stopped, the cells taken apart, and the cake of sediment having been removed, the cells are returned to the press frame, to' be again put into operation. The filtrate is caught in a trough.

Fig. 8a.

In another form of press, instead of the central opening, there is a hole in the cornel' of each plate and distance frame in such a position that, when placed in the press, the holes form a continuous channel through the corner of the whole series of cells. A small hole drilled on the inside of each distance frame, at right angles to the direction of the channel, admits the liquid into each cell. The filter is a piece of cloth hung over the distance frame in such a way that both sides of the frame are covered. A frame so covered is put between each pair of grooved plates. Small holes are cut in the cloth to correspond to the channel in the corners of the cells. The method of filtration is the same as in the central feed machines.

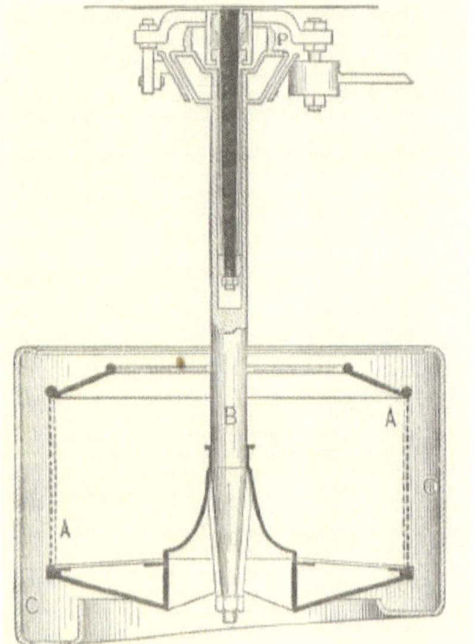

Fig. 9.

The pressure obtained by the force pump may be only a few pounds, or it may rise to several hundred pounds per square inch. The filter press may contain from a dozen to fifty or more cells, and these cells may be as large as four feet in diameter. For many purposes the press is surrounded by coils or jackets, through which steam or refrigerating solutions may be circulated, according as hot or cold filtration is desired. The filter press is very rapid in its action and is extensively employed in industrial chemical work. For use with acid or corrosive liquids, the plates and distance frames are often covered with lead or some alloy which is not easily corroded.

The centrifugal machine is, to a great extent, replacing the filter press and other filters, especially when crystals are to be removed. This furnishes the most rapid method and leaves the substance almost dry. The centrifugal machine is a cylindrical box or basket (A) of wire gauze or perforated sheet metal, fixed 011 a vertical shaft (B), which rotates at a very high speed. The contents of the box are driven to the outer wall by the centrifugal force, the solid matter being retained by the gauze or screen. The liquid passes through and is caught in a fixed shell (C), surrounding the rotating basket. These machines are of various sizes from 12 to 60 inches diameter, and 8 to 36 inches, depth of basket. Two general forms are in use: the over-driven type, in which the driving pulley (P) is fixed at the upper end of the shaft, above the basket; and the under-driven type, in which the basket is placed on the upper end of the shaft, and the pulley below. In the over-driven type it is frequently customary to suspend the shaft in flexible bearings. Thus the basket is enabled to adjust itself to any change in the centre of gravity, caused by unequal loading, and runs without vibration.

Sand filters are sometimes used for work on a large scale. These are made as follows; Into a box having a perforated bottom, is put a layer of coarse gravel; this is covered with finer pebbles; these by sand, and a jute or canvas cloth covers the whole. A wooden or iron grating is added to protect the filter, when the sediment is shoveled out. The filter is often placed over a receptacle from which the air may be exhausted, thus affording pressure filtration if necessary.

References

- Kleine, Christopher; Wotruba, Hermann and Robben, Mathilde: A new tool for mining engineers – The sensor-based sorter. Sustainable Development in the Minerals Industry (SDIMI) Conference 2011, Aachen, Germany, ISBN 978-3-86797-119-5
- Concept-of-theoretical-plates-in-column-chromatography: lab-training.com, Retrieved 21 April 2018
- Aqueous-twophase-extraction-advances-for-bioseparation-21559821: omicsonline.org, Retrieved 29 March 2018
- Department of Processing and Recycling. "Sensor-Based Sorting 2014". RWTH Aachen University. Retrieved 4 March 2016.
- Stripping-chemical-engineering-660700: accessscience.com, Retrieved 19 July 2018
- What-is-electrowinning: wisegeek.com, Retrieved 20 June 2018
- Centrifugation, chemical-engineering: sciencedirect.com, Retrieved 10 July 2018
- Robben C, Characteristics of sensor-based sorting technology and implementation in mining [Book]. - Shaker Verlag GmbH, in der Schriftenreihe zur Aufbereitung und Veredlung, ISBN 978-3-8440-2498-2
- Chromatography, science: britannica.com, Retrieved 31 March 2018
- Decantation, chemistry: physics-chemistry-class.com, Retrieved 29 May 2018
- What-is-distillation-601964: thoughtco.com, Retrieved 26 March 2018

2

Electrophoresis

Electrophoresis refers to the motion of dispersed particles with respect to a fluid when acted by a uniform electric field. The diverse types of electrophoresis, such as capillary electrophoresis, di-electrophoresis, free-flow electrophoresis, gel electrophoresis, moving-boundary electrophoresis, etc. and their applications have been closely examined in this chapter.

Electrophoresis is the term used to describe the motion of particles in a gel or fluid within a relatively uniform electric field. Electrophoresis may be used to separate molecules based on charge, size, and binding affinity. The technique is mainly applied to separate and analyze biomolecules, such as DNA, RNA, proteins, nucleic acids, plasmids, and fragments of these macromolecules. Electrophoresis is one of the techniques used to identify source DNA, as in paternity testing and forensic science.

Electrophoresis of anions or negatively charged particles is called anaphoresis. Electrophoresis of cations or positively charged particles is called cataphoresis.

Electrophoresis was first observed in 1807 by Ferdinand Frederic Reuss of Moscow State University, who noticed clay particles migrated in water subjected to a continuous electric field.

Working of Electrophoresis

In electrophoresis, there are two primary factors that control how quickly a particle can move and in what direction. First, the charge on the sample matters. Negatively charged species are attracted to the positive pole of an electric field, while positively charged species are attracted to the negative end. A neutral species may be ionized if the field is strong enough. Otherwise, it doesn't tend to be affected.

The other factor is particle size. Small ions and molecules can move through a gel or liquid much more quickly than larger ones.

While a charged particle is attracted to an opposite charge in an electric field, there are other forces that affect how a molecule moves. Friction and the electrostatic retardation force slow the progress of particles through the fluid or gel. In the case of gel electrophoresis, the concentration of the gel can be controlled to determine the pore size of the gel matrix, which influences mobility. A liquid buffer is also present, which controls the pH of the environment.

As molecules are pulled through a liquid or gel, the medium heats up. This can denature the molecules as well as affect the rate of movement. The voltage is controlled to try to minimize the time required to separate molecules, while maintaining a good separation and keeping the chemical species intact. Sometimes electrophoresis is performed in a refrigerator to help compensate for the heat.

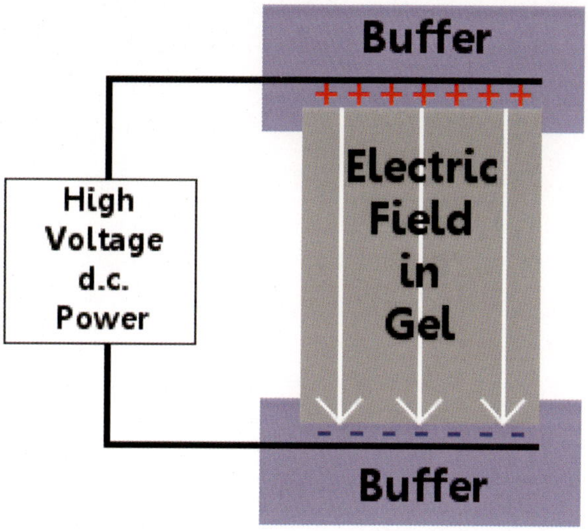

Types of Electrophoresis

Electrophoresis encompasses several related analytical techniques. Examples include:

- Affinity electrophoresis - Affinity electrophoresis is a type of electrophoresis in which particles are separated based on complex formation or biospecific interaction.
- Capillary electrophoresis - Capillary electrophoresis is a type of electrophoresis used to separate ions depending mainly on the atomic radius, charge, and viscosity. As the name suggests, this technique is commonly performed in a glass tube. It yields quick results and a high resolution separation.
- Gel electrophoresis - Gel electrophoresis is a widely used type of electrophoresis in which molecules are separated by movement through a porous gel under the influence of an electrical field. The two main gel materials are agarose and polyacrylamide. Gel electrophoresis is used to separate nucleic acids (DNA and RNA), nucleic acid fragments, and proteins.
- Immunoelectrophoresis - Immunoelectrophoresis is the general name given to a variety of electrophoretic techniques used to characterize and separate proteins based on their reaction to antibodies.
- Electroblotting - Electroblotting is a technique used to recover nucleic acids or proteins following electrophoresis by transferring them onto a membrane. The polymers polyvinylidene fluoride (PVDF) or nitrocellulose are commonly used. Once the specimen has been recovered, it can be further analyzed using stains or probes. A western blot is one form of electroblotting used to detect specific proteins using artificial antibodies.
- Pulsed-field gel electrophoresis - Pulsed-field electrophoresis is used to separate macromolecules, such as DNA, by periodically changing the direction of the electric field applied to a gel matrix. The reason the electric field is changed is because traditional gel electrophoresis is unable to efficiently separate very large molecules that all tend to migrate together. Changing the direction of the electric field gives the molecules additional

directions to travel, so they have a path through the gel. The voltage is generally switched between three directions: one running along the axis of the gel and two at 60 degrees to either side. Although the process takes longer than traditional gel electrophoresis, it's better at separating large pieces of DNA.

- Isoelectric focusing - Isoelectric focusing (IEF or electrofocusing) is a form of electrophoresis that separates molecules based on different isoelectric points. IEF is most often performed on proteins because their electrical charge depends on pH.

Electrophoresis Theory

The surface adsorbed sample strongly affects suspended particles by applying electric surface charge, on which an external electric field exerts an electrostatic coulomb force. According to the double layer theory, all surface charges in fluids are screened by a diffuse layer of ions, which has the same absolute charge but opposite sign with respect to that of the surface charge. The electric field also exerts a force on the ions in the diffuse layer which has direction opposite to that acting on the surface charge This force is not actually applied to the particle, but to the ions in the diffuse layer located at some distance from the particle surface, and part of it is transferred all the way to the particle surface through viscous stress. This part of the force is also called electrophoretic retardation force. When the electric field is applied and the charged particle to be analyzed is at steady movement through the diffuse layer, the total resulting force is zero:

$$F_{toto} = 0 = F_{el} + F_f + F_{ret}$$

Considering the drag force on the moving particles due to the viscosity of the dispersant, in the case of low turbulence and moderate electric charge strength E, the velocity of a dispersed particle v is simply proportional to the applied field, which leaves the electrophoretic mobility μ_e defined as:

$$\mu_e = \frac{v}{E}$$

The most known and widely used theory of electrophoresis was developed in 1903 by Smoluchowsky

$$\mu_e = \frac{\varepsilon_r \varepsilon_0 \zeta}{\eta},$$

where ε_r is the dielectric constant of the dispersion, ε_0 is the permittivity of free space ($C^2 N^{-11} m^{-2}$), η is dynamic viscosity of the dispersion medium (Pa s), and ζ is zeta potential (i.e., the electrokinetic potential of the slipping plane in the double layer).

The Smoluchowski theory is very powerful because it works for dispersed particles of any shape at any concentration. Unfortunately, it has limitations on its validity. It follows, for instance, from the fact that it does not include Debye length κ^{-1}. However, Debye length must be important for electrophoresis, as follows immediately from the Figure on the right. Increasing thickness of the double layer (DL) leads to removing point of retardation force further from the particle surface. The thicker DL, the smaller retardation force must be.

Detailed theoretical analysis proved that the Smoluchowski theory is valid only for sufficiently thin DL, when particle radius a is much greater than the Debye length:

$$a\kappa >> 1$$

This model of "thin Double Layer" offers tremendous simplifications not only for electrophoresis theory but for many other electrokinetic theories. This model is valid for most aqueous systems because the Debye length is only a few nanometers there. It breaks only for nano-colloids in solution with ionic strength close to water.

The Smoluchowski theory also neglects contribution of surface conductivity. This is expressed in modern theory as condition of small Dukhin number.

$$Du << 1$$

In the effort of expanding the range of validity of electrophoretic theories, the opposite asymptotic case was considered, when Debye length is larger than particle radius:

$$a\kappa < 1$$

Under this condition of a "thick Double Layer", Huckel predicted the following relation for electrophoretic mobility:

$$\mu_e = \frac{2\varepsilon_r\varepsilon_0\zeta}{3\eta}$$

This model can be useful for some nanoparticles and non-polar fluids, where Debye length is much larger than in the usual cases. There are several analytical theories that incorporate surface conductivity and eliminate the restriction of a small Dukhin number pioneered by Overbeek and Booth. Modern, rigorous theories valid for any zeta potential and often any $a\kappa$ stem mostly from DukhinSemenikhin theory. In the thin Double Layer limit, these theories confirm the numerical solution to the problem provided by O'Brien and White.

Capillary Electrochromatography

Capillary electrochromatography (CEC) is a miniaturized, hybrid separation technique that combines micro-high-performance liquid chromatography (μHPLC) and capillary electrophoresis (CE). The driving force in CEC is electroosmotic flow (EOF), which is generated by the application of an electric field to a capillary that is usually packed with typical HPLC stationary phase particles. Column efficiency in CEC is usually much higher than that in HPLC with an identical column, since EOF generates a plug-like flow profile in contrast to the parabolic flow profile in HPLC. Therefore, CEC has both the high selectivity and peak capacity of HPLC and the high efficiency and resolution of CE. In 1974, Pretorius successfully used EOF as the driving force through a glass tube packed with a chromatographic stationary phase, which provided column efficiency of about 31,000 N/m. However, CEC had not received much attention until Jorgenson separated nonionic aromatic compounds by applying a voltage across a packed capillary column, and Knox proved high efficiency and other advantages of CEC in theory and practice.

However, CEC has not become a mainstream separation technique because of the problems and difficulties associated with the lack of proper instrumentation and in-house column fabrication. In addition, the high applied voltage across the column in "pure" CEC format often causes Joule heating, which leads to bubble formation that can result in dryout of the column and disruption of the current. Therefore, pressurized CEC (pCEC), or electrokinetic high-performance liquid chromatography (eHPLC), with EOF combining hydraulic pressure as its driving force has gained more attention.

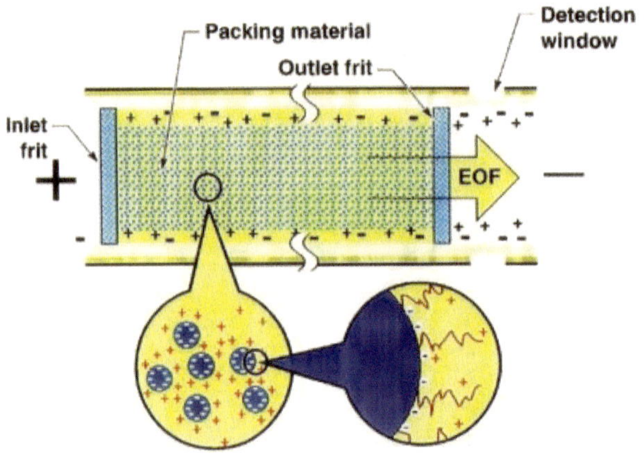

Figure- Schematic of a packed capillary column in CEC.

In pCEC, the retention mechanism for neutral compounds is essentially based on chromatographic partition. However, for charged compounds, both chromatographic partition and electrophoretic mobility contribute to the separation mechanism. Therefore, high efficiency, high resolution, high peak capacity, and fast speed separations with pCEC can be achieved. The dual mechanism of pCEC makes it suitable for both neutral and charged compounds and dramatically enhances the separation selectivity compared to a standalone HPLC or CE system. In addition, it is possible to perform pCEC, μHPLC, and CE independently on the same system. Unlike in "pure" CEC, the binary solvent delivery pump provides the capability of gradient elution, and the rotary-type injector in pCEC permits quantitative sample injection. Furthermore, pCEC can be readily coupled with on-column detection schemes such as UV, laser-induced fluorescence (LIF), MS, and electrochemical (EC) detection. Due to its micro-nanoscale feature, pCEC is also economically attractive and environmentally friendly.

Experimental

Apparatus

All pCEC experiments were carried out using a TriSep™-2100 capillary electrochromatography system, which consisted of a binary solvent gradient module, high-voltage power supply (±30 kV); microfluidic control module including a six-port injector; workstation; and on-column variable-wavelength UV-VIS or LIF detector. As illustrated in figure, the mixed mobile phases, delivered by the binary solvent delivery system, flow through the six-port injection valve. Sample is injected into the sample loop and is then carried by the mobile phase to the splitting cross, where the grounding electrode is connected. Depending on the split ratio, about 1% of the sample and mobile phase flows into the capillary column. A capillary restrictor or backpressure regulator connected with the splitting cross can control the flow speed in the column and maintain a constant pressure at the inlet of the capillary column to suppress bubble formation. The waste vial contains both the capillary column outlet and the positive or negative electrode. The capillary columns with detection windows were installed into the UV-VIS or LIF detector, which transmitted signals to the chromatographic workstation.

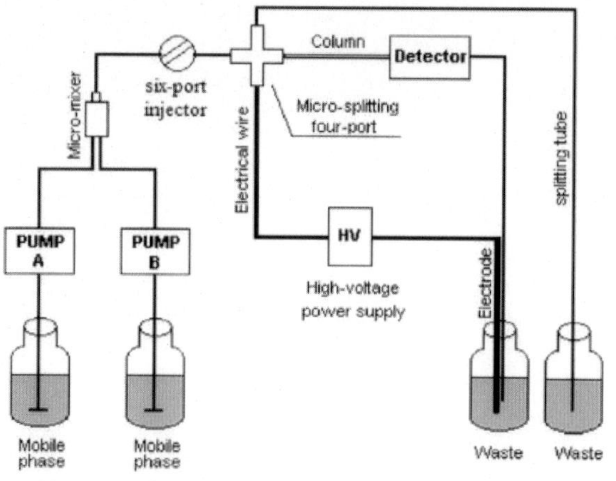

Figure- Schematic of TriSep-2100 pCEC system.

Reagents and Materials

Chromatographic-grade methanol, acetonitrile, and other analytical-grade chemicals were used. All solutions were filtered using 0.22-μm filters.

Column

Capillaries with 75~150 μm i.d. and 365 μm o.d. were packed electrokinetically with, in most cases, 3–5 μm porous octadecylsilica (ODS) particles. Detection windows were generated by burning off about 2 mm polyimide coating. Some capillary columns used in this work were packed with 1.5-μm nonporous C18 particles to increase column efficiency and resolution. Exotic stationary materials, such as gold microspheres, can be packed into the capillary column.

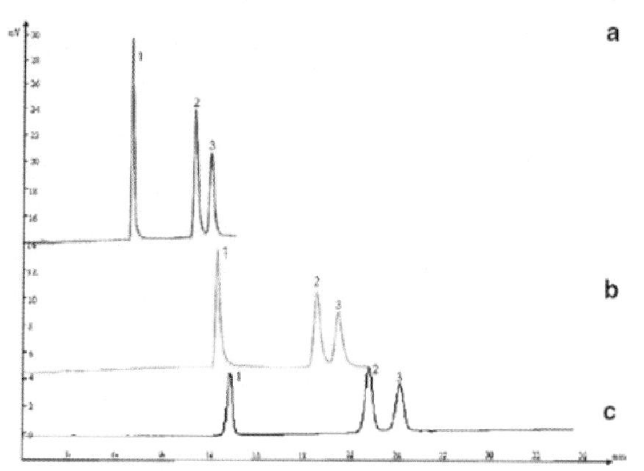

Figure- Separation of three flavone compounds by a) pCEC, b) μHPLC, and c) HPLC.

Column—a) and b): 50 cm (effective length 25 cm) × 150 μm i.d. capillary packed with 5 μm C18, and c): 250 mm × 4.6 mm i.d, 5 μm C18. Mobile phase: methanol—15 mM Na2HPO4 (55:45, v/v). Flow rate: a) 500 nL/min, b) 400 nL/min, and c) 1 mL/min. Voltage: a) 10 kV, b) and c) 0 kV. Detector: UV at 360 nm. Peak identification: 1) quercetin, 2) isorhamnetin, and 3) kaempferol.

Results and Discussion

High-speed Separation and High Column Efficiency

Flavone compounds	Theoretical plate number (N/m)			Retention time (min)		
	pCEC	μHPLC	HPLC	pCEC	μHPLC	HPLC
Quercetin	105,030	60,936	32,108	7.135	12.983	13.679
Isorhamnetin	96,986	49,898	31,328	11.568	19.371	22.634
Kaempferol	82,904	35,892	28,654	13.872	20.790	24.401

Table- Comparison of column efficiency and retention time between pCEC, μHPLC, and HPLC

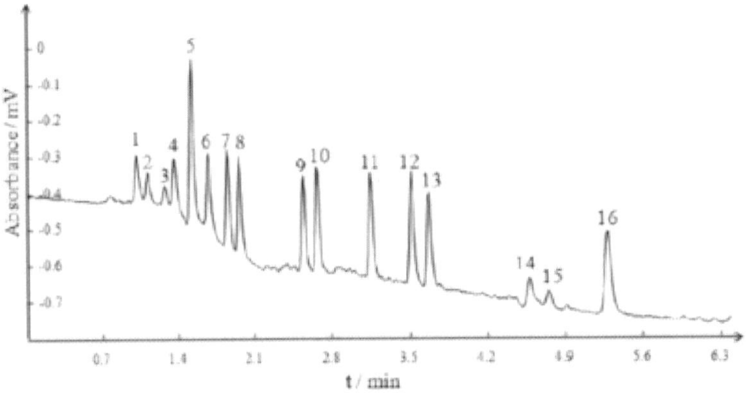

Figure- Separation of 16 PAHs in pCEC. Column: 35 cm (effective length 10 cm) × 100 μm i.d. capillary packed with 1.5 μm NPS C18.

Mobile phase A: 2 mM Tris H2O, B: 2 mM Tris acetonitrile. Gradient: 0~2 min, 50~75%B. Flow rate: 400 nL/min. Voltage: 30 kV. Detector: UV at 254 nm. Peak identification: 1) naphthalene, 2) acenaphthylene, 3) acenaphthene, 4) fluorene, 5) phenanthrene, 6) anthracene, 7) fluoranthene, 8) pyrene, 9) benz[a]anthracene, 10) chrysene, 11) benzo[b]fluoranthene, 12) benzo[k]fluoranthene, 13) benzo[a]pyrene, 14) dibenz[a,h]anthracene, 15) benzo[ghi]perylene, 16) indeno-[1,2,3-cd] pyrene.

The major advantage of pCEC is its high efficiency and fast speed due to its combined driving force, i.e., EOF and pressure. Flavone compounds widely exist in many traditional Chinese medicines. Although they can be separated and detected by HPLC, μHPLC, and pCEC, as shown in figure, a much higher theoretical plate number was obtained using pCEC rather than μHPLC and HPLC. In addition, the retention time of flavone compounds separated by pCEC was nearly only half of the retention time in HPLC under similar pressure. The column efficiency and retention time in the three separation modes are listed in Table. The pCEC system makes the separation speed much faster with application of the dual mechanism and small particles. For example, 16 polycyclic aromatic hydrocarbon (PAH) components were baseline separated within 6 min, as demonstrated in figure.

Capillary Electrophoresis

Capillary electrophoresis is an analytical technique that separates ions based on their electrophoretic mobility with the use of an applied voltage. The electrophoretic mobility is dependent upon the charge of the molecule, the viscosity, and the atom's radius. The rate at which the particle moves is directly proportional to the applied electric field--the greater the field strength, the faster the mobility. Neutral species are not affected, only ions move with the electric field. If two ions are the same size, the one with greater charge will move the fastest. For ions of the same charge, the smaller particle has less friction and overall faster migration rate. Capillary electrophoresis is used most predominately because it gives faster results and provides high resolution separation. It is a useful technique because there is a large range of detection methods available.

Instrumental Setup

A typical capillary electrophoresis system consists of a high-voltage power supply, a sample introduction system, a capillary tube, a detector and an output device. Some instruments include a temperature control device to ensure reproducible results. This is because the separation of the sample depends on the electrophoretic mobility and the viscosity of the solutions decreases as the column temperature rises. Each side of the high voltage power supply is connected to an electrode. These electrodes help to induce an electric field to initiate the migration of the sample from the anode to the cathode through the capillary tube. The capillary is made of fused silica and is sometimes coated with polyimide. Each side of the capillary tube is dipped in a vial containing the electrode and an electrolytic solution, or aqueous buffer. Before the sample is introduced to the column, the capillary must be flushed with the desired buffer solution. There is usually a small window near the cathodic end of the capillary which allows UV-VIS light to pass through the analyte and measure the absorbance. A photomultiplier tube is also connected at the cathodic end of the capillary, which enables the construction of a mass spectrum, providing information about the mass to charge ratio of the ionic species.

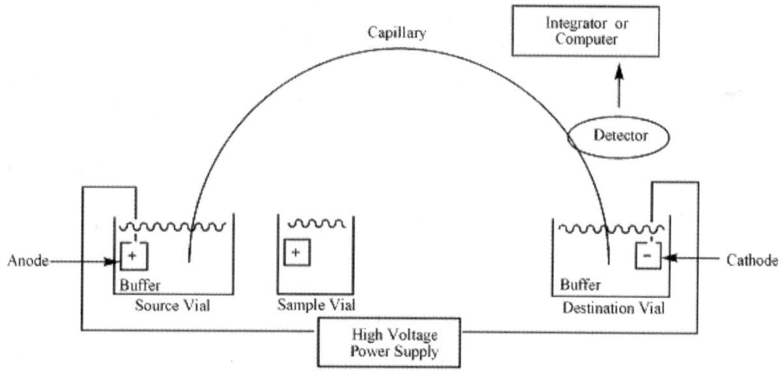

Figure: Instrumental Setup.

Theory

Electrophoretic Mobility

Electrophoresis is the process in which sample ions move under the influence of an applied voltage. The ion undergoes a force that is equal to the product of the net charge and the electric field strength. It is also affected by a drag force that is equal to the product of *f*, the translational friction coefficient, and the velocity. This leads to the expression for electrophoretic mobility:

$$\mu_{EP} = \frac{q}{f} = \frac{q}{6\pi\eta r}$$

where f for a spherical particle is given by the Stokes' law; η is the viscosity of the solvent, and r is the radius of the atom. The rate at which these ions migrate is dictated by the charge to mass ratio. The actual velocity of the ions is directly proportional to E, the magnitude of the electrical field and can be determined by the following equation:

$$\upsilon = \mu_{EP} E$$

This relationship shows that a greater voltage will quicken the migration of the ionic species.

Electroosmotic Flow

The electroosmotic flow (EOF) is caused by applying high-voltage to an electrolyte-filled capillary. This flow occurs when the buffer running through the silica capillary has a pH greater than 3 and the SiOH groups lose a proton to become SiO^- ions. The capillary wall then has a negative charge, which develops a double layer of cations attracted to it. The inner cation layer is stationary, while the outer layer is free to move along the capillary. The applied electric field causes the free cations to move toward the cathode creating a powerful bulk flow. The rate of the electroosmotic flow is governed by the following equation:

$$\mu_{EOF} = \frac{\epsilon}{4\pi\eta} E\zeta$$

where ε is the dielectric constant of the solution, η is the viscosity of the solution, E is the field strength, and ζ is the zeta potential. Because the electrophoretic mobility is greater than the electroosmotic flow, negatively charged particles, which are naturally attracted to the positively charged anode, will separate out as well. The EOF works best with a large zeta potential between the cation layers, a large diffuse layer of cations to drag more molecules towards the cathode, low resistance from the surrounding solution, and buffer with pH of 9 so that all the SiOH groups are ionized.

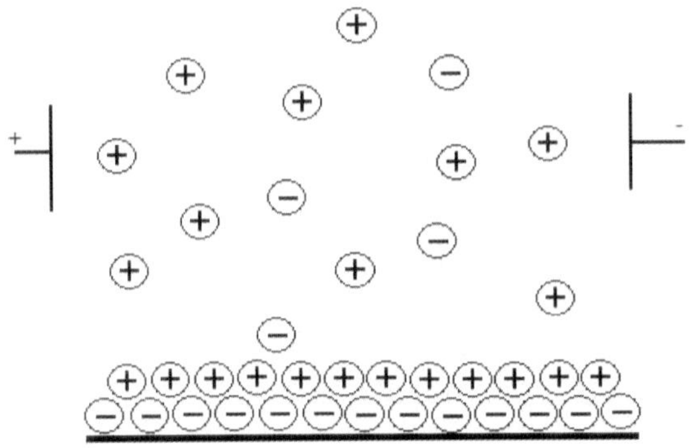

Electroosmotic Flow (EOF)

Figure: Electroosmotic Flow due to Applied Voltage

Capillary Electroseparation Methods

There are six types of capillary electroseparation available: capillary zone electrophoresis (CZE), capillary gel electrophoresis (CGE), micellar electrokinetic capillary chromatography (MEKC), capillary electrochromatography (CEC), capillary isoelectric focusing (CIEF), and capillary isotachophoresis (CITP). They can be classified into continuous and discontinuous systems as shown in figure. A continuous system has a background electrolyte acting throughout the capillary as a buffer. This can be broken down into kinetic (constant electrolyte composition) and steady-state (varying electrolyte composition) processes. A discontinuous system keeps the sample in distinct zones separated by two different electrolytes.

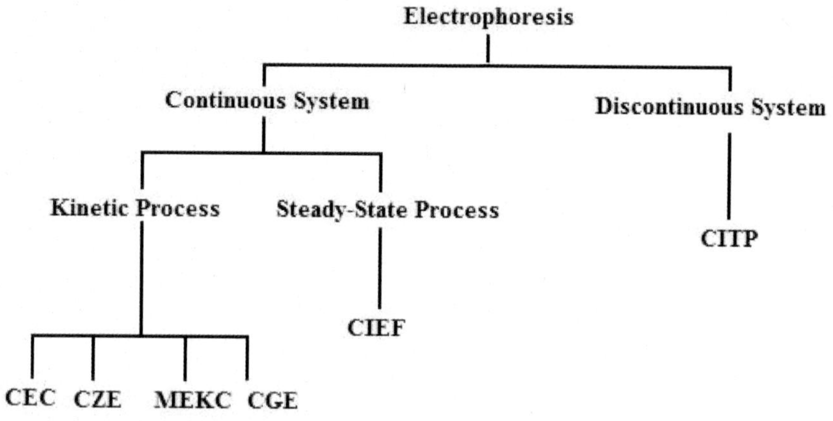

Figure: *Categorization of Electrophoresis Techniques*

Capillary Zone Electrophoresis (CZE)

Capillary Zone Electrophoresis (CZE), also known as free solution capillary electrophoresis, it is the most commonly used technique of the six methods.A mixture in a solution can be separated into its individual components quickly and easily.The separation is based on the differences in electrophoretic mobility, which is directed proportional to the charge on the molecule, and inversely proportional to the viscosity of the solvent and radius of the atom.The velocity at which the ion moves is directly proportional to the electrophoretic mobility and the magnitude of the electric field.

The fused silica capillaries have silanol groups that become ionized in the buffer. The negatively charged SiO^- ions attract positively charged cations, which form two layers—a stationary and diffuse cation layer. In the presence of an applied electric field, the diffuse layer migrates towards the negatively charged cathode creating an electrophoretic flow (μep) that drags bulk solvent along with it. Anions in solution are attracted to the positively charged anode, but get swept to the cathode as well. Cations with the largest charge-to-mass ratios separate out first, followed by cations with reduced ratios, neutral species, anions with smaller charge-to-mass ratios, and finally anions with greater ratios. The electroosmotic velocity can be adjusted by altering pH, the viscosity of the solvent, ionic strength, voltage, and the dielectric constant of the buffer.

Capillary Gel Electrophoresis (CGE)

CGE uses separation based on the difference in solute size as the particles migrate through the gel. Gels are useful because they minimize solute diffusion that causes zone broadening, prevent the capillary walls from absorbing the solute, and limit the heat transfer by slowing down the molecules. A commonly used gel apparatus for the separation of proteins is capillary SDS-PAGE. It is a highly sensitive system and only requires a small amount of sample.

Micellar Electrokinetic Capillary Chromatography (MEKC)

MEKC is a separation technique that is based on solutes partitioning between micelles and the solvent. Micelles are aggregates of surfactant molecules that form when a surfactant is added to a solution above the critical micelle concentration. The aggregates have polar negatively charged surfaces and are naturally attracted to the positively charged anode. Because of the electroosmotic flow toward the cathode, the micelles are pulled to the cathode as well, but at a slower rate. Hydrophobic molecules will spend the majority of their time in the micelle, while hydrophilic molecules will migrate quicker through the solvent. When micelles are not present, neutral molecules will migrate with the electroosmotic flow and no separation will occur. The presence of micelles results in a retention time to where the solute has little micelle interaction and retention time tmc where the solute strongly interacts. Neutral molecules will be separated at a time between to and tmc. Factors that affect the electroosmotic flow in MEKC are: pH, surfactant concentration, additives, and polymer coatings of the capillary wall.

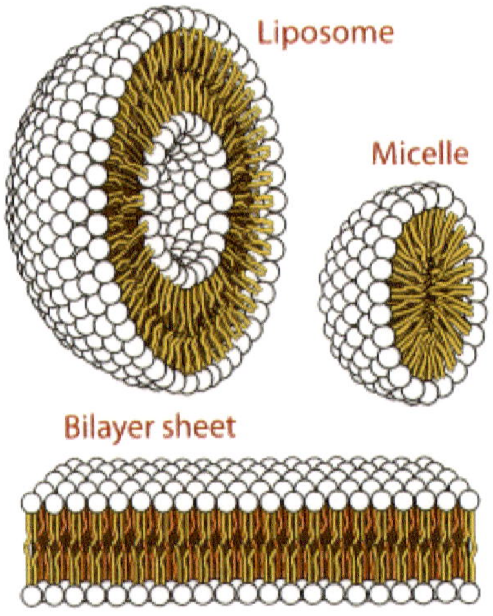

Figure: Micelles

Capillary Electrochromatography (CEC)

The separation mechanism is a packed column similar to chromatography. The mobile liquid passes over the silica wall and the particles. An electroosmosis flow occurs because of the charges on the stationary surface. CEC is similar to CZE in that they both have a plug-type flow compared to the pumped parabolic flow that increases band broadening.

Capillary Isoelectric Focusing (CIEF)

CIEF is a technique commonly used to separate peptides and proteins. These molecules are called zwitterionic compounds because they contain both positive and negative charges. The charge depends on the functional groups attached to the main chain and the surrounding pH of the environment. In addition, each molecule has a specific isoelectric point (pI). When the surrounding pH is equal to this pI, the molecule carries no net charge. To be clear, it is not the pH value where a protein has all bases deprotonated and all acids protonated, but rather the value where positive and negative charges cancel out to zero. At a pH below the pI, the molecule is positive, and then negative when the pH is above the pI. Because the charge changes with pH, a pH gradient can be used to separate molecules in a mixture. During a CIEF separation, the capillary is filled with the sample in solution and typically no EOF is used (EOF is removed by using a coated capillary). When the voltage is applied, the ions will migrate to a region where they become neutral (pH=pI). The anodic end of the capillary sits in acidic solution (low pH), while the cathodic end sits in basic solution (high pH). Compounds of equal isoelectric points are "focused" into sharp segments and remain in their specific zone, which allows for their distinct detection.

Calculating pI

An amino acid with n ionizable groups with their respective pKa values pK_1, pK_2, ... pk_n will have the pI equal to the average of the group pkas: $pI = (pK_1+pK_2+...+pk_n)/n$. Most proteins have many ionizable sidechains in addition to their amino- and carboxy- terminal groups. The pI is different for each protein and it can be theoretically calculated according to the Henderson-Hasselbalch approximation, if we know amino acids composition of protein. In order to experimentally determine a protein's pI 2-Dimensional Electrophoresis (2-DE) can be used. The proteins of a cell lysate are applied to a pH immobilized gradient strip, upon electrophoresis the proteins migrate to their pI within the strip. The second dimension of 2-DE is the separation of proteins by MW using a SDS-gel.

Dielectrophoresis

Dielectrophoresis is a phenomenon that occurs in a non-uniform electric field as a result of polarization, as shown in Fig. DEP force is generated by the interaction between the induced dipoles and a non-uniform electric field. The magnitude and direction of the DEP force is determined by the intrinsic dielectric properties of the particles and the medium. As mentioned, the characteristics of DEP force differ from those of other cell sorters that require an antibody antigen reaction. The magnitude and direction of the DEP force is expressed with the below equation.

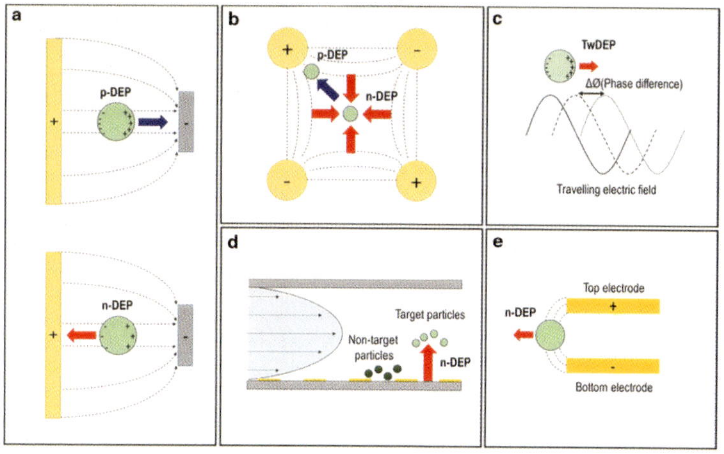

Figure: Schematic diagram of dielectrophoresis.

(a): A particle in a non-uniform electric field experiencing positive DEP force and negative DEP force, (b): a particle experiencing DEP trapping, (c): a particle in an electric field with a phase gradient (traveling wave dielectrophoresis), (d): target and non-target particles experiencing DEP-FFF, (e): a particle experiencing DEP barrier.

$$F_{DEP} = 2\pi r^3 \varepsilon_m \varepsilon_o \mathrm{Re}[f_{CM}] \nabla |E_{rms}|^2$$

where r indicates the radius of the target particle, ε_m and ε_o are the permittivity of the medium and the vacuum state, the term Re[f_{CM}] indicates the real parts of f_{CM}. f_{CM} is the Clausius–Mossotti (CM) factor, and $\boldsymbol{E}_{rms}$ is the root-mean-square of the electric field. Researchers transform equation $F_{DEP} = 2\pi r^3 \varepsilon_m \varepsilon_o \mathrm{Re}[f_{CM}] \nabla |E_{rms}|^2$ as necessary to various forms according to technique. For example, in the case of TwDEP force (Figure c), the magnitude of the acting force on a particle can be expressed as:

$$F_{TWD} = -\frac{4\pi\varepsilon_m \varepsilon_o r^3 \mathrm{Im}(f_{CM}) E^2}{4(w+g)}$$

where the term Im[f_{CM}] indicates the imaginary part, w is the width of the electrode, and g is the spacing between each electrode. Also, the force of the DEP barrier (Figure e), which is generated between top and bottom electrodes, can be approximated as:

$$F_{DEP} = \frac{27}{32}\pi^2 \varepsilon_m \mathrm{Re}[f_{CM}] r^3 \frac{U^2}{a^3}\left[1 + 0\left(\frac{r^2}{a^2}\right)\right]$$

where U is the applied root-mean-square voltage, and a is the channel height, which is the same as the spacing between the top and bottom electrode pair.

The direction of DEP force depends on the sign of the CM factor in equation $F_{DEP} = 2\pi r^3 \varepsilon_m \varepsilon_o \mathrm{Re}[f_{CM}] \nabla |E_{rms}|^2$, and it is defined as:

$$f_{CM} = \frac{\varepsilon_p^* - \varepsilon_m^*}{\varepsilon_p^* + 2\varepsilon_m^*}$$

$$\varepsilon_i^* = \varepsilon_0 \varepsilon_i - j\frac{\sigma_i}{\omega}, \text{ where, } i = m, p$$

where ε_p^* and ε_m^* denote the complex permittivity of the particle and the medium, j is the imaginary vector ($j = \sqrt{-1}$), σ is the conductivity of i, and ω is the angular frequency ($\omega = 2\pi f$) of the applied AC electric field. In the case of Re[f_{CM}] > 0, particles in a non-uniform AC electric field move toward the higher electric field, and this is called p-DEP force. The blue arrow in figure (a) indicates its direction. In the case of Re[f_{CM}] < 0, particles in an electric field are repelled from the higher electric field, and this is called n-DEP force. The red arrow in figure (a) indicates its direction. Hence, the input frequency is an important factor. In particular, the condition in which the frequency at the CM factor is zero is called cross-over frequency. This is expressed as:

$$f_{CO} \approx \frac{\sigma_m}{\sqrt{2}\pi r C_{mem}}$$

where σ m is the medium conductivity, and C mem is the capacitance per unit area of the cell plasma membrane. In other words, when an application employs n-DEP force, the input frequency should be selected properly to ensure that $\mathrm{Re}[f_{CM}] < 0$. Consequently, DEP force-based cell manipulating tools in the microfluidics can be designed based on the above equations.

DEP Trapping

Dielectrophoretic trapping is an application suitable for dealing with submicron particles such as bacteria and viruses. Its concept is shown in figure. In order to separate submicron particles, DEP trapping generally harnesses p-DEP force in a still fluid. Suehiro implemented a DEP impedance measurement (DEPIM) employing p-DEP to trap suspended biological cells [e.g. Escherichia coli (E. coli) and Serratia marcescens (Serratia)]. They trapped target cells within a still fluid and washed the non-target cells after agglutinating the target cells by antigen–antibody reaction. Also, they employed interdigitated patterned electrode arrays and trapped target cells spaced between electrodes deposited in parallel. Lapizco-Encinas presented insulator-based (electrodeless) dielectrophoresis (iDEP) to separate live and dead E. coli. In their paper, p-DEP and n-DEP force were selectively utilized. Since dead cells show much lower DEP mobility than live cells, dielectrophoresis is a more appropriate technique than those using other electrokinetic phenomena. Their DACS was designed to have numerous micron-sized posts, as shown in figure (i.e. extruded electrodes). Although DEP trapping is a useful tool to differentiate target particles, it has limitations. Rather than separating biological particles continuously, DEP tapping is a one-shot method. In addition, DEP trapping requires flushing to collect trapped target cells (or wash un-trapped non-target cells). These limitations cause low throughput and recovery rates in DEP tapping-based DACSes. In order to overcome these limitations, transporting techniques (TwDEP force) and particle sorting in a continuous flow (e.g. DEP-FFF) have been proposed.

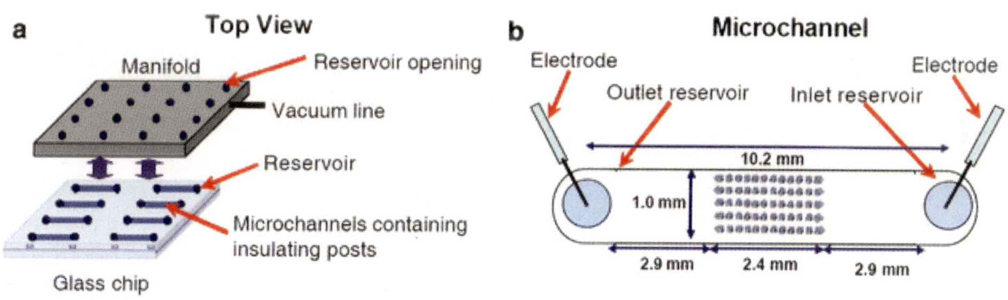

Traveling Wave Dielectrophoresis

Unlike other DACS techniques, traveling wave dielectrophoresis utilizes an AC electric field with a phase gradient. It is suitable for separating micron-sized particles such as blood cells. The phase shifts induce particle transportation along or against the direction of the travelling wave. For cell separation, it is a more adequate method than DEP trapping because the phase shifts enable particle motion even in a still fluid. Morgan introduced TwDEP force-based cell sorting by employing a multilayered electrode on a large area to achieve high throughput. Although they separated the components of whole blood, the large area of the configuration caused a resistance increase at the microelectrode. Therefore, Choi proposed a unique multilayered bus bar design to deliver an improved TwDEP force-based DACS, as shown in figure. They exploited a quadrature phase (0°, 90°, 180° and 270°) with four microelectrode tracks. In addition, in order to implement size-based particle sorting, a microelectrode track was designed with gradually increasing gaps from 10 to

40 μm. Consequently, they demonstrated separation feasibility by fractionating four different diametric latex particles (i.e. 3, 6, 10 and 20 μm) successfully. In order to break through the limitations of conventional DACSes, Cheng proposed a transformed cell sorting device employing two different electrode arrays that generated TwDEP force, as shown in figure. They achieved an approximate fourfold velocity increase by using a focusing mechanism that employed TwDEP force within a continuous flow. Conclusively, they showed that a TwDEP system could achieve a higher throughput than DEP trapping. However, since it required complicated electrode structures, they had difficulty in improving the recovery rate and reproducibility.

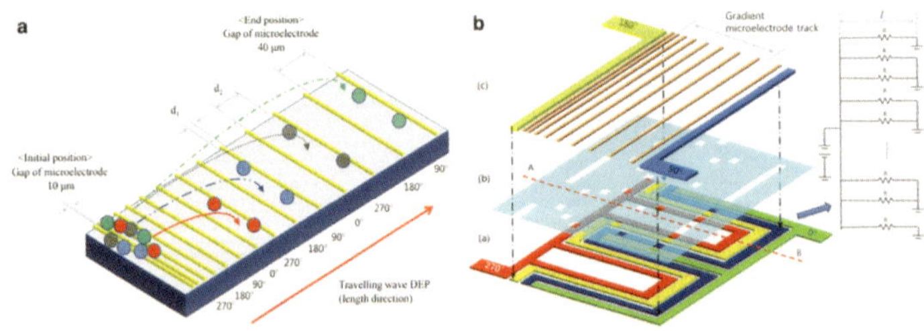

Figure: The proposed electrode with three layers by Choi (a) Schematic view of the high-throughput gradient TwDEP device, (b) three layers of the TwDEP device

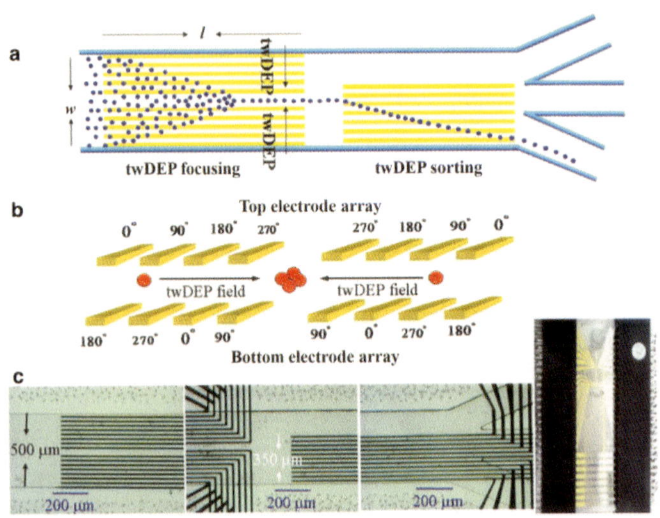

Figure: Schematic diagram of TwDEP force-based cell sorter by Cheng (a) Top view of electrode arrangement, (b) side view of electrode structure, (c) images of TwDEP focusing electrodes, TwDEP sorting electrodes and packaged chip

DEP Field Flow Fractionation

The working principle of the DEP field-flow fractionation is shown in figure. DEP-FFF is an outstanding technique because it was originally proposed to separate biological particles within a continuous fluidic flow. In other words, a DACS that harnesses DEP-FFF is able to perform continuous cell separation with the unceasing injection of target particles. Particle movement in the microchannel depends on three acting forces—DEP force, hydrodynamic force, and gravity. Particles passing through the main channel can experience sedimentation and levitation due to the gravity and n-DEP force, respectively. When the DEP force is stronger than gravity, the particles levitate diagonally rather than vertically due to the hydrodynamic force of the fluidic flow. Since DEP force depends on dielectric properties and the radius of the particles, multiple particle types with different sizes or dielectric properties can be fractionated. A detailed analysis concerning the acting force for different particles with DEP-FFF was implemented by Gascoyne and Wang Lewpiriyawong fabricated a PDMS-based micro-device utilizing DEP-FFF, and they demonstrated simulation studies. They selected an AgPDMS composite as their electrode material. In order to select the proper input voltage, they implemented a separation test under various voltage conditions (0, 25, 50 and 55 V) with latex polystyrene microspheres (5, 10 and 15 μm), as shown in figure. Since the AgPDMS caused voltage drops, they were able to select a higher voltage than in the simulation analysis. Consequently, they demonstrated size-dependent particle sorting within a continuous microfluidic flow. Piacentini performed the size-based separation of platelets from red blood cells, as shown in figure. In the same manner as that used by Lewpiriyawong they investigated the trajectory in the microchannel through a numerical simulation. Their device employed a "liquid electrode" defined by Demierre et al. which is a kind of planar electrode deposited at the bottom of the empty zone positioned vertically from the main channel. They achieved striking results, with a recovery rate of over 98 %. On the other hand, Gupta developed a device named ApoStream® to separate circulating tumor cells (CTCs) from normal and healthy blood cells in a microfluidic flow. They not only separated various CTCs (e.g. SKOV3 and MDA-MB-231), but also confirmed the viability of the completely separated cells. As a result, the viability of the cancer cells fractionated by Apostream® was greater than 97.1 %, and the throughput was reported as 5000 cells/s. Also, Apostream® has come close to being commercialized, and the device has been utilized in research fields. Recent news regarding its commercialization is continually being reported on their website (ApoCell).

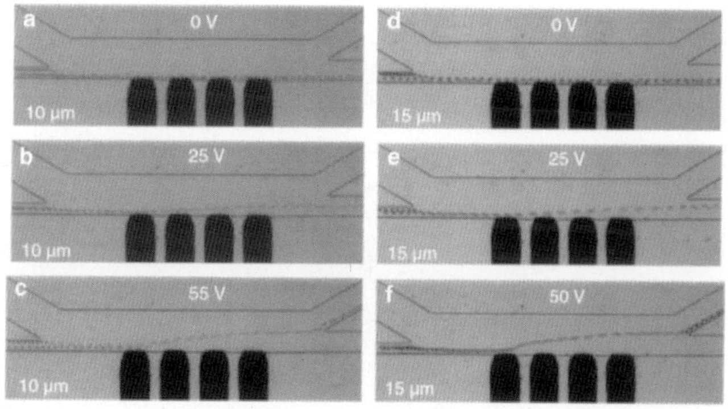

Figure: Video clip of particle levitation according to input voltage (10 μm particle at (a) 0 V, (b) 25 V, (c) 55 V, and 15 μm particle at (d) 0 V, (e) 25 V and (f) 50 V)

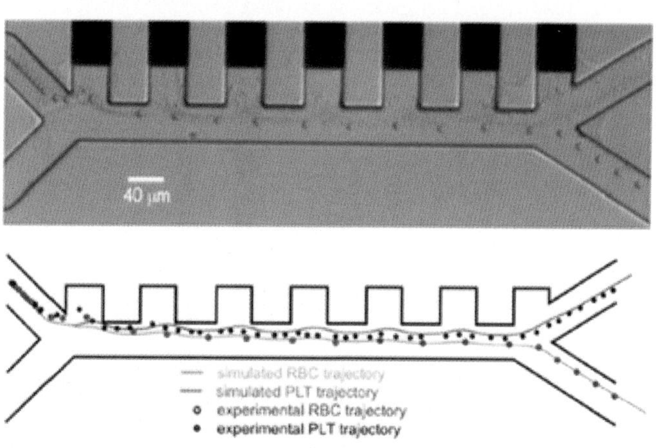

Figure: Trajectory of RBCs and PLTs in the device and graphical representation of trajectory between simulation and experiment

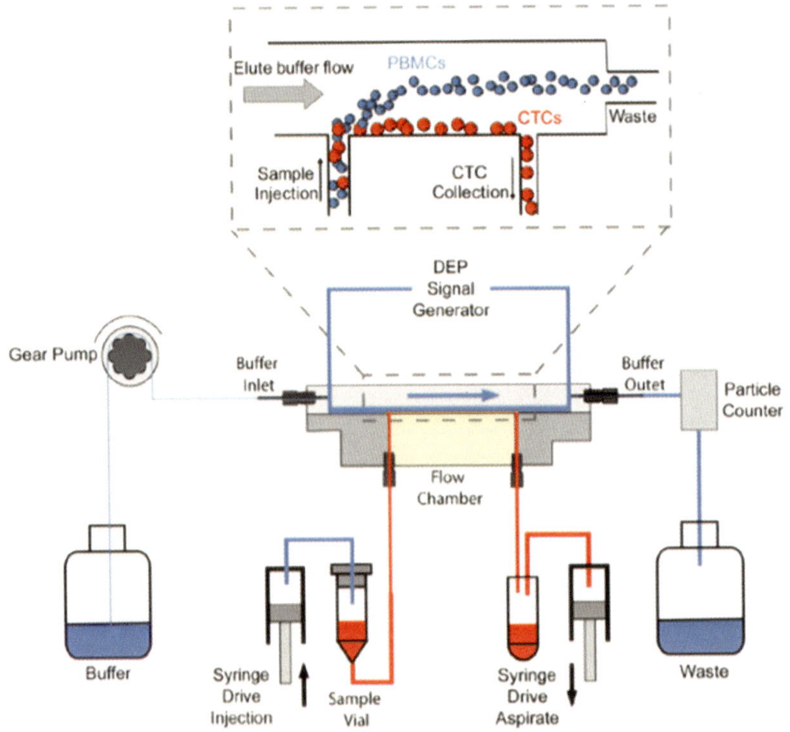

Figure: Schematic diagram of the Apostream

DEP Barrier

Another way to exploit a continuous flow is to generate a dielectrophoretic barrier to hinder cell movement at the main channel. Generally, top–bottom patterned electrodes are deposited to generate n-DEP force and consequently deflect target particles toward a target area. When the DEP force between the top and bottom electrodes (electrode pair) is sufficient, the angled electrode pair enables the deflection of the target particles along the electrode pair. DACSes harnessing DEP barrier always install repeated electrode pairs to guarantee high separation efficiency. This technique is convenient since it does not forcibly translocate but just switches the direction of the particles in the channel, though not against the fluidic flow. In addition, the DEP barrier technique has advantages in terms of size-based target particle selection because changes in the channel depth (distance between top and bottom electrodes) can control the magnitude of the DEP barrier under the same input conditions (input voltage and frequency), as shown in equation. Durr introduced a DACS that harnessed DEP barrier, as shown in figure. They reported an electrode pair as a 3-D deflector array. They analyzed the threshold velocity for particle size according to voltage variation, as well as an electric field in a numerical simulation. As shown in figure, Kim presented a multi-target DACS (MT-DACS), and they separated multiple biological particles (MC1061 strain of *E. coli*) harnessing the same mechanism. They confirmed their separation results using flow cytometry. Since three kinds of particles were selected as target cells, three outlets and two kinds of angled electrode groups with different angles were fabricated. Compared to single pass DACSes, they achieved a 1000-fold overall enrichment of the separated target cells. In addition, they found that there was no detectable cross-contamination near the outlet channels. Kim proposed a novel DACS with a vertically arrayed micro channel to harness gravity with cantilever type electrode arrays. Since the proposed DACS harnessed gravity to shunt the injected cells, they did not require an additional flow generator such as micro syringe pump (or other micro-device), which is utilized by almost DACSes. Also, for high throughput, which is a weak point in current DACSes, they designed a meso-sized channel (L × W×H = 2 mm × 0.5 mm × 25 mm) that differed from conventional micro-devices. They injected a mixture including three kinds of particles (10, 25 and 50 μm), and they investigated the sedimentation patterns of each colony. Consequently, they achieved a high separation efficiency of 94.7 %. Lee designed a simplified DACS to obtain high reliability. Their DACS consisted of a meso-sized channel (15 mm × 40 mm × 0.2 mm) encompassed by two plates that were as a patterned top–bottom electrode to generate DEP barrier as shown in figure. In the case of conventional DACSes, the large number of components complicates the assembly procedure. Hence they insisted that complex structures could cause a leaking phenomenon between components and reduce reproducibility and reliability. They determined the proper input condition based on a numerical simulation and demonstrated a separation test with K562 cells (cancer cells found in bone marrow). In the experiments, the DACS achieved a separation efficiency of 94.74 %, a throughput of 17,000 cells/min and a recovery rate of 49.42 %.

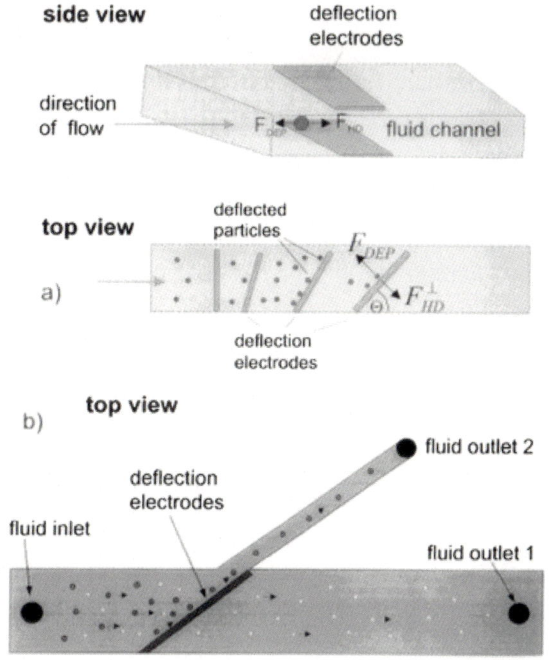

Figure: Schematic diagram of DACS harnessing DEP barrier

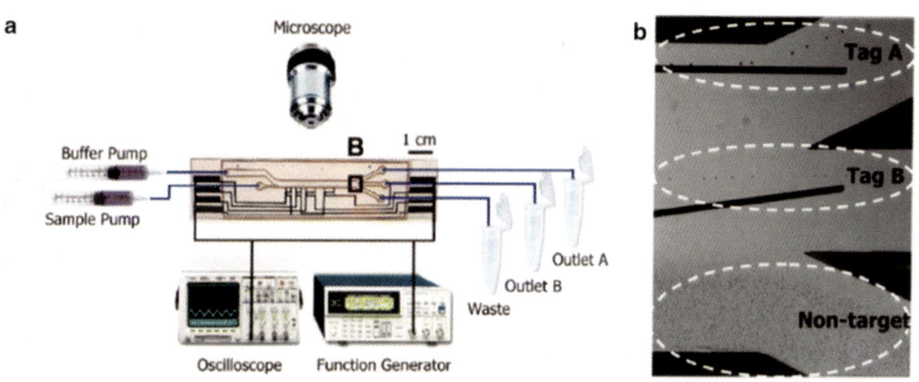

Figure: Experimental setup (a) and optically captured images (b) at the crossroads

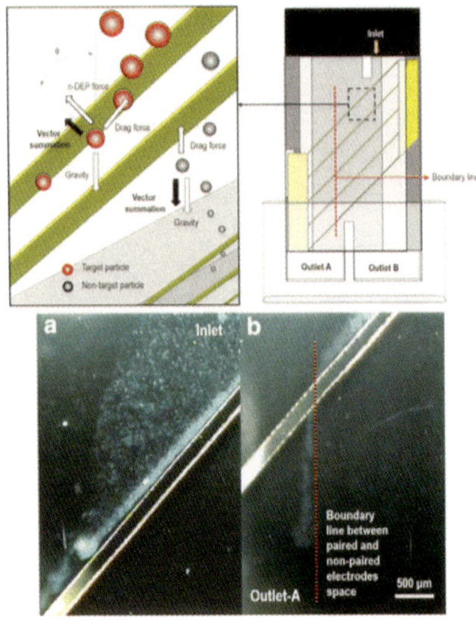

Figure: Simplified DACS configuration and video clip of separation process (a) Deflected K562 cells above the DEP barrier, (b) K562 cells sinks beyond boundary line

Free-flow Electrophoresis

Free Flow Electrophoresis (FFE) is an electrophoresis procedure working continuously in the absence of a stationary phase (or solid support material such as a gel). It separates preparatively charged particles ranging in size from molecular to cellular dimensions according to their electrophoretic mobilities (EPMs) or isoelectric points (pIs). Samples are injected continuously into a thin buffer film, which may be segmented or uniform, flowing through a chamber formed by two narrowly spaced glass plates. Perpendicularly to the electrolyte and sample flow, current may be applied while the fluid is flowing (continuous FFE) or while the fluid flow is transiently stopped (interval FFE). In any case the applied electric field leads to movement of charged sample components towards the respective counterelectrode according to their electrophoretic mobilities or isoelectric points. The sample and the electrolyte used for a separation enter the separation chamber at one end and the electrolyte containing different sample components as separated bands is fractionated at the other side.

Gel Electrophoresis

Gel Electrophoresis is a procedure used in molecular biology to separate and identify molecules (such as DNA, RNA, protein, complexes) by size. The separation of these molecules is achieved by placing them in a gel made up of small pores and setting an electric field across the gel. The molecules will move based on their inherent electric charge (i.e., negatively charged molecules move away from the negative pole) and smaller molecules will move faster than larger molecules; thus, a size separation is achieved within the pool of molecules running through the gel. The gel works in a similar manner to a sieve separating particles by size; the electrophoresis works to move the particles, using their inherent electric charge, through the sieve.

Purpose of Gel Electrophoresis

The purpose of gel electrophoresis is to visualize, identify and distinguish molecules that have been processed by a previous method such as PCR, enzymatic digestion or an experimental condition. Often, mixtures of nucleic acids or proteins that are collected from a previous experiment/method are run through gel electrophoresis to determine identity or differentiate molecules.

Types of Gel Electrophoresis

There are two types of gel electrophoresis: native and denaturing. A native gel electrophoresis usually attempts to keep RNA or protein in its native structure while running it through the gel. A denaturing gel electrophoresis attempts to reduce the RNA or protein into its most linear structure before or during gel electrophoresis. The denaturation of the RNA or protein is accomplished by adding a reducing agent to the sample, gel and/or buffer. The reducing agent separates bonds within the RNA or protein molecule and thereby reduces its secondary structure. The secondary structure of a protein or RNA will influence, in a non-linear manner, how fast it migrates through a gel. A denatured, linear form of RNA or protein, however, will migrate proportionally to its linear size (base pairs or kilo Daltons). Denaturing gel electrophoresis is often more accurate for size identification, whereas native gel electrophoresis is usually used to identify protein complexes.

Examples of Gel Electrophoresis

- TAE Agarose Gel Electrophoresis is most commonly used for DNA.
- TBE and Denaturing PAGE (polyacrylamide gel electrophoresis) are common for RNA separation.
- SDS PAGE is a denaturing gel electrophoresis commonly used for protein identification and separation.

Gel Electrophoresis Steps

The broad steps involved in a common DNA gel electrophoresis protocol:

1. Preparing the samples for running:

   The DNA is isolated and preprocessed (e.g. PCR, enzymatic digestion) and made up in solutionwith some basic blue dye to help visualize the movement of the sample through the gel.

2. An agarose TAE gel solution is prepared:

   TAE buffer provides a source of ions for setting up the electric field during electrophoresis. A weight to volume concentration of agarose in TAE buffer is used to prepare the solution. For example, if a 1% agarose gel is required, 1g of agarose is added to 100mL of TAE. The agarose percentage used is determined by how big or small the DNA is expected to be. If one is looking at separating a pool of smaller size DNA bands (<500bp), a higher percentage agarose gel (>1%) is prepared. The higher percentage agarose creates a denser sieve to increase separation of small DNA length differences. The agarose-TAE solution is heated to dissolve the agarose.

3. Casting the gel:

   The agarose TAE solution is poured into a casting tray that, once the gel solution has cooled down and solidified, creates a gel slab with a row of wells at the top.

4. Setting up the electrophoresis chamber:

   The solid gel is placed into a chamber filled with TAE buffer. The gel is positioned so that the chamber wells are closest to the negative electrode of the chamber.

5. Loading the gel:

   The gel chamber wells are loaded with the DNA samples and usually a DNA ladder is also loaded as reference for sizes.

6. Electrophoresis:

   The negative and positive leads are connected to the chamber and to a power supply where voltage is set. Turning on the power supply sets up the electric field and the negatively charged DNA samples will start to migrate through the gel and away from the negative electrode towards the positive.

7. Stopping electrophoresis and visualizing the DNA:

   Once the blue dye in the DNA samples has migrated through the gel far enough, the power supply is turned off and the gel is removed and placed into an ethidium bromide solution. Ethidium bromide intercalates between DNA and is visible in UV light. Sometimes ethidium bromide is added directly to the agarose gel solution in step 2. The ethidium bromide stained gel is then exposed to UV light and a picture is taken. DNA bands are visualized in from each lane corresponding to a chamber well. The DNA ladder that was loaded is also visualized and the length of the DNA bands can be estimated. An example is given in the figure below.

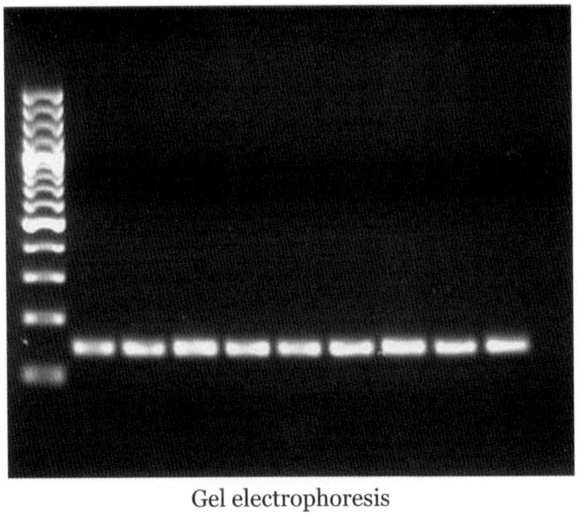

Gel electrophoresis

Differencial Gel Electrophoresis

DIGE – Differential Gel Electrophoresis – is a variant of gel electrophoresis that allows for simultaneous separation of up to three samples on one gel, bringing a new level of statistical confidence and reliability to 2D gel electrophoresis.

The system uses fluorescent cyanine dyes covalently bound to the proteins prior the 2D gel electrophoresis process. The dyes (Cy™2, Cy3 and Cy5) are mass and charge-matched, but have unique fluorescent properties. Therefore the labeled proteins migrate simultaneously on the 2-DE gel but still produce distinct excitation and emission spectra.

This means multiple protein samples labeled with different dyes can be co-migrated on the same gel. The gel is scanned using an adapted fluorescence scanner or camera system to generate overlaid, multi-channel images for each gel.

There are two different CyDye™ DIGE Fluor dyes available.

- The minimal dyes are NHS ester dyes that label lysine residues, and are recommended for normal applications. Multiplexing up to three samples and labeling of 50 µg protein is possible.
- The saturation dyes are maleimide dyes that label cysteine residues, and should be used when samples are precious or available in very small amounts. Multiplexing up to two samples and labeling of 5 µg protein is possible.

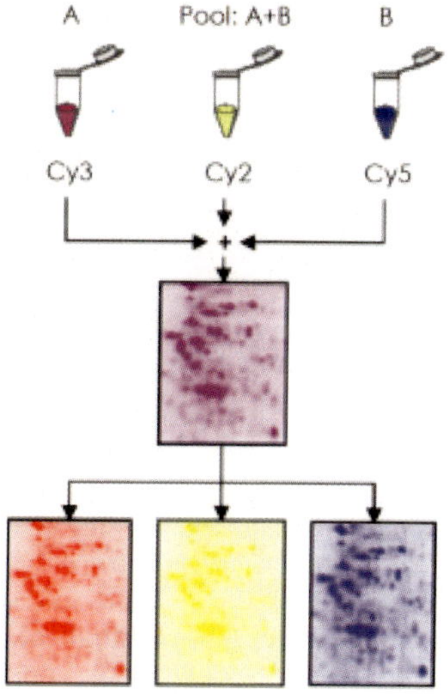

Internal standard

This type of methodology allows the inclusion of an internal standard that ideally is a pool of all the samples within the experiment. The internal standard is labeled with one of the cyanine dyes (typically Cy2 for minimal dyes and Cy3 for saturation dyes) and run on each gel in the experiment. This creates an image that is the average of all experimental samples, with all proteins in the experiment represented. The presence of the internal standard in every DIGE gel provides an inherent link between samples that can be exploited to match and normalize the protein quantities across samples.

Advantages

The advantages of the DIGE technology using an internal standard are:

- Matching between gels is more straightforward. Matching between images of the same gel is implicit because running differences are eliminated for the corresponding samples. Matching between gels can be performed between internal standard images, for which variations in spot patterns are due only to electrophoretic differences.
- The use of the internal standard virtually eliminates experimental gel-to-gel variation and enables accurate quantification of induced biological change between samples. Quantitative comparisons of protein between samples are made based on the relative change of each protein spot to its own in-gel internal standard.
- No technical replicates need to be run to confirm differences in protein abundance, thus reducing the number of gels required per experiment as well as analysis time and cost.

Two-dimensional gel electrophoresis

Two-dimensional gel electrophoresis, abbreviated as 2-DE, is a form of gel electrophoresis commonly used to analyze mixtures of proteins by two properties in two dimensions respectively. In two-dimensional gel electrophoresis, proteins are first separated by their pI in isoelectric focusing and then further separated by molecular weight through SDS-PAGE, thus the sample proteins are distributed across the two-dimensional gel profile. This technique expands the number of proteins that can be identified, and provides more efficient data and detailed information for proteomics analysis.

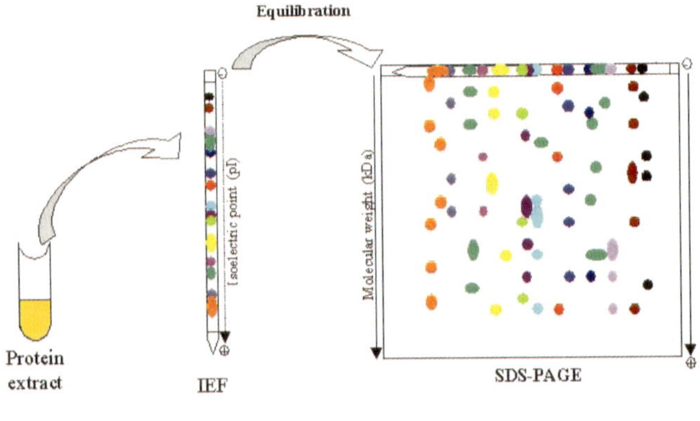

2-D electrophoresis begins with 1-D electrophoresis and then separates the molecules by a second property in a direction 90 degrees from the first. In 1-D electrophoresis, the proteins separated in one dimension will lie along a lane, and then the molecules are spread out across in the 2-D gel. Generally, it is unlikely that two molecules will be similar in both two distinct properties, so molecules are more effectively separated in 2-D electrophoresis than in 1-D electrophoresis.

2DE uses these four core steps: sample solubilization, isoelectric focusing (IEF), SDS-PAGE, and – whew. – analysis.

Step 1: Sample Solubilization

Much like regular ol' SDS-PAGE, those tissue or blood samples for 2DE need to be processed and solubilized before they can be loaded into the IEF gel. In an ideal world, all proteins would solubilize immediately with no qualitative or quantitative changes. In the real world, not all proteins are equal, and this first step immediately tilts 2DE towards detecting highly soluble and abundant proteins. Don't reach for the Tween and Triton, however; solubilization requires IEF-compatible lysis reagents, such as electrically neutral detergents (e.g. CHAPS) and chaotropes like urea. Obnoxiously enough, urea can carbamylate proteins, while those pesky detergents can interfere with downstream steps. As a result, this step requires a great deal of sample-specific optimization to avoid bias and contamination.

Step 2: Isoelectric Focusing

At last, your sample is solubilized. Time to load it on an IEF gel where, similar to SDS-PAGE, the proteins will be pushed through the acrylamide gel by an electric field. Where IEF gets more exciting is that the gel incorporates a pH gradient, and each protein moves only until it reaches its isoelectric point (pI). The pI is the pH where a protein has no net charge, meaning the field has no effect and the protein stays put, focusing tightly into a band within 0.01 pH unit of its pI. As straightforward as this sounds, IEF throws several wrenches into the scientific works. First, proteins become less soluble and can even precipitate out as they move closer to their pI, especially in low-salt, IEF-friendly buffers. Second, IEF gels and buffers interfere with sample prep for mass spectrometry (MS) and can be difficult to stain for analysis. This means IEF must almost always be done first so that SDS-PAGE can make the sample MS-compatible. Finally, it's very easy to contaminate your sample with keratin, so this step requires gloves, diligent depilation, and working behind a "sneeze shield."

Step 3: SDS-PAGE

Finally, the second dimension! But first, those beautifully focused proteins need to be solubilized again in SDS before they can be separated by their molecular weights on an orthogonal second axis. Other than this equilibration and some additional care with timing and voltages, this step uses molecular weights to separate proteins through the tried-and-true SDS-PAGE method. At the end, the gel will have the proteins aligned along two axes: isoelectric point vs. molecular weight.

Step 4: Analysis

This final step depends on your particular experimental endpoints: are you comparing protein expression? and/or identifying proteins? When comparing protein expression across different experimental samples, the gels are typically stained with silver or Coomassie blue for total protein. Various image analysis platforms are then used to scan and compare the location and intensities of the separated proteins. Naturally, analysis is not as clean and simple as slapping a gel into the scanner. To improve reproducibility between samples, gels should be run in parallel, which can lead to logistic and diplomatic challenges in a lab with only one power supply and an imminent lab meeting. Furthermore, interpreting that Morse-like pattern of dots and splotches requires protein identification and annotation beforehand, either through in-house efforts or via sweet, sweet annotated databases of similar samples.

There are two general options for those in the protein identification phase. The first is to run a western blot, transferring the proteins from the SDS-PAGE gel. The second approach – which is more powerful, laborious and expensive – is to excise the proteins in the gel, digest them, and send them out for identification by MS. While this can yield buckets and buckets of high-resolution information – for example, the type of post-translational modifications or disease-related adducts to proteins, amazingly enough – there are stringent requirements for a sample to be MS-worthy. Clean-up steps are typically required to remove the detergents and protein stains, and some stains – notably silver, which often requires development with a formaldehyde solution – interfere with MS sample preparation. Furthermore, MS requires typically several hundred nanograms of protein per sample, which means your protein of interest must be visible by the less-sensitive Coomassie blue stain.

Isoelectric Focusing

Isoelectric focusing (IEF) also known simply as electrofocusing, is a technique for separating charged molecules, usually proteins or peptides, on the basis of their isoelectric point (pI), i.e., the pH at which the molecule has no charge. IEF works because in an electric field molecules in a pH gradient will migrate towards their pI.

In most cases, a commercially available immobilized pH gradient (IPG) strip is used. The IPG strip consists of an acrylamide gel that contains wide pores to prevent a sieving effect based on protein mass, with a pH gradient. Various gradients are available, with wider gradients, such as pH 3-10 that are used for whole proteome analysis, and narrower ranges, such as pH 5-8 that are used for more specialist applications.

The sample is usually combined with carrier ampholytes to assist in migration. Ampholytes are a mixture of charged molecules with a range of pIs that matches the pI range of the IPG strip. The migration of the ampholytes encourages the sample molecules to move along the pH gradient. Ampholyte mixtures of a variety of pI ranges are commercially available.

After separation across the pH gradient, the sample is further separated (in 2D-PAGE) or analyzed.

Applications of IEF

IEF is "traditionally" used as first stage separation for 2D-PAGE, separating proteins by charge prior to second dimension separation by SDS-PAGE. This additional separation allows resolution of a couple of thousand proteins on a 2D-PAGE gel–enough for the entire proteome of an organelle or bacterium. It can also be used to examine post-translation modifications of proteins.

Another use for IEF is for fractionation of proteins or peptides prior to mass spec. Previously, it was difficult to recover molecules separated by IEF. However, with the Agilent OFFGEL system, the proteins or peptides remain in solution rather than being trapped in the gel as with standard IEF. Electrofocusing of proteins in this way provides a convenient alternative to SDS-PAGE for sample fractionation prior to mass spec.

Using IEF to separate peptides also provides an alternative to strong cation exchange (SCX) fractionation, which is one of the most popular techniques for the separation of peptides prior to mass spectrometry. IEF separation of peptides has been shown to result in more peptide identifications from whole proteome samples than with SCX for some samples.

Get Specialized for Practical Applications.

A more specialist version of IEF is capillary IEF (CIEF). This is more challenging but can be a useful way to accurately determine a protein's pI. It can also be useful for a variety of practical applications, such as accurate analysis of erythropoietin, QC of monoclonal antibodies, and for analysis of hemoglobin charge variants as an indicator of the severity of diabetes.

How it Works in Practice

Setting up an IEF experiment is probably easier than understanding the theory behind what's actually happening.

1. The IPG strips are rehydrated (face down) in a denaturing buffer (>6 M urea) with detergent, usually including carrier ampholytes (it's a case of trial and error, but for many samples, it works better to add the ampholytes with the sample).
2. There are 2 ways to actually load the sample–you either include it in the rehydration buffer or you load it in a small plastic cup at the end of strip (again, some samples work better with one or other way of loading–give both a try and see what works best for you).
3. The strips are then placed in the IEF apparatus, filter paper wicks are placed over the ends of the gel (these collect salts and proteins/peptides that are outside of the pI range of the IEF strip).
4. Electrodes are placed on top—you're ready to run.

Voltages used differ greatly for different systems, so it is advisable to do your homework in advance and look it up in the manual (or ask a lab mate)—but make sure you use the right voltage for the length strip you are using.

Moving-boundary Electrophoresis

In moving boundary electrophoresis the ions move through a simple liquid. Think of a beaker containing buffer and two electrodes. If an electric field is induced and charged molecules are added to the buffer solution, those molecules will move.

While there are specialized applications for moving boundary electrophoresis (e.g., capillary electrophoresis), it is not a common lab technique for two major reasons.

1. Foremost, the resolution between different populations of molecules (bands) can be lost due to fluid motion if special precautions are not taken. Fluid motion can occur through physical disturbance (e.g., waves induced by mechanical vibrations) or through thermal convection (Joule heating).

2. The time and electrophoretic distance needed to separate molecules with similar charges and masses may be impractical.

Kinetic Capillary Electrophoresis

Kinetic capillary electrophoresis (KCE) constitutes a toolset of homogeneous kinetic affinity methods for measuring rate constants of formation (k_+) and dissociation (k_-) of non-covalent biomolecular complexes, C, formed from two binding partners, A and B. A parameter-based approach of extracting k_+ and k_- from KCE electropherograms relies on a small number of experimental parameters found from the electropherograms and used in explicit expressions for k_+ and k_- derived from approximate solutions to mass transfer equations. Deriving the explicit expressions for k_+ and k_- is challenging but it is justified as the parameter-based approach is the simplest way of finding k_+ and k_- from KCE electropherograms. Here, we introduce a unique approximate analytical solution of mass transfer equations in KCE termed a "two-peak approximation" and a corresponding parameter-based method for finding k_+ and k_-. The two-peak approximation is applicable to any KCE method in which: (i) A* binds B to form C* (the asterisk denotes a detectable label on A), (ii) two peaks can be identified in a KCE electropherogram and (iii) the concentration of B remains constant. The last condition holds if B is present in access to A* and C* throughout the capillary. In the two-peak approximation, the labeling of A serves only for detection of A and C and, therefore, is not required if A (and thus C) can be observed with a label-free detection technique. The proposed two-peak approximation was studied, in particular, its accuracy, by using the simulated propagation patterns built with the earlier-developed exact solution of the mass-transfer equations for A* and C*. The results prove that the obtained approximate solution of mass transfer equations is correct. They also show that the two-peak approximation facilitates finding k_+ and k_- with a relative error of less than 10% if two peaks can be identified on a KCE electropherogram. Importantly, the condition of constant concentration of B is always satisfied in macroscopic approach to studying kinetics at equilibrium (MASKE) whether or not B is in excess to A* and C*, and, thus, the two-peak approximation is applicable to MASKE. It completes a toolset of fitting-free methods for processing MASKE data and makes MASKE a simple practical method for finding k_+ and k_- of "fast", "slow", and "intermediate-rate" non-covalent interactions.

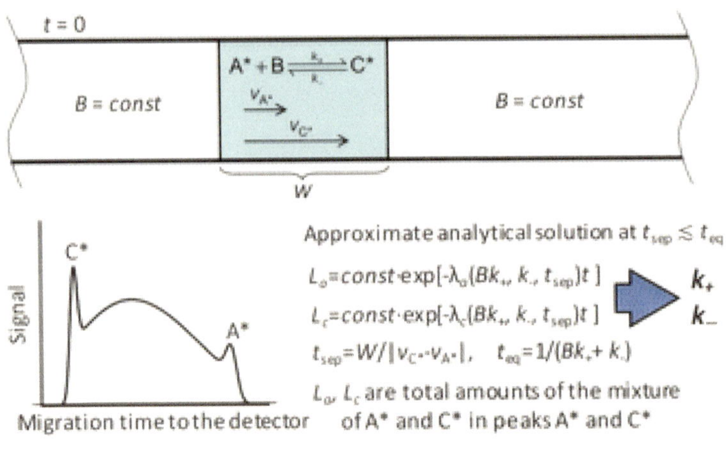

References

- Electrophoresis: chemicool.com, Retrieved 21 June 2018
- Gel-electrophoresis: biologydictionary.net, Retrieved 24 March 2018
- 2d-electrophoresis, services: creative-proteomics.com, Retrieved 16 May 2018
- How-does-2d-gel-electrophoresis-work-9606: bitesizebio.com, Retrieved 20 June 2018
- Isoelectric-focusing-separation-proteins-peptides-26811: bitesizebio.com, Retrieved 30 April 2018

Chromatography

Chromatography is a technique that is used for the separation of a mixture into its constituents. It can either be analytical or preparative. Some of the diverse topics covered in this chapter include high-performance liquid chromatography, electrochromatography, centrifugal partition chromatography, supercritical fluid chromatography, etc. which will provide an overview of the different chromatographic techniques.

'Chromatography' is an analytical technique commonly used for separating a mixture of chemical substances into its individual components, so that the individual components can be thoroughly analyzed. There are many types of chromatography e.g., liquid chromatography, gas chromatography, ion-exchange chromatography, affinity chromatography, but all of these employ the same basic principles.

Principles of Chromatography

Let's first familiarize ourselves with some terms that are commonly used in the context of chromatography:

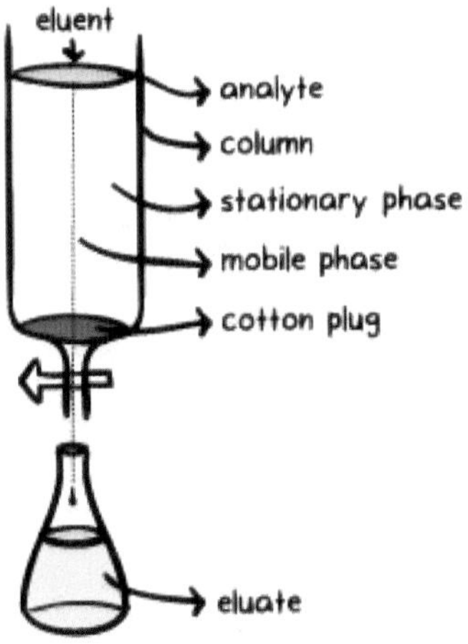

Term	Definition
Mobile phase or carrier	solvent moving through the column
Stationary phase or adsorbent	substance that stays fixed inside the column
Eluent	fluid entering the column
Eluate	fluid exiting the column (that is collected in flasks)
Elution	the process of washing out a compound through a column using a suitable solvent
Analyte	mixture whose individual components have to be separated and analyzed

Now let's try to understand the principle of chromatography. Let us draw a pictorial representation of a column chromatographic separation set up.

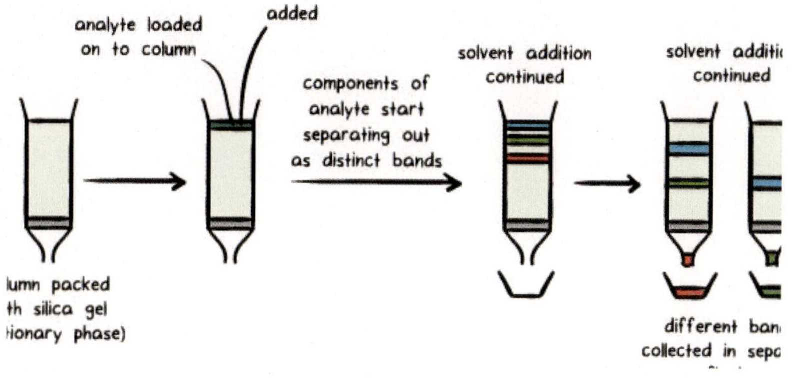

As depicted above, the analyte is loaded over the silica bed (packed in the column) and allowed to adhere to the silica. Here, silica acts as the stationary phase. Solvent (mobile phase) is then made to flow through the silica bed (under gravity or pressure). The different components of the analyte exhibit varying degrees of adhesion to the silica and as a result they travel at different speeds through the stationary phase as the solvent flows through it, indicated by the separation of the different bands. The components that adhere more strongly to the stationary phase travel more slowly compared to those with a weaker adhesion. Analytical chromatography can be used to purify compounds ranging from milligram to gram scale.

Before we move on, let's conduct a simple experiment to exemplify the power of a chromatographic separation.

1. Take a few leaves and crush them in a mortar.
2. Spot a drop of the leaf extract on a strip of chromatographic paper ~ 0.5 cm above the edge of the paper. Chromatographic paper is made of cellulose and is quite polar in nature.
3. Place the strip of paper in a jar that contains a small volume of propanone (acetone). There should be just enough propanone that the edge of the paper dips in it comfortably. Place a lid on the jar to avoid any evaporation of the solvent.
4. Let the solvent rise up the paper by capillary action. Remove the paper strip from the jar once the solvent has reached the 'solvent front' level. 5) What do you think you will notice?

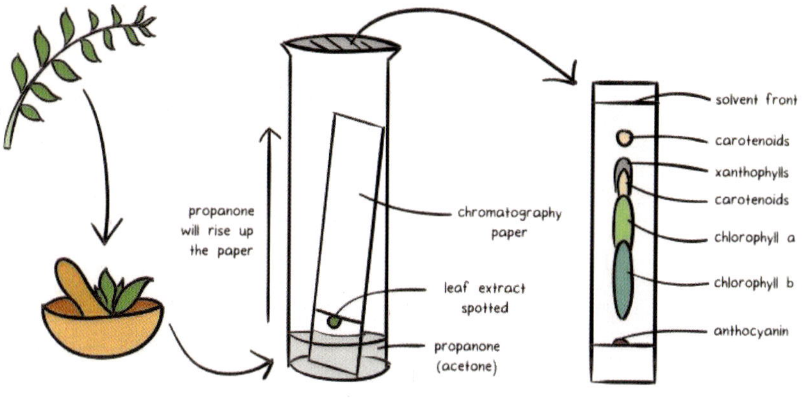

The various components of the leaf pigment separate out. Could you have ever imagined that a leaf pigment was made up of so many compounds?

Principle of separation of different components: Differential affinities (strength of adhesion) of the various components of the analyte towards the stationary and mobile phase results in the differential separation of the components. Affinity, in turn, is dictated by two properties of the molecule: 'Adsorption' and 'Solubility'.

We can define adsorption as the property of how well a component of the mixture sticks to the stationary phase, while solubility is the property of how well a component of the mixture dissolves in the mobile phase.

- Higher the adsorption to the stationary phase, the slower the molecule will move through the column.
- Higher the solubility in the mobile phase, the faster the molecule will move through the column.

So, the interplay between the above two factors determines the differential rates at which the different components of the analyte will move through the column. Adsorption and solubility of a molecule can be manipulated by choosing the appropriate stationary phase and mobile phase.

Now, the question arises why do different compounds possess different affinities towards the stationary and mobile phases? "Polarity" of the compounds dictates their affinities towards the stationary and mobile phases. Let's understand this through an example.

Suppose we have a mixture of two molecules A and B, where 'A' is a protein and 'B' is a lipid. Our column is packed with silica, which is polar in nature; our mobile phase is hexane, which is non-polar in nature. What do you think will happen when we load this mixture of A and B onto this column?

'A', being polar in nature, will adsorb on to the polar stationary phase (silica). 'B' being non-polar in nature, will readily dissolve in the non-polar mobile phase (hexane) without adhering to silica, and will thus elute out of the column with hexane. Once B is eluted out, the mobile phase will be changed to something polar like acetonitrile. By doing so we will now force A to detach from the silica and dissolve in the polar solvent, acetonitrile, and get eluted out of the column with acetonitrile. This is illustrated in the diagram below.

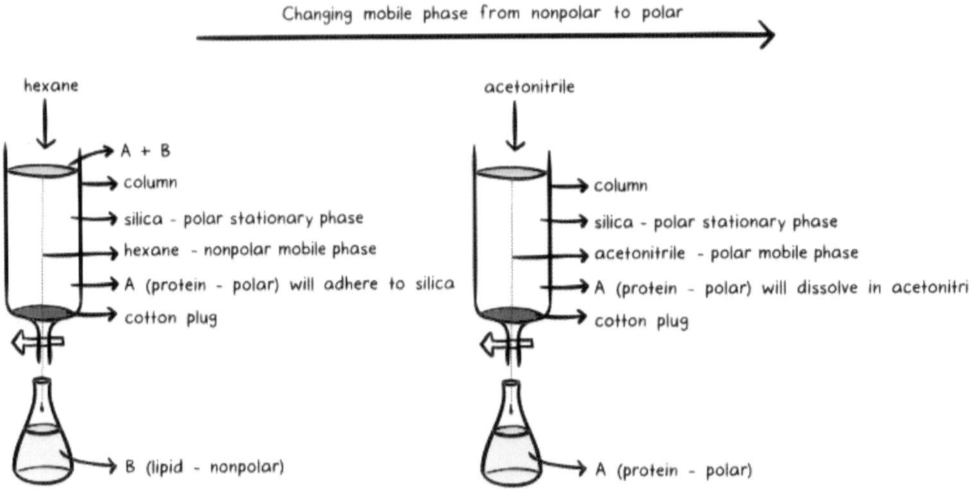

High-performance Liquid Chromatography

High-performance liquid chromatography or high-pressure liquid chromatography (HPLC) is a chromatographic method that is used to separate a mixture of compounds in analytical chemistry and biochemistry so as to identify, quantify or purify the individual components of the mixture.

Reversed phase HPLC/UHPLC chromatography is a commonly used separation mode. It provides dynamic retention of compounds possessing hydrophobic and organic functionality. A combination of hydrophobic and van der Waals type interactions between all the target compound and both the stationary and mobile phases enables the retention of these compounds by reversed phase.

Working of HPLC

In very small amounts, the sample mixture to be separated and tested is sent into a stream of mobile phase percolating via the column. There are different types of columns available with sorbents of varying particle sizes and surfaces.

The mixture moves through the column at varying velocities and interact with the sorbent, also known as the stationary phase. The velocity of each component in the mixture depends on its chemical nature, the nature of the column and the composition of the mobile phase. The time at which a specific analyte emerges from the column is termed as its retention time. The retention time is measured under specific conditions and considered as the identifying characteristic of a given analyte.

Sorbent particles might be hydrophobic or polar in nature. The commonly used mobile phases include any miscible combination of water and organic solvents such as acetonitrile and methanol. Water-free mobile phases can also be used.

The aqueous component of the mobile phase might contain acids like formic, phosphoric or trifluoroacetic acid or salts to enable the separation of the sample components. The composition of the mobile phase is either maintained as a constant or as varied during the chromatographic analysis. The constant approach is effective for the separation of the sample components that are not very dissimilar in their affinity for the stationary phase. In the varied approach, the composition of the mobile phase differs from low to high eluting strength. The eluting strength of the mobile phase is reflected by analyte retention times where high eluting strength produces fast elution.

The composition of the mobile phase is chosen based on the intensity of interactions between several sample components and the stationary phase.

The HPLC partitioning process is quite similar to the liquid-liquid extraction process except that the former is a continuous process unlike the latter which is a step-wise process. It is recommended that trial partitioning processes be performed to determine the exact HPLC method that would provide adequate separation.

Benefits of HPLC

The key benefits of HPLC systems are as follows:

- Controls and automates chromatography instrumentation.

- Provides data management, security features, and reporting and instrument validation.
- Powerful and adaptable.
- Increases productivity by managing all the areas of analysis - from sample to instrument, and from separation to reporting results.
- Affordable.

Applications of HPLC

HPLCs can be used in the following applications:

- Water purification.
- Preconcentration of trace components.
- Ligand-exchange chromatography.
- Ion-exchange chromatography of proteins.
- High-pH anion-exchange chromatography of carbohydrates and oligosaccharides.

Partition Chromatography

The chromatographic technique that involves the separation of components between two liquid phases; original solvent and film of solvent used in column is called as partition chromatography. In this technique the molecules get preferential separated between two liquid phases. One of the liquid phases is stationary phase whereas another is mobile phase in liquid state.

The components of mixture get dispersed into both phases preferentially. Remember polar molecules get partitioned into polar phase and non-polar molecules get attracted towards non-polar phase only. This method can be applied to Liquid-liquid, liquid-gas chromatography only not on solid-gas type as solid phase cannot move.

Partition Chromatography Technique

Partition chromatography is a type of liquid chromatography which is used to separate components of a given sample. The separation of components is based on the interactions of the particles of sample with the mobile and stationary phases.

In partition chromatography, both phases are in liquid state.

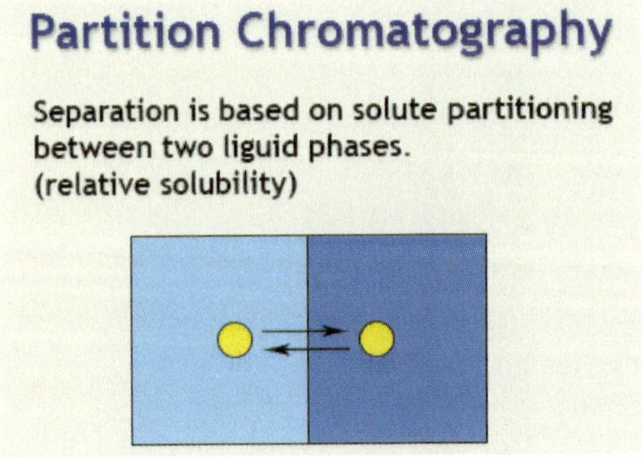

Partition Chromatography Principle

The principle of partition chromatography is that the separation of components of given sample occurs due to partition of components between two liquid phases. In this process, the stationary phase is coated with a liquid surface. The solid surface must be immiscible in the mobile phase. Due to partition of component between both phases reduces some components compare to others.

It results the separation of components. Here the stationary phase immobilizes the liquid surface and makes it stationary phase. The mobile phase passes over the stationary phase and separate out. The separation depends on the relative solubility in the stationary liquid layer. Because of different partition coefficient, different component of sample are separated.

Partition Chromatography Procedure

In the partition chromatography, the separation of components of the mixture occurs due to distribution of components between two liquid phases. The separation of components occurs on the basis of partition coefficients of constituents of mixture. The flow of components or mobile phase separates all components over stationary phase.

The polar phase attracts polar components of sample and non-polar components are attracted by non-polar phase of the system. Partition chromatography can be done with different phases like Liquid-liquid and liquid-gas chromatography.

Partition Chromatography Applications

Partition chromatography is one of best chromatographic method to get high performance and pure components. Partition Chromatography is used for final purification natural extracts, synthetic mixtures and biological matrices. It is also used for fractionization of complex crude extracts.

Normal–phase Chromatography

Normal phase chromatography is one where in the stationary phase is polar in nature and the mobile phase is non-polar in nature. Chromatography is a separation process which employs two phases i.e. stationary phase and a mobile phase.

The stationary phase is water soluble and polar in nature. While the mobile phase is water insoluble and less polar in nature. In the initial phases of discovery and development of chromatography, silica was widely used as the stationary phase.

This silica gel is polar in nature and hence this type of chromatography is normal type. While the other ones which do not employ silica gel or any polar stationary phase are considered as reverse phase type.

The mobile phase used in this type are some organic solvents which are non polar. But interestingly this method of chromatography finds less use in research and industry. Instead the reverse phase chromatography is widely used. This is because of the wide advantages that method offers.

Normal Phase Chromatography Principle

In routine chromatography, sand or silica gel which is polar in nature is used as stationary phase. The mobile phase used is mostly non polar and organic in nature.

When compounds like plant extracts are to be separated, those with polar nature tend to bind to the stationary phase. While the compounds with non-polar nature tend move fast with the non-polar mobile phase as they have less affinity to silica phase Hence the non-polar compounds are eluted out first leaving the polar ones inside. These compounds retained can be seen as colored bands called as chromatograms.

In this method, solvents like water, alcohol and other highly polar ones are avoided from being used as mobile phases. Instead, non-polar solvents like petroleum ether, hexane etc are widely used.

Applications of Normal Phase Chromatography

Tough this method is less widely used, it is still the preferred method for:

1. This is the most preferred method for column chromatography.
2. It is used for separation of non-polar compounds in organic chemistry.
3. It is also used to separate plant extracts as the plant material has lot of pigments, lipids etc. which are non-polar in nature.
4. When the components of the mixture are damaged by water, they can be separated by this method.
5. Also isomers and compound which are too hydrophyllic and very hydrophobic can be separated well.

Displacement Chromatography

More efficient chromatographic methods is an increasing need for purified proteins in basic life science research as in the pharmaceutical industry. In all cases, target proteins must be separated from components of the host cell system or from other sources, like tissue. The well-established gradient chromatography is associated with the disadvantage that the binding capacity of the column cannot be fully explored when a high resolution is desired. Displacement chromatography offers an alternative here, as up to 80% of the binding capacity of a column can be utilized. Displacement chromatography (DC) is, along with isocratic and gradient chromatography, one of three basic modes of liquid chromatography. With DC, the sample is diluted in a sample application buffer, termed carrier, which promotes a high affinity of the target molecules to bind to the stationary phase of the chromatographic material. During sample application, the sample molecules will distribute themselves along the column according to their affinity to the stationary phase. As a result, the molecules with the highest affinity will be located at the top of the column and the molecules with lower affinity will be bound further away from the top of the column. Pumping continuously, a displacer containing eluent through the column will elute the adsorbed sample molecules. This step starts directly after sample application is finished. The displacer molecule, which should have a higher affinity towards the stationary phase than any other component of the sample, affects the displacement of the sample molecules as it passes down the column, thereby causing the sample molecules to arrange themselves in sample bands (displacement train) and elute in order of increasing affinities towards the stationary phase. The DC completes when the displacer totally saturates the stationary phase and thus elutes. DC is beneficial for the preparative purifi cation of products from chemical synthesis but can also be applied for the separation of molecules in complex mixtures, therefore used for proteomics. Recently, DC has been utilized for the separation of protein species, which is one of the most challenging issues in protein separation and a big problem in the production of therapeutic proteins. (According to Jungblut et al. Protein species are defined as proteins, which are coded by a single unique gene but differ in their exact chemical composition). During the production of therapeutic proteins, different species of the same protein are generated. These species are often nonfunctional or even harmful when administrated to patients. Zhang et al. demonstrated that a separation of charge variants of a defined mAb is possible with DC.

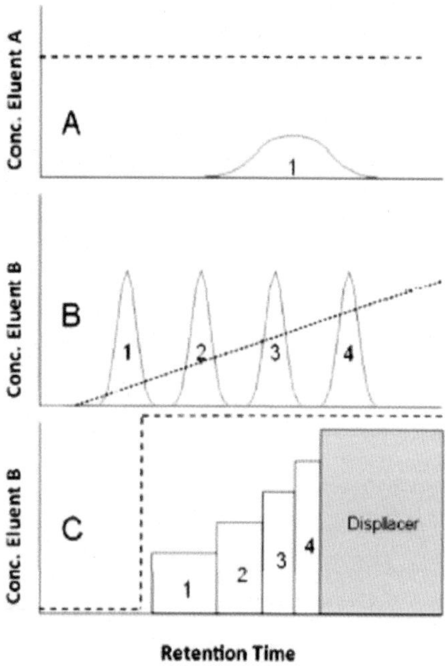

Figure: Scheme of the three basic modes of liquid chromatography: a) isocratic mode; b) gradient mode; c) displacement mode.

Dashed lines represent time course of the relative eluent concentration. The solid line represents the time course of the relative concentration of the molecules 1-4, eluting from the column. In this idealized example, it is assumed that all four diff erent molecules are present in equal concentrations. The affinity of the molecules follows the order 1 < 2 < 3 < 4. It is assumed that the molecules 2-4 did not elute from the column under the isocratic mode (a) because of their higher affinity towards the stationary phase. Note, that under the mode displacement (c) molecules elute in bands (rectangular peak shape) from the column.

Sample Displacement Chromatography (SDC)

DC requires the careful selection of an appropriate displacer and the displacer may co-elute with the target protein thus making an additional separation step necessary. Sample displacement chromatography (SDC) circumvents the displacer problem, because SDC does not need a displacer. When separating a mixture of molecules with SDC, the dissolved sample is pumped onto several columns connected in series. During sample loading, the molecules compete for the binding sites at the stationary phase. Molecules with the highest affinity bind to the first column, whereas molecules with lower affinities bind to the subsequent columns. After sample loading has been finished, the columns are disconnected and the adsorbed molecules are eluted from each individual column segment. With SDC, complex mixtures of proteins have already been separated.

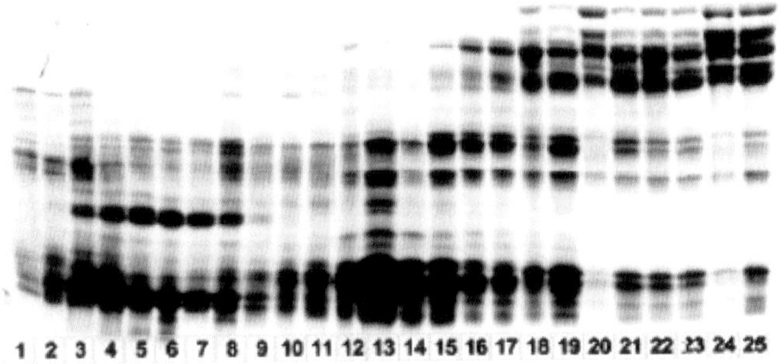

Figure . SDS-PAGE analysis of the eluates of an SDBC of a protein extract from renal tissue. The numbers represent the numbers of the chromatographic material segments. Tube number 1 was initially incubated with the sample. The last tube (number 25) contained the proteins with the lowest affinity.

Sample Displacement Batch Chromatography (SDBC)

A drawback in SDC is that the binding capacity of the stationary phase in relationship to the sample must be determined because the sample size and the total binding capacity of the columns connected in series should be harmonized. If the sample size in relation to the total binding capacity is too small, this will result in an insufficient separation. In this case, most of the molecules will bind to the fi rst column. If the sample size in relation to the total binding capacity is too large, a part of the sample will not be separated. Furthermore, the available column size may not be optimal in relation to the sample amount. For sample displacement chromatography performed in the batch mode (SDBC), this problem is less critical. Here the chromatographic material is handled in a buffer suspension allowing fl exible segment sizes. In addition, with the batch approach, even the chromatography of crude extracts is possible, which cannot be processed with columns due to the risk of column blocking. For SDBC, the total binding capacity is roughly estimated or experimentally determined yielding the total amount of chromatographic material, required for SDBC. An example for a typical calculation of the required amount of chromatographic material for SDBC is demonstrated. After equilibration of the calculated total amount of chromatographic material in a sample application buffer, the suspension is distributed to reaction vials or tubes. After sedimentation of the chromatographic material, the supernatant is removed. The sample is applied to the first tube, incubated while continuously rotating the tube for keeping the sample, buffer, and chromatographic material in suspension. After settling down the chromatographic material, the supernatant is moved to the second tube. After incubation, the supernatant of the second tube is transferred to the third tube. Sample loading will be fi nished when the sample has passed the last tube. The chromatographic materials in the individual tubes are washed and thereafter incubated individually with an elution buffer for desorption of the bound molecules. In figure, an SDBC separation result of a complex mixture of a protein extract from renal tissue is shown. The distribution of the pattern of protein bands demonstrates the high quality of the SDBC separation. In this protocol, it is demonstrated that even a complex protein sample can be separated with SDBC. Furthermore, scaling-up an SDBC separation is shown to be very simple. All parameters simply must be multiplied with the up-scaling factor. Additionally, SDBC is a very inexpensive approach for the separation of proteins, which does not require sophisticated HPLC equipment. In order to achieve optimal results with SDBC, a screening of parameters is recommended.

Reversed-phase Chromatography (RPC)

Reversed phase chromatography has found both analytical and preparative applications in the area of biochemical separation and purification. Molecules that possess some degree of hydrophobic character, such as proteins, peptides and nucleic acids, can be separated by reversed phase chromatography with excellent recovery and resolution. In addition, the use of ion pairing modifiers in the mobile phase allows reversed phase chromatography of charged solutes such as fully deprotected oligonucleotides and hydrophilic peptides. Preparative reversed phase chromatography has found applications ranging from micropurification of protein fragments for sequencing to process scale purification of recombinant protein products.

Theory of Reversed Phase Chromatography

The separation mechanism in reversed phase chromatography depends on the hydrophobic binding interaction between the solute molecule in the mobile phase and the immobilised hydrophobic ligand, i.e. the stationary phase. The actual nature of the hydrophobic binding interaction itself is a matter of heated debate but the conventional wisdom assumes the binding interaction to be the result of a favourable entropy effect. The initial mobile phase binding conditions used in reversed phase chromatography are primarily aqueous which indicates a high degree of organised water structure surrounding both the solute molecule and the immobilised ligand. As solute binds to the immobilised hydrophobic ligand, the hydrophobic area exposed to the solvent is minimised. Therefore, the degree of organised water structure is diminished with a corresponding favourable increase in system entropy. In this way, it is advantageous from an energy point of view for the hydrophobic moieties, i.e. solute and ligand, to associate.

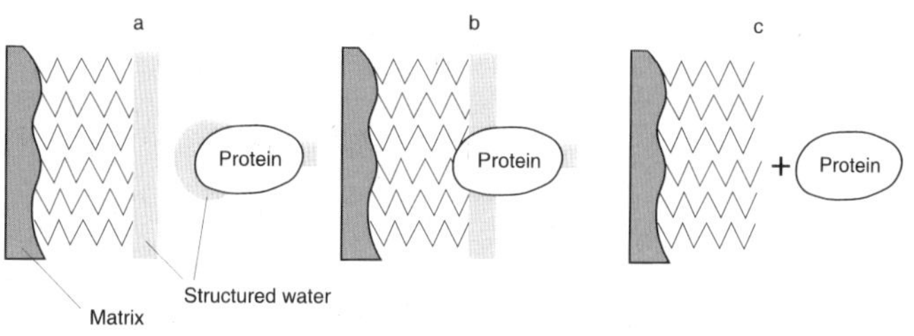

Figure: Interaction of a solute with a typical reversed phase medium. Water adjacent to hydrophobic regions is postulated to be more highly ordered than the bulk water. Part of this 'structured' water is displaced when the hydrophobic regions interact leading to an increase in the overall entropy of the system.

Reversed phase chromatography is an adsorptive process by experimental design, which relies on a partitioning mechanism to effect separation. The solute molecules partition (i.e. an equilibrium is established) between the mobile phase and the stationary phase. The distribution of the solute between the two phases depends on the binding properties of the medium, the hydrophobicity of the solute and the composition of the mobile phase. Initially, experimental conditions are designed to favour adsorption of the solute from the mobile phase to the stationary phase. Subsequently, the mobile phase composition is modified to favour desorption of the solute from the stationary phase back into the mobile phase. In this case, adsorption is considered the extreme equilibrium state where the distribution of solute molecules is essentially 100% in the stationary phase. Conversely, desorption is an extreme equilibrium state where the solute is essentially 100% distributed in the mobile phase.

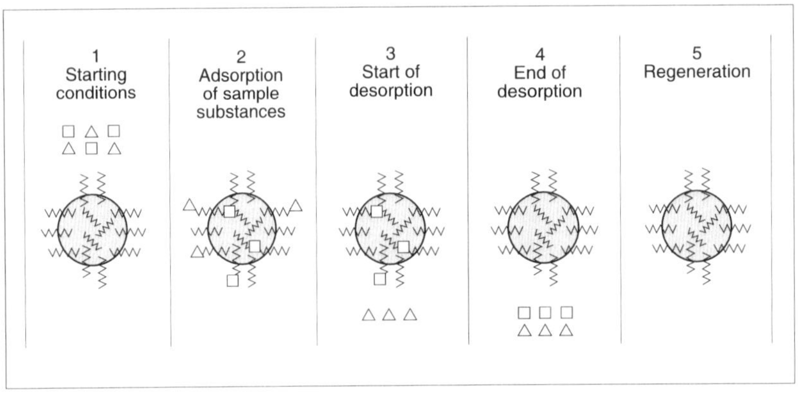

Fig: Principle of reversed phase chromatography with gradient elution.

Reversed phase chromatography of biomolecules generally uses gradient elution instead of isocratic elution. While biomolecules strongly adsorb to the surface of a reversed phase matrix under aqueous conditions, they desorb from the matrix within a very narrow window of organic modifier concentration. Along with these high molecular weight biomolecules with their unique adsorption properties, the typical biological sample usually contains a broad mixture of biomolecules with a correspondingly diverse range of adsorption affinities. The only practical method for reversed phase separation of complex biological samples, therefore, is gradient elution.

Separations in reversed phase chromatography depend on the reversible adsorption/desorption of solute molecules with varying degrees of hydrophobicity to a hydrophobic stationary phase. The majority of reversed phase separation experiments are performed in several fundamental steps as illustrated in figure.

- The first step in the chromatographic process is to equilibrate the column packed with the reversed phase medium under suitable initial mobile phase conditions of pH, ionic strength and polarity (mobile phase hydrophobicity). The polarity of the mobile phase is controlled by adding organic modifiers such as acetonitrile. Ion-pairing agents, such as trifluoroacetic acid, may also be appropriate. The polarity of the initial mobile phase (usually referred to as mobile phase A) must be low enough to dissolve the partially hydrophobic solute yet high enough to ensure binding of the solute to the reversed phase chromatographic matrix.
- In the second step, the sample containing the solutes to be separated is applied. Ideally, the sample is dissolved in the same mobile phase used to equilibrate the chromatographic bed.

The sample is applied to the column at a flow rate where optimum binding will occur. Once the sample is applied, the chromatographic bed is washed further with mobile phase A in order to remove any unbound and unwanted solute molecules.

- In the third step, bound solutes are next desorbed from the reversed phase medium by adjusting the polarity of the mobile phase so that the bound solute molecules will sequentially desorb and elute from the column. In reversed phase chromatography this usually involves decreasing the polarity of the mobile phase by increasing the percentage of organic modifier in the mobile phase. This is accomplished by maintaining a high concentration of organic modifier in the final mobile phase (mobile phase B). Generally, the pH of the initial and final mobile phase solutions remains the same. The gradual decrease in mobile phase polarity (increasing mobile phase hydrophobicity) is achieved by an increasing linear gradient from 100% initial mobile phase A containing little or no organic modifier to 100% (or less) mobile phase B containing a higher concentration of organic modifier. The bound solutes desorb from the reversed phase medium according to their individual hydrophobicities.
- The fourth step in the process involves removing substances not previously desorbed. This is generally accomplished by changing mobile phase B to near 100% organic modifier in order to ensure complete removal of all bound substances prior to re-using the column.
- The fifth step is re-equilibration of the chromatographic medium from 100% mobile phase B back to the initial mobile phase conditions.

Separation in reversed phase chromatography is due to the different binding properties of the solutes present in the sample as a result of the differences in their hydrophobic properties. The degree of solute molecule binding to the reversed phase medium can be controlled by manipulating the hydrophobic properties of the initial mobile phase. Although the hydrophobicity of a solute molecule is difficult to quantitate, the separation of solutes that vary only slightly in their hydrophobic properties is readily achieved. Because of its excellent resolving power, reversed phase chromatography is an indispensable technique for the high performance separation of complex biomolecules.

Typically, a reversed phase separation is initially achieved using a broad range gradient from 100% mobile phase A to 100% mobile phase B. The amount of organic modifier in both the initial and final mobile phases can also vary greatly. However, routine percentages of organic modifier are 5% or less in mobile phase A and 95% or more in mobile phase B.

The technique of reversed phase chromatography allows great flexibility in separation conditions so that the researcher can choose to bind the solute of interest, allowing the contaminants to pass unretarded through the column, or to bind the contaminants, allowing the desired solute to pass freely. Generally, it is more appropriate to bind the solute of interest because the desorbed solute elutes from the chromatographic medium in a concentrated state. Additionally, since binding under the initial mobile phase conditions is complete, the starting concentration of desired solute in the sample solution is not critical allowing dilute samples to be applied to the column.

Ionic binding may sometimes occur due to ionically charged impurities immobilised on the reversed phase chromatographic medium. The combination of hydrophobic and ionic binding effects is referred to as mixed-mode retention behaviour. Ionic interactions can be minimised by judiciously selecting mobile phase conditions and by choosing reversed phase media which are commercially produced with high batch-to-batch reproducibility and stringent quality control methods.

Ion-exchange Chromatography

Ion exchange chromatography (or ion chromatography) is a process that allows the separation of ions and polar molecules based on their affinity to the ion exchanger. It can be used for almost any kind of charged molecule including large proteins, small nucleotides and amino acids.

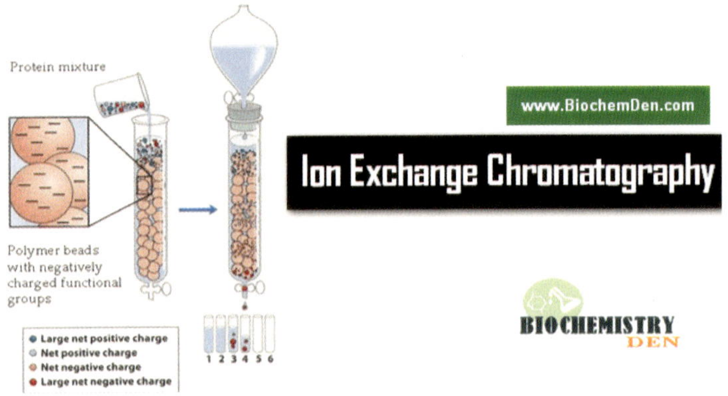

Principle

Exchange of ions is the basic principle in this type of Chromatography. In this process two types of exchangers i.e., cationic and anionic exchangers can be used.

- Cationic exchangers possess negatively charged group, and these will attract positively charged cations. These exchangers are also called "Acidic ion exchange" materials, because their negative charges result from the ionization of acidic group.

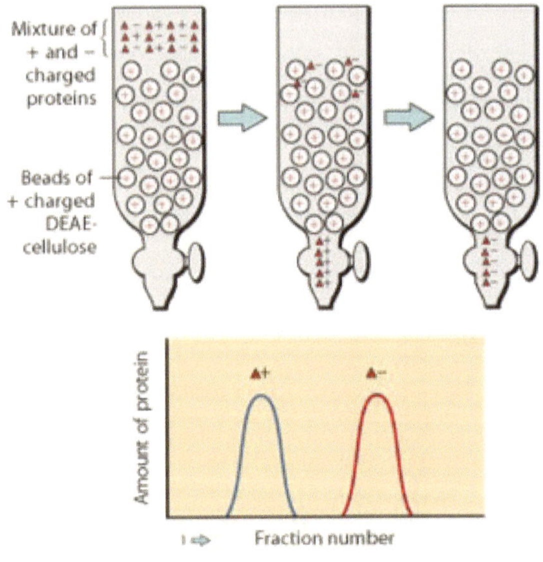

- Anionic exchangers have positively charged groups that will attract negatively charged anions. These are also called "Basic ion exchange" materials.

→**Cationic exchanger:**

$$RSO_3^- \ldots\ldots\ldots\ldots Na^+ \; + \; N^+H_3R' \;\; \leftrightarrow \;\; RSO_3 \ldots N^+H_3R' \; + \; Na^+$$

Exchanger — counter ion — Charged molecule to be exchanged — Bound molecular ion — Exchanged ion

→**Anionic exchanger:**

$$R_4N^+ \ldots Cl^- \; + \; {}^-OOC\,R' \;\; \leftrightarrow \;\; (R^+)_4N \ldots {}^-OOC\,R' \; + \; Cl^-$$

Types of Ion Exchange Resins

- Two main groups of materials are used to prepare ion exchange resins: Polystyrene and Cellulose.
- Resins made from both of these materials differ in their flow properties, ion accessibility and chemicals and mechanical stability.
- Polystyrene resins are proposed by polymerization reaction of styrene and divinyl benzene.
- A higher concentration of divinyl benzene produces higher cross linkages.
- Polystyrene resins are very useful for separating small molecular weight compounds.
- Increasing the cross linkage increases the rigidity, reduces swelling, reduces porosity & reduces the solubility of the polymeric structure.
- Sulfonic acids are strong acids with good proton dissociation ability. By sulfonation process, acidic functional groups are easily attached to nearly every aromatic nucleus.
- Resins substituted with sulfonic acid groups are strong cationic exchangers.
- To prepare weekly acidic exchanger, carbohydrate groups can be attached to the aromatic rings instead of sulfonic acid group.
- If basic functional groups are introduced, the resin can exchange anions rather than cations. Strong anion exchangers are prepared with a tertiary amine, yielding a strongly basic quaternary ammonium group. Weak anionic exchangers can be prepared with secondary amines, yielding a weakly basic tertiary amine.

Cellulose resins have much greater permeability to macromolecular polyelectrolytes and possess a much lower charge density as compared to polystyrene exchangers.

- Carboxymethyl cellulose *(CM-cellulose) – Cationic exchanger*
- DEAE cellulose – Anionic exchanger

Preparation of the Exchange Medium

There are three steps are of absolute importance:

1) Swelling of medium: (Pre-cycling)

Swelling makes the functional groups to be exposed for ion exchange.

- Swellimg of anion exchangers is usually carried out by treating it. first with an acid (0.5N HCl) and then with base (0.5N NaOH).
- Exactly the reverse is the case with cationic exchangers. The matrix can be treated with EDTA for impurity eliminations.

2) Removal of very small particles:

These fines will decrease flow rate and unsatisfactory reaction. To remove fines, the exchanger is repeatedly suspended in a large volume of water and after the larger polymers have settle down, the slow sedimenting materials decanted.

3) Equilibration with counter ions:

This is accomplished by washing the exchanger with different reagents depending upon the desired counter ion to be introduced.

- NaOH –> counter ion to be introduced is "Na^+"
- HCl –> counter ion to be introduced is "H^+"
- NaNO3 –> counter ion to be introduced is "NO_3"

Choice of Buffers

- Anionic exchange Chromatography should be carried out with cationic buffers.
- Cationic exchange Chromatography should be carried out with anionic buffers.
- The pK of the buffer should be as near as possible to the pH at which the system is buffered. This results in high buffer capacity, which can with stand the local changes of pH in the column easily.

Buffers	PH range
Ammonium acetate	4 to 6
Ammonium formate	3 to 5
Pyridinium formate	3 to 6
Pyridinium acetate	4 to 6
Ammonium carbonate	8 to 10

Practical Procedure

- Ion exchange separations are carried out mainly in columns packed with an ion-exchanger. These ionic exchangers are commercially available. They are made up of styrene and divinyl benzene.
- DEAE-cellulose is an anionic exchanger, CM-cellulose is a cationic exchanger. The choice of the exchanger depends upon the charge of particle to be separated. To separate anions "Anionic exchanger" is used, to separate cations "Cationic exchanger" is used.
- First the column is filled with ion exchanger then the sample is applied followed by the buffer. The tris-buffer, pyridine buffer, acetate buffer, citrate and phosphate buffers are widely used.
- The particles which have high affinity for ion exchanger will come down the column along with buffers. In next step using corresponding buffer separates the tightly bound particles.
- Then these particles are analyzed spectroscopically.

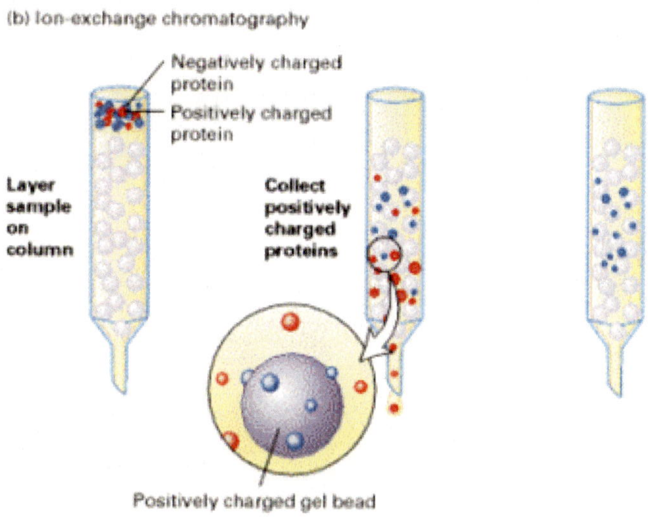

Applications of Ion Exchange Chromatography

1. It is extremely used in the analysis of amino acids. The amino acid "Autoanalyzer" is based on in exchange principle.
2. To determine the base composition of nucleic acids. Chargaff used this technique for established the equivalence of Adenine and Thymine; Guanine and Cytosine.
3. This is most effective method for water purification. Complete deionization of water (or) a non-electrolyte solution is performed by exchanging solute cations for hydrogen ions and solute anions for hydroxyl ions. This is usually achieved by method is used for softening of drinking water.
4. Proteins are also successfully separated by this technique.
5. It is also used for the separation of many vitamins, other biological amines, and organic acids and bases.

Affinity Chromatography

Affinity chromatography is a separation method based on a specific binding interaction between an immobilized ligand and its binding partner. Examples include antibody/antigen, enzyme/substrate, and enzyme/inhibitor interactions. The degree of purification can be quite high depending on the specificity of the interaction and, consequently, it is generally the first step, if not the only step, in a purification strategy.

Working of Affinity Chromatography

Affinity chromatography generally involves the following steps:

1. Incubate crude sample (e.g., cell lysate or serum) with the affinity support to allow the target molecule in the sample to bind to the immobilized ligand.
2. Wash away non-bound sample components from the support using appropriate buffers that maintain the binding interaction between target and ligand.
3. Elute (dissociate and recover) the target molecule from the immobilized ligand by altering the buffer conditions so that the binding interaction no longer occurs.

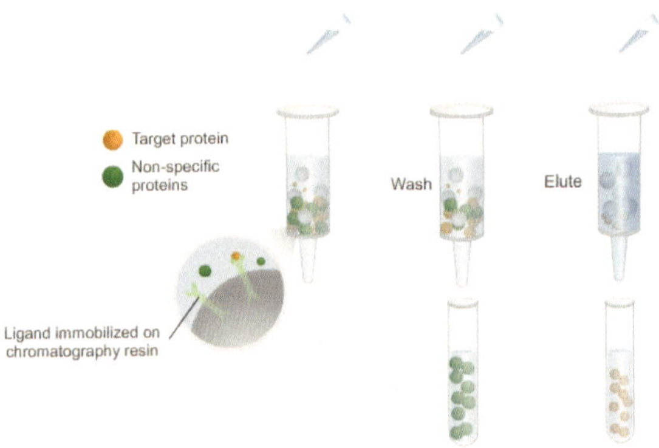

Small scale affinity chromatography using an antibody immobilized to a solid support. Chromatography has three main components: the mobile phase or solvent containing proteins, the stationary or solid phase also called the medium or resin (which may be agarose or other porous resin) and the chromatography column. Affinity chromatography is very selective and provides high resolution with an intermediate to high sample loading capacity. The protein of interest is tightly bound to the resin under conditions that favor specific binding to the ligand, and unbound contaminants are washed off. The bound protein is then recovered in a highly purified form by changing conditions to favor elution. Elution conditions may be specific, such as a competitive ligand, or nonspecific, such as changing pH, ionic strength, or polarity. The target protein is eluted in a purified and concentrated form.

A single pass of a serum or cell-lysate sample through an affinity column can achieve greater than 1000-fold purification of a specific protein so that only a single band is detected after gel electrophoresis (e.g., SDS-PAGE) analysis.

Most commonly, ligands are immobilized or “coupled” directly to solid support material by formation of covalent chemical bonds between particular functional groups on the ligand (e.g., primary amines, sulfhydryls, carboxylic acids, aldehydes) and reactive groups on the support. However, indirect coupling approaches are also possible. For example, a glutathione S-transferase (GST)-tagged fusion protein can be first captured to a glutathione support via the glutathione-GST affinity interaction and then secondarily chemically crosslinked to immobilize it. The immobilized GST-tagged fusion protein can then be used to affinity purify binding partner(s) of the fusion protein.

Ligands that bind to general classes of proteins (e.g., antibodies) or commonly used fusion protein tags (e.g., Histidine tag, or His-tag) are commercially available in pre-immobilized forms ready to use for affinity chromatography. Alternatively, more specialized ligands such as specific antibodies or antigens of interest can be immobilized using one of several commercially available activated affinity supports; for example, a peptide antigen can be immobilized to a support and used to purify antibodies that recognize the peptide.

Binding and Elution Buffers for Affinity Chromatography

Most affinity chromatography procedures involving protein:ligand interactions use binding buffers at physiologic pH and ionic strength, such as phosphate buffered saline (PBS). This is especially true when antibody:antigen or native protein:protein interactions are the basis for the affinity chromatography. Once the binding interaction occurs, the support is washed with additional buffer to remove nonbound components of the sample. Nonspecific (e.g., simple ionic) binding interactions can be minimized by adding low levels of detergent or by moderate adjustments to salt concentration in the binding and/or wash buffer. Finally, elution buffer is added to break the binding interaction and release the target molecule, which is then collected in its purified form.

Elution buffers dissociate binding partners by extremes of pH (low or high), high salt (ionic strength), the use of detergents or chaotropic agents that denature one or both of the molecules, removal of a binding factor or competition with a counter ligand. In most cases, subsequent dialysis or desalting is required to exchange the purified protein from elution buffer into a more suitable buffer for storage or downstream analysis.

The most widely used elution buffer for affinity chromatography based on protein interactions is 0.1 M glycine•HCl, pH 2.5-3.0. This buffer effectively dissociates most protein:protein and antibody:antigen binding interactions without permanently affecting protein structure. However, some antibodies and proteins are damaged by low pH, so eluted protein fractions are best neutralized immediately by addition of 1/10th volume of alkaline buffer such as 1 M Tris•HCl, pH 8.5.

Aqueous Normal-phase Chromatography

Aqueous Normal-Phase (ANP) Chromatography, frequently referred to as Hydrophilic Interaction Liquid Chromatography (HILIC), is a variation of normal-phase chromatography with the distinction that one of the major components of the mobile phase is water. Typical eluents of ANP consists of 50-95% acetonitrile in water or an aqueous buffer. The high volatility of the eluent makes ANP LC-MS friendly where one can realize a dramatic increase in sensitivity compared to reversed-phase chromatography. A requirement for ANP is a polar stationary phase. Common phases include bare silica, fluorinated phases such as Discovery HS F5, cyano, diol, and even certain phenyl phases such as the Ascentis Phenyl.

Many classes of polar compounds can be retained in ANP. These include polar neutrals, polar acids, and polar and non-polar basic amines. Both polar and ionic interactions appear to contribute to retention and selectivity in this mode of chromatography. HILIC retention can be simply described as the partitioning of a polar analyte into and out of an absorbed water enriched layer on the surface of the column. On the other hand, charged polar analytes can also undergo ion-exchange with the silanol groups on the surface. It is important to understand that either mechanism can dominate depending on the analyte and the mobile phase conditions.

Benefits of ANP/HILIC Separation

- Retain polar compounds that are too water soluble to be retained by reversed-phase chromatography.
- Achieve complimentary selectivity to reversed-phase chromatography.
- Gain enhanced sensitivity in mass spectrometry.
- Facilitate recovery of preparative chromatography fractions due to high-volatility mobile phase.

Biogenic amines is one class of compounds that can benefit from the use of ANP chromatography. Shown in figure is the analysis of four biogenic amines utilizing the Ascentis Si column under ANP conditions. These small, polar compounds can be diffi cult to analyze by traditional reversed-phase chromatography utilizing C18 or other hydrophobic columns.

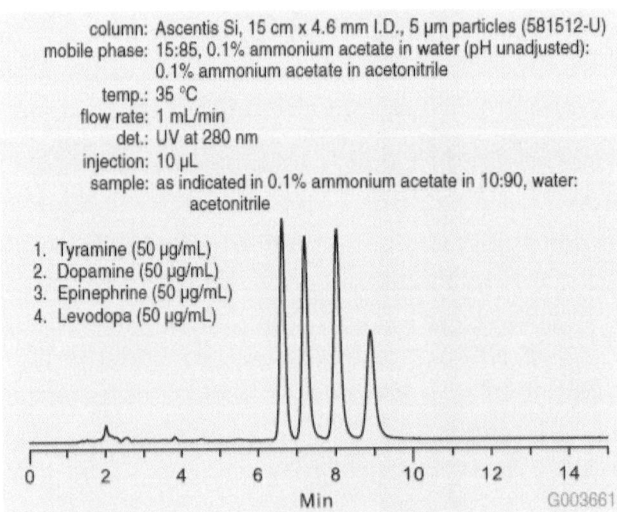

Figure: Separation of Biogenic Amines on Ascentis Si

Method Considerations

A consideration when using ANP chromatography is that many features are the opposite of reversed-phase chromatography. For example, add more organic solvent to the mobile phase to increase retention. Run gradients from low aqueous content to high aqueous content. Dilute samples with the organic solvent, not water. Finally, ionexchange can be a dominant mechanism, so adjusting buffer pH and ionic strength can have a dramatic effect on retention.

Techniques Based on Chromatographic Bed Shape

Column Chromatography

Column chromatography is a chromatography technique used to separate mixture of chemical substances into its individual compounds. Column chromatography is a widely used method for the purification or separation of chemical compound mixture in lab.

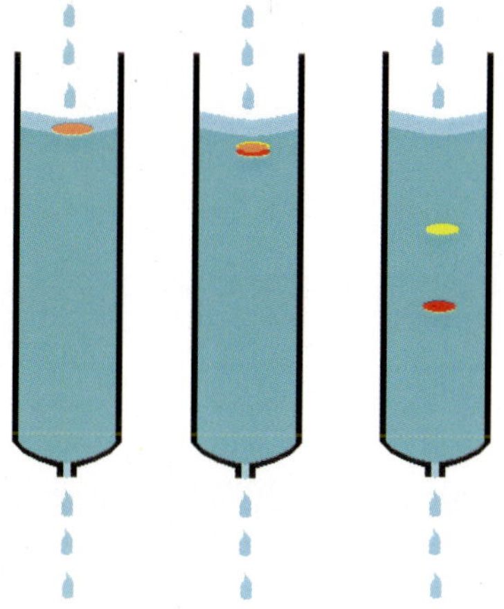

Principles of Column Chromatography

Column Chromatography consists of two phases: one mobile phase and one contiguous stationery phase. The stationery phase is solid and the mobile phase is liquid. The compound mixture moves along with the mobile phase through stationery phase and separates depending on the different degree of adhesion (to the silica) of each component in the sample or the compound mixture.

The stationery phase

A glass tube with a circle large inlet and a small outlet with a plug or tap, named as column is used for this column chromatography. The column is placed vertically with a stand where the outlet is downward.

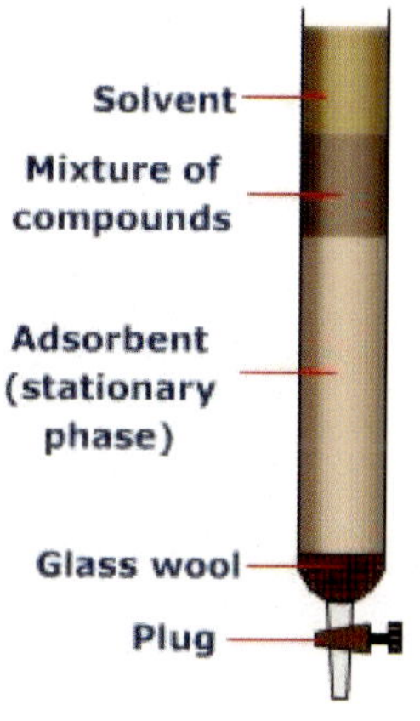

A piece of cotton wool is entered into the outlet and placed over the plug if there are no glass wool present to stop escaping the stationery phase from the column. There are two procedure to prepare the column by packing with silica or alumina:

- Dry method: In dry method at first the column is filled with dry powdered silica. Then the mobile phase, a suitable solvent is flushed through it until all the silica are wet and settled. From this point till the end always the column need to keep wet with solvent.
- Wet method: In wet method firstly a slurry of silica and solvent is prepared and then poured onto the column using a funnel. More solvent must be used until the silica is settled into it.

Process

Column chromatography works in few steps:

Step 1: The mobile phase or eluent is either solvent or solvent mixture. The upper level of mobile phase should be same as the stationery phase. That means the stationery phase should be wet with the solvent. On this stage the compound mixture what need to be separated, are added from the top of the column in such a way that the top level of it is not disturbed. By turning on the tap below it is allowed to adsorbed on the surface of the silica.

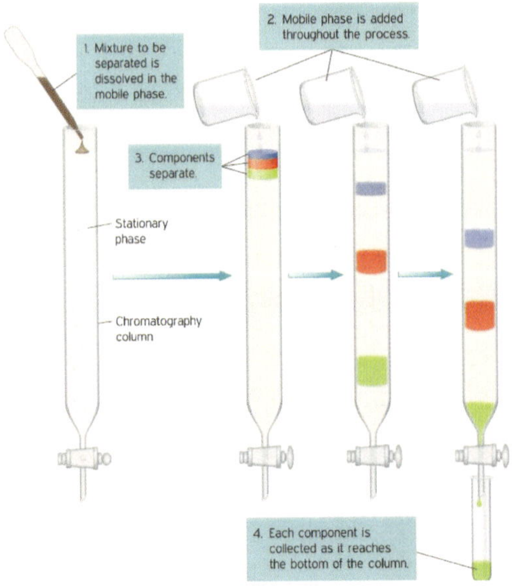

Step 2: Then the solvent or a suitable solvent mixture is added at first touching the side of the glass column slowly and carefully so that the top level of the stationery phase is not disturbed. The solvent is repeatedly added as many times as needed throughout the process.

Step 3: When the tap, is on the compounds in the compound mixture move along with the eluent depending on the polarity of the sample molecule. The non polar components travel faster than the polar component.

Suppose if any compound mixture contains three compounds blue, red and green. According to polarity the order of these compounds are blue>red>green. That means blue is the most polar compound and thus will have less tendency to move along with the mobile phase.

Step 4: The green colored compound will travel first as it is less polar that other two. When it is near end of the column a clean test tube is taken to collect the green sample. After this the red and at last the most polar blue compound is collected, all in separate test tubes.

Thus a compound mixture is separated or purified by using column chromatography.

Applications

Column chromatography is one of the most convenient and widely used methods for purifying compounds. Often, synthetic reactions will produce multiple products and column chromatography can be used to isolate each of the compounds for further examination. Column chromatography is extremely valuable when synthesizing or isolating novel compounds, as very little needs to be known about a compound and its' physical properties prior to the purification process.

The pharmaceutical industry routinely uses column chromatography to purify compounds as part of its early stage drug development process. Often in these preliminary stages researchers will construct libraries of compounds around a lead compound, then subsequently use column chromatography to purify the newly synthesized compounds. The extensive use and versatility of this purification technique has prompted educators to incorporate the technique into the undergraduate curriculum.

Planar Chromatography

Biomolecules are widespread throughout the world. It is an organic molecule pro-duced by a living organism, including large polymeric molecules such as proteins, polysaccharides and nucleic acids as well as small molecules such as a primary metabolites, secondary metabolites, and natural products. Many sample preparation techniques are used in biomolecule analysis; the method selected depends on the complexity of the sample, the nature of the matrix and the analytes, and the analytical techniques available. Planar chromatography is an important analytical method with other chromatographic techniques such as gas chromatography (GC), high perfor-mance liquid chromatography (HPLC), and supercritical fluid chromatography (SFC). Thin-layer chromatography (TLC), although less sensitive and efficient than some other separation methods, has many advantages. Planar chromatography is most effective for the low-cost analysis of samples requiring minimal sample clean-up. Planar chromatography is also selected for biomolecular step analysis because:

- Single use of stationary phase minimizes sample preparation requirements.
- Parallel separation of numerous samples enhances high throughput.
- Ease of postchromatographic derivatization improves method selectivity and specificity.
- Detection and/or quantitation steps can easily be repeated under different con-ditions.
- All chromatographic information is stored on the plate and can be re-evaluated if required.
- Several screening protocols for different analytes can be carried out simultaneously.
- Selective derivatizing reagents can be used for individual or group identification of the analytes.

- Detection of the separated spots with specific and sensitive color reagents
- Visual detection of UV-absorbing compounds is possible in field analyses by use of a UV lamp.
- Detection by contact with X-ray film, digital bio- and autoradiography, and even quantitative assay by use of enzymes is possible.
- TLC plates can be documented by video scans or photographs.
- Planar chromatography combined with modern video scanning and densitometry enables quantitative analysis.
- Planar chromatography coupled with densitometry enables detection of the spotsor zones through scanning of the chromatograms with UV–Vis light in thetransmission, reflectance, or fluorescence mode.
- With multi-wavelength scanning of the chromatograms, spectral data of theanalytes can be directly acquired from the TLC plates and can further be compared with the spectra of the analytes from software library.
- Additional information for structural elucidation can be obtained by planarchromatography combined with MS (fast atom bombardment (FAB) and liquid secondary mass spectrometry (SIMS)).
- The whole procedure of chromatographic development can be followed visually, so any distortion of the solvent front, etc., can be observed directly.
- The chromatogram can be developed simply by dipping the plate into a mobile phase.
- The possibility of two-dimensional development with a single adsorbent.
- The possibility of two-dimensional development on, for example, silica – octa- decyl silica coupled layers (Multi-K SC5 and CS5 dual phase).
- Planar chromatography is also the easiest technique, which performs multidi-mensional separation (e.g., by graft chromatography or multidimensional chro- matography).

Summing up, planar chromatography is one of principal separation techniques, which plays the key role also in biomolecule analysis.

There are two main branches of planar chromatography: paper chromatography and thin-layer chromatography (TLC). Although both practices follow very similar principles and methods, they use different materials as beds or planes for the stationary phase of the process.

Paper Chromatography

Paper chromatography has proved to be very successful in the analysis of chemical compound and lipid sample in particular. In paper chromatography, the sample mixture is applied to a piece of filter paper, the edge of the paper is immersed in a solvent, and the solvent moves up the paper by capillary action.

Components of the mixture are carried along with the solvent up the paper to varying degrees, depending on the compound's preference to be adsorbed onto the paper versus being carried along with the solvent. The paper is composed of cellulose to which polar water molecules are adsorbed, while the solvent is less polar, usually consisting of a mixture of water and an organic liquid.

The paper is called the stationary phase while the solvent is referred to as the mobile phase. In order to obtain a measure of the extent of movement of a component in a paper chromatography experiment, we can calculate an "Rf value" for each separated component in the developed chromatogram. An Rf value is a number that is defined as distance traveled by the component from application point.

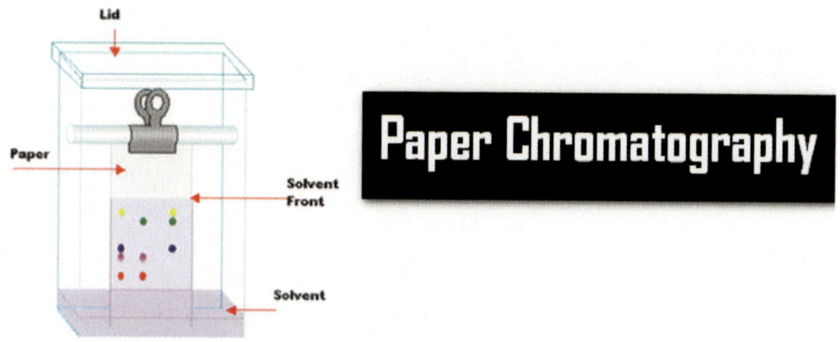

Nature of the Paper

The paper commonly used consists of highly purified cellulose. Cellulose, a homopolysaccharide of glucose. Contains several thousand anhydro-glucose units-linked through oxygen atoms. The paper exhibits weak ion exchange and adsorptive properties. Modified forms of paper have been produced in which the paper has been impregnated with alumina, silica gel, and ion-exchange resin etc.

The chemical composition of whatmann filter paper no: 1 is: a-cellulose (98 to 99%), b-cellulose (0.3 to 1%), Pentosans (0.4 to 0.8%), Ash (0.07 to 0.1%) & ether soluble matter (0.015 to 0.1%).

Apparatus

The apparatus required for paper chromatography are:

1. Support for paper
2. Solvent trough
3. Airtight chamber
4. Whattmann filter paper number 1
5. Capillary tubes
6. Samples – Amino acids (or) Pigments
7. Solvents
8. Platinum loop

Paper Development

There are two main techniques, which may be employed for the development of paper chromatograms:

1. Ascending techniques.
2. Descending techniques.
3. Radial development.
4. Two-dimensional chromatography.

Ascending Techniques

The filter paper is then dried and equilibrated by putting it into an airtight cylindrical jar, which contains an aqueous solution of a solvent. The most widely applicable solvent mixture is n-Butanol: Acetic acid: Water (4:1:5), which is abbreviated as BAW.

The sheet of paper is supported on a frame with the button edge in contact with a trough with solvent. The arrangement is contained in an airtight tank lined with paper saturated with the solvent to prove a constant atmosphere and separations are carried out in a constant temperature room. Thus, the solvent will ascend into the paper this process is, therefore, termed "Ascending Chromatography".

Descending Techniques

The end of the filter paper may be put into the solvent mixture contained in a narrow trough mounted near the top of the container. In this chromatography, the solvent will descend into the paper and this process is then termed "Descending Chromatography".

This method is convenient for compounds, which have similar Rf values since the solvent drips off the bottom of the paper, thus giving a wider separation.

Two-dimensional Chromatography (2D)

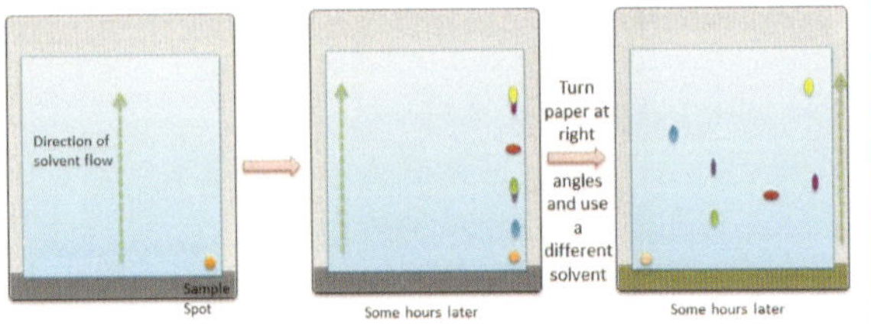

2-Dimensional Paper Chromatography

The mixture is separated then the first solvent, which should be volatile: then after drying, the paper is turned through 900 and separation is carried out in the second solvent.

After locating the migrated unknown sample along with standard known sample, a map is obtained and comparing their position with a map of known compounds can identify compounds.

Locating the Compounds

The strip is removed when the solvent has migrated over most of the available space. The distance to which the solvent has run is marked. In most cases, the completed Chromatogram is colorless with no indication of the presence of any compounds. Such a chromatogram is said as "Undeveloped" for locating the various compounds.

Large non-polar side chain amino acids

- →Leucine, Isoleucine
- →Phenylalanine, Tryptophan
- →Valine, Methionine
- →Lysine

} Migrate faster in **n-butanol: Acetic acid: Water (4:1:5)**

Shorter non-polar side chains

- →Proline, Alanine
- →Glycine

} Migrate faster than large non-polar side chain amino acids

Polar side chains

- →Threonine, Glutamic acid,
- →Serine,Arginine,
- →Aspartic acid, Histidine
- →Lysine,Cysteine

} Migrate faster than large non-polar side chain amino acids

The filter paper strip is first dried, then sprayed with 0.5% Ninhydrin in acetone and at least heated for a few minutes at 80 to 100° C. the reaction occurs and the colored spots appear at the sites of the amino acids, such as Chromatogram is now called "Developed".

In paper chromatography, the stationary cellulose phase is more polar than the mobile organic phase.

Identifying the Compounds

The ratio of the distance traveled by a component (i.e. amino acid) to that traveled by the solvent front, both measured from the marked point of the application of the mixture, is called the "Resolution front (Rf)" value for that component.

$$Rf = \frac{\text{Distance from origin run by the compound}}{\text{Distance from origin run by the solvent}}$$

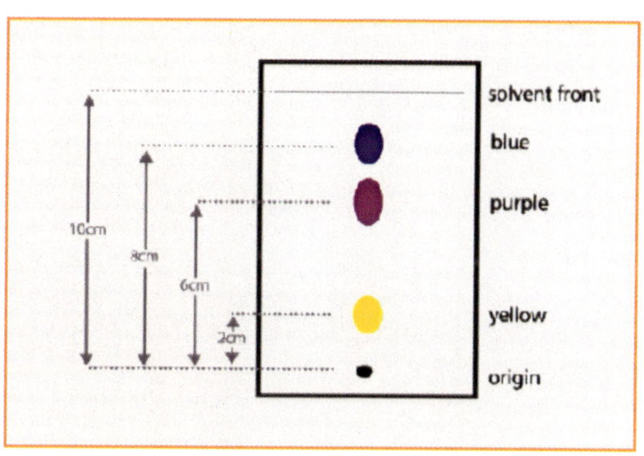

Detection

The filter paper strip may be sprayed with ninhydrin and heated so that the colored spots indicating the location of amino acids may develop. The color densities of these spots may be measured with a recording(or) reflectance photometer device.

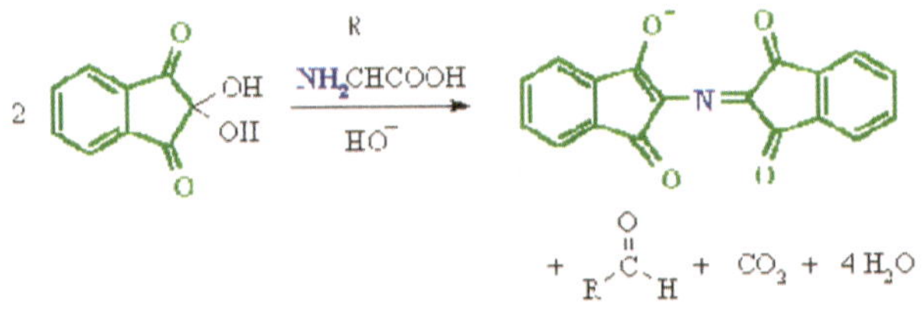

Ninhydrin Test

- Amines (including α-amino acids) react with ninhydrin to give a colored product.
- It can be used qualitatively (e.g. for chromatographic visualization) or quantitatively (e.g. for peptide sequencing).
- The α-amino acids typically give a blue-purple product.
- Proline, a secondary amine, gives a yellow-orange product.
- The test is sensitive enough that ninhydrin can be used for the visualization of fingerprints.

Applications of Paper Chromatography

- To check the control of purity of pharmaceuticals,
- To the detection of adulterants,
- To detect the contaminants in foods and drinks,
- To the study of ripening and fermentation,
- For the detection of drugs and dopes in animals & humans
- To the analysis of cosmetics,
- To the analysis of the reaction mixtures in biochemical labs.

Thin-layer chromatography (TLC)

TLC is a type of planar chromatography.

- It is routinely used by researchers in the field of phyto-chemicals, biochemistry, and so forth, to identify the components in a compound mixture, like alkaloids, phospholipids, and amino acids.
- It is a semi quantitative method consisting of analysis.
- High performance thin layer chromatography (HPTLC) is the more sophisticated or more precise quantitative version.

Principle

Similar to other chromatographic methods, thin layer chromatography is also based on the principle of separation.

- The separation depends on the relative affinity of compounds towards stationary and the mobile phase.
- The compounds under the influence of the mobile phase (driven by capillary action) travel over the surface of the stationary phase. During this movement, the compounds with higher affinity to stationary phase travel slowly while the others travel faster. Thus, separation of components in the mixture is achieved.
- Once separation occurs, the individual components are visualized as spots at a respective level of travel on the plate. Their nature or character are identified by means of suitable detection techniques.

System Components

TLC system components consists of:

- TLC plates, preferably ready made with a stationary phase: These are stable and chemically inert plates, where a thin layer of stationary phase is applied on its whole surface layer. The stationary phase on the plates is of uniform thickness and is in a fine particle size.
- TLC chamber: This is used for the development of TLC plate. The chamber maintains a uniform environment inside for proper development of spots. It also prevents the evaporation of solvents, and keeps the process dust free.
- Mobile phase: This comprises of a solvent or solvent mixture The mobile phase used should be particulate-free and of the highest purity for proper development of TLC spots. The solvents recommended are chemically inert with the sample, a stationary phase.
- A filter paper.: This is moistened in the mobile phase, to be placed inside the chamber. This helps develop a uniform rise in a mobile phase over the length of the stationary phase.

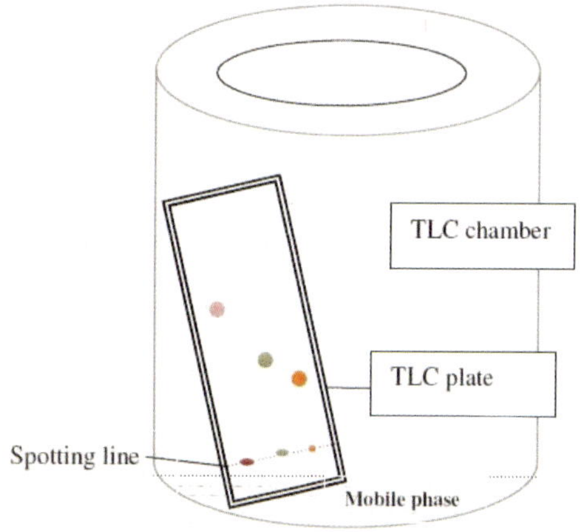

Procedure

The stationary phase is applied onto the plate uniformly and then allowed to dry and stabilize. These days, however, ready-made plates are preferred:

- With a pencil, a thin mark is made at the bottom of the plate to apply the sample spots.
- Then, samples solutions are applied on the spots marked on the line in equal distances.
- The mobile phase is poured into the TLC chamber to a leveled few centimeters above the chamber bottom. A moistened filter paper in mobile phase is placed on the inner wall of the chamber to maintain equal humidity (and also thereby avoids edge effect this way).
- Now, the plate prepared with sample spotting is placed in TLC chamber so that the side of the plate with the sample line is facing the mobile phase. Then the chamber is closed with a lid.
- The plate is then immersed, such that the sample spots are well above the level of mobile phase (but not immersed in the solvent — as shown in the picture) for development.
- Allow sufficient time for the development of spots. Then remove the plates and allow them to dry. The sample spots can now be seen in a suitable UV light chamber, or any other methods as recommended for the said sample.

Advantages

- It is a simple process with a short development time.
- It helps with the visualization of separated compound spots easily.
- The method helps to identify the individual compounds.
- It helps in isolating of most of the compounds.
- The separation process is faster and the selectivity for compounds is higher (even small differences in chemistry is enough for clear separation).
- The purity standards of the given sample can be assessed easily.
- It is a cheaper chromatographic technique.

Applications

- To check the purity of given samples.
- Identification of compounds like acids, alcohols, proteins, alkaloids, amines, antibiotics, and more.
- To evaluate the reaction process by assessment of intermediates, reaction course, and so forth.
- To purify samples, i.e for the purification process.
- To keep a check on the performance of other separation processes.

Being a semi quantitative technique, TLC is used more for rapid qualitative measurements than for quantitative purposes. But due its rapidity of results, easy handling and inexpensive procedure, it finds its application as one of the most widely used chromatography techniques.

Techniques Based on Physical State

Gas Chromatography

Gas chromatography is a term used to describe the group of analytical separation techniques used to analyze volatile substances in the gas phase. In gas chromatography, the components of a sample are dissolved in a solvent and vaporized in order to separate the analytes by distributing the sample between two phases: a stationary phase and a mobile phase. The mobile phase is a chemically inert gas that serves to carry the molecules of the analyte through the heated column. Gas chromatography is one of the sole forms of chromatography that does not utilize the mobile phase for interacting with the analyte. The stationary phase is either a solid adsorbant, termed gas-solid chromatography (GSC), or a liquid on an inert support, termed gas-liquid chromatography (GLC).

Sample Injection

A sample port is necessary for introducing the sample at the head of the column. Modern injection techniques often employ the use of heated sample ports through which the sample can be injected and vaporized in a near simultaneous fashion. A calibrated microsyringe is used to deliver a sample volume in the range of a few microliters through a rubber septum and into the vaporization chamber. Most separations require only a small fraction of the initial sample volume and a sample splitter is used to direct excess sample to waste. Commercial gas chromatographs often allow for both split and splitless injections when alternating between packed columns and capillary columns. The vaporization chamber is typically heated 50° C above the lowest boiling point of the sample and subsequently mixed with the carrier gas to transport the sample into the column.

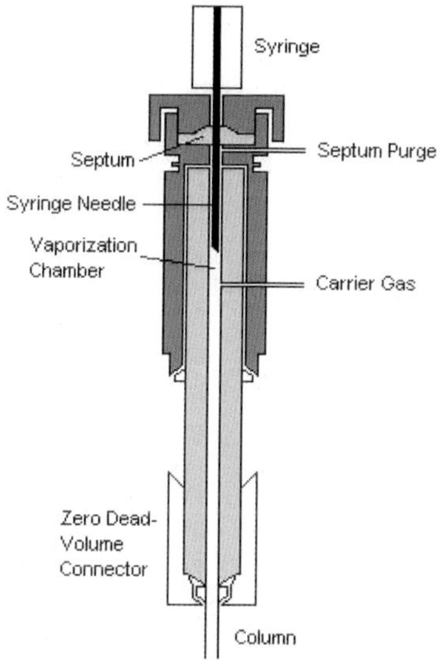

Figure: A cross-sectional view of a microflash vaporizer direct injector.

Carrier Gas

The carrier gas plays an important role, and varies in the GC used. Carrier gas must be dry, free of oxygen and chemically inert mobile-phase employed in gas chromatography. Helium is most commonly used because it is safer than, but comprable to hydrogen in efficiency, has a larger range of flow rates and is compatible with many detectors. Nitrogen, argon, and hydrogen are also used depending upon the desired performance and the detector being used. Both hydrogen and helium, which are commonly used on most traditional detectors such as Flame Ionization(FID), thermal conductivity (TCD) and Electron capture (ECD), provide a shorter analysis time and lower elution temperatures of the sample due to higher flow rates and low molecular weight. For instance, hydrogen or helium as the carrier gas gives the highest sensitivity with TCD because the difference in thermal conductivity between the organic vapor and hydrogen/helium is greater than other carrier gas. Other detectors such as mass spectroscopy, uses nitrogen or argon which has a much better advantage than hydrogen or helium due to their higher molecular weights, in which improve vacuum pump efficiency.

All carrier gasses are available in pressurized tanks and pressure regulators, gages and flow meters are used to meticulously control the flow rate of the gas. Most gas supplies used should fall between 99.995% - 99.9995% purity range and contain a low levels (< 0.5 ppm) of oxygen and total hydrocarbons in the tank. The carrier gas system contains a molecular sieve to remove water and other impurities. Traps are another option to keep the system pure and optimum sensitive and removal traces of water and other contaminants. A two stage pressure regulation is required to use to minimize the pressure surges and to monitor the flow rate of the gas. To monitor the flow rate of the gas a flow or pressure regulator was also require onto both tank and chromatograph gas inlet. This applies different gas type will use different type of regulator. The carrier gas is preheated and filtered with a molecular sieve to remove impurities and water prior to being introduced to the vaporization chamber. A carrier gas is typically required in GC system to flow through the injector and push the gaseous components of the sample onto the GC column, which leads to the detector.

Gas Recommendations for Capillary Columns

Detector	Carrier gas	Preferred makeup gas	Second choice	Detector, anode purge, or reference gas
Electron Capture	Hydrogen	Argon/Methane	Nitrogen	Anode purge must be same as makeup
	Helium	Argon/Methane	Nitrogen	
	Nitrogen	Nitrogen	Argon/Methane	
	Argon/Methane	Argon/Methane	Nitrogen	
Flame Ionization	Hydrogen	Nitrogen	Helium	Hydrogen and air for detector
	Helium	Nitrogen	Helium	
	Nitrogen	Nitrogen	Helium	
Flame Photometric	Hydrogen	Nitrogen		Hydrogen and air for detector
	Helium	Nitrogen		
	Nitrogen	Nitrogen		
	Argon	Nitrogen		
Nitrogen-Phosphorus	Helium	Nitrogen	Helium**	Hydrogen and air for detector
	Nitrogen	Nitrogen	Helium**	
Thermal Conductivity	Hydrogen*	Must be same as carrier and reference gas	Must be same as carrier and reference gas	Reference must be same as carrier and makeup
	Helium			
	Nitrogen			

* When using hydrogen with a thermal conductivity detector, vent the detector exhaust to a fume hood or a dedicated exhaust to avoid buildup of hydrogen gas.

** Helium is not recommended as a makeup gas at flow rates > 5 mL/min. Flow rates above 5 mL/min shorten detector life.

Figure: Gas Recommendations for Capillary Columns

Gas Recommendations for Packed Columns

Detector	Carrier gas	Comments	Detector, anode purge, or reference gas
Electron Capture	Nitrogen	Maximum sensitivity	Nitrogen
	Argon/Methane	Maximum dynamic range	Argon/Methane
Flame Ionization	Nitrogen	Maximum sensitivity	Hydrogen and air for detector
	Helium	Acceptable alternative	
Flame Photometric	Hydrogen		Hydrogen and air for detector
	Helium		
	Nitrogen		
	Argon		
Nitrogen-Phosphorus	Helium	Optimum performance	Hydrogen and air for detector
	Nitrogen	Acceptable alternative	
Thermal Conductivity	Helium	General use	Reference must be same as carrier
	Hydrogen	Maximum sensitivity (Note A)	
	Nitrogen	Hydrogen detection (Note B)	
	Argon	Maximum hydrogen sensitivity (Note B)	

Note A: Slightly greater sensitivity than helium. Incompatible with some compounds.
Note B: For analysis of hydrogen or helium. Greatly reduces sensitivity for other compounds.

Figure: Gas Recommendations for Packed Columns

Column Oven

The thermostatted oven serves to control the temperature of the column within a few tenths of a degree to conduct precise work. The oven can be operated in two manners: isothermal programming or temperature programming. In isothermal programming, the temperature of the column is held constant throughout the entire separation. The optimum column temperature for isothermal operation is about the middle point of the boiling range of the sample. However, isothermal programming works best only if the boiling point range of the sample is narrow. If a low isothermal column temperature is used with a wide boiling point range, the low boiling fractions are well resolved but the high boiling fractions are slow to elute with extensive band broadening. If the temperature is increased closer to the boiling points of the higher boiling components, the higher boiling components elute as sharp peaks but the lower boiling components elute so quickly there is no separation.

In the temperature programming method, the column temperature is either increased continuously or in steps as the separation progresses. This method is well suited to separating a mixture with a broad boiling point range. The analysis begins at a low temperature to resolve the low boiling components and increases during the separation to resolve the less volatile, high boiling components of the sample. Rates of 5-7° C/minute are typical for temperature programming separations.

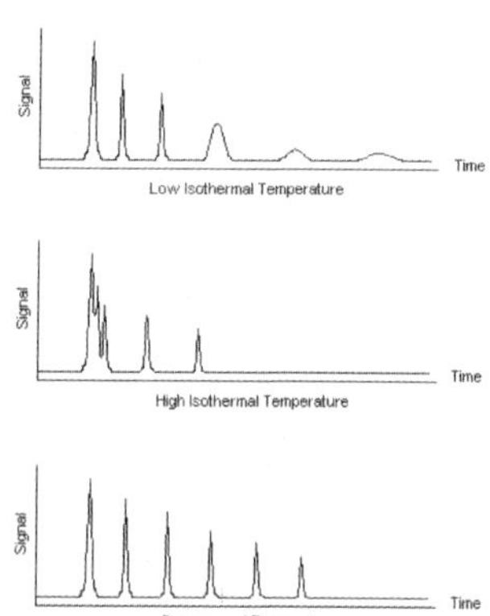

Figure: The effect of column temperature on the shape of the peaks.

Open Tubular Columns and Packed Columns

Open tubular columns, which are also known as capillary columns, come in two basic forms. The first is a wall-coated open tubular (WCOT) column and the second type is a support-coated open tubular (SCOT) column. WCOT columns are capillary tubes that have a thin later of the stationary phase coated along the column walls. In SCOT columns, the column walls are first coated with a thin layer (about 30 micrometers thick) of adsorbant solid, such as diatomaceous earth, a material which consists of single-celled, sea-plant skeletons. The adsorbant solid is then treated with the liquid stationary phase. While SCOT columns are capable of holding a greater volume of stationary phase than a WCOT column due to its greater sample capacity, WCOT columns still have greater column efficiencies.

Most modern WCOT columns are made of glass, but T316 stainless steel, aluminum, copper and plastics have also been used. Each material has its own relative merits depending upon the application. Glass WCOT columns have the distinct advantage of chemical etching, which is usually achieved by gaseous or concentrated hydrochloric acid treatment. The etching process gives the glass a rough surface and allows the bonded stationary phase to adhere more tightly to the column surface.

One of the most popular types of capillary columns is a special WCOT column called the fused-silica wall-coated (FSWC) open tubular column. The walls of the fused-silica columns are drawn from purified silica containing minimal metal oxides. These columns are much thinner than glass columns, with diameters as small as 0.1 mm and lengths as long as 100 m. To protect the column, a polyimide coating is applied to the outside of the tubing and bent into coils to fit inside the thermostatted oven of the gas chromatography unit. The FSWC columns are commercially available and currently replacing older columns due to increased chemical inertness, greater column efficiency and smaller sampling size requirements. It is possible to achieve up to 400,000 theoretical plates with a 100 m WCOT column, yet the world record for the largest number of theoretical plates is over 2 million plates for 1.3 km section of column.

Packed columns are made of a glass or a metal tubing which is densely packed with a solid support like diatomaceous earth. Due to the difficulty of packing the tubing uniformly, these types of columns have a larger diameter than open tubular columns and have a limited range of length. As a result, packed columns can only achieve about 50% of the efficiency of a comparable WCOT column. Furthermore, the diatomaceous earth packing is deactivated over time due to the semi-permanent adsorption of impurities within the column. In contrast, FSWC open tubular columns are manufactured to be virtually free of these adsorption problems.

	Type of Column			
	FSWC	WCOT	SCOT	Packed
Length	10 to 1000 m	10 to 1000 m	10 to 100 m	1 to 6 m
Inner Diameter	0.1 to 0.3 mm	0.25 to 0.75 mm	0.5 mm	2 to 4 mm
Efficiency (plates/m)	2000 to 4000	1000 to 4000	600 to 1200	500 to 1000
Sample Size	10 to 75 ng	10 to 1000 ng	10 to 1000 ng	10 to 10^6 ng
Pressure	Low	Low	Low	High
Speed	Fast	Fast	Fast	Slow
Inertness	Best	Good	Fair	Poor

Figure: Properties of gas chromatography columns.

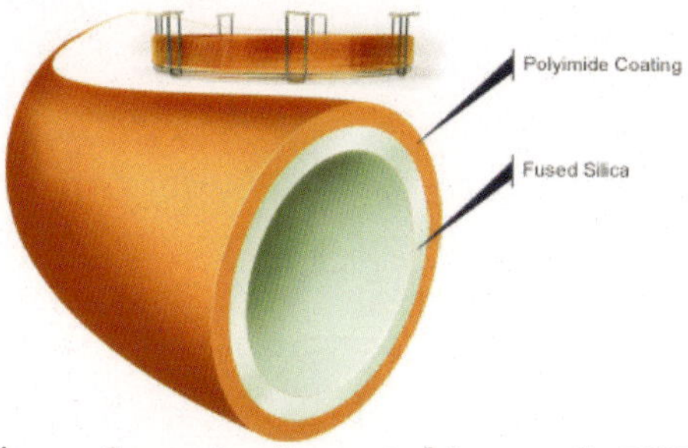

Figure: Computer Generated Image of a FSWC column (specialized for measuring BAC levels)

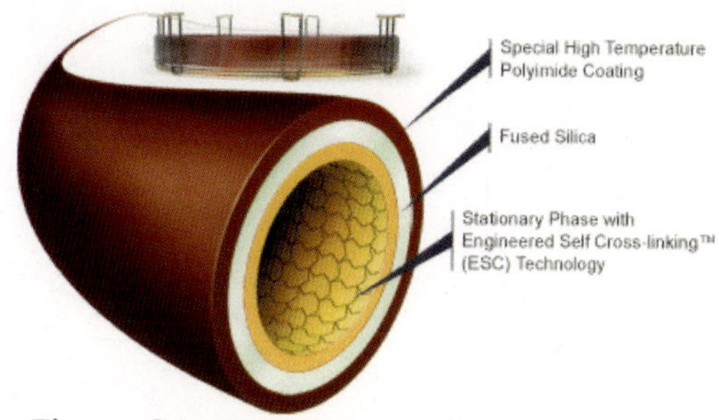

Figure: Computer Generated Image of a FSWC column (specialized to withstand extreme heat)

Different types of columns can be applied for different fields. Depending on the type of sample, some GC columns are better than the others. For example, the FSWC column shown in figure is designed specially for blood alcohol analysis. It produces fast run times with baseline resolution of key components in under 3 minutes. Moreover, it displays enhanced resolutions of ethanol and acetone peaks, which helps with determining the BAC levels. This particular column is known as Zebron-BAC and it made with polyimide coating on the outside and the inner layer is made of fused silica and the inner diameter ranges from .18 mm to .25 mm. There are also many other Zebron brand columns designed for other purposes.

Another example of a Zebron GC column is known as the Zebron-inferno. Its outer layer is coated with a special type of polyimide that is designed to withstand high temperatures. As shown in figure, it contains an extra layer inside. It can withstand up to 430° C to be exact and it is designed to provide true boiling point separation of hydrocarbons distillation methods. Moreover, it is also used for acidic and basic samples.

Detection Systems

The detector is the device located at the end of the column which provides a quantitative measurement of the components of the mixture as they elute in combination with the carrier gas. In theory, any property of the gaseous mixture that is different from the carrier gas can be used as a detection method. These detection properties fall into two categories: bulk properties and specific properties. Bulk properties, which are also known as general properties, are properties that both the carrier gas and analyte possess but to different degrees. Specific properties, such as detectors that measure nitrogen-phosphorous content, have limited applications but compensate for this by their increased sensitivity.

Each detector has two main parts that when used together they serve as transducers to convert the detected property changes into an electrical signal that is recorded as a chromatogram. The first part of the detector is the sensor which is placed as close the the column exit as possible in order to optimize detection. The second is the electronic equipment used to digitize the analog signal so that a computer may analyze the acquired chromatogram. The sooner the analog signal is converted into a digital signal, the greater the signal-to-noise ratio becomes, as analog signal are easily susceptible to many types of interferences.

An ideal GC detector is distinguished by several characteristics. The first requirement is adequate sensitivity to provide a high resolution signal for all components in the mixture. This is clearly an idealized statement as such a sample would approach zero volume and the detector would need infinite sensitivity to detect it. In modern instruments, the sensitivities of the detectors are in the range of 10-8 to 10-15 g of solute per second. Furthermore, the quantity of sample must be reproducible and many columns will distort peaks if enough sample is not injected. An ideal column will also be chemically inert and and should not alter the sample in any way. Optimized columns will be able to withstand temperatures in the range of -200° C to at least 400° C. In addition, such a column would have a short linear response time that is independent of flow rate and extends for several orders of magnitude. Moreover, the detector should be reliable, predictable and easy to operate.

Understandably, it is not possible for a detector meet all of these requirements.

Type of Detector	Applicable Samples	Detection Limit
Mass Spectrometer (MS)	Tunable for any sample	.25 to 100 pg
Flame Ionization (FID)	Hydrocarbons	1 pg/s
Thermal Conductivity (TCD)	Universal	500 pg/ml
Electron-Capture (ECD)	Halogenated hydrocarbons	5 fg/s
Atomic Emission (AED)	Element-selective	1 pg
Chemiluminescence (CS)	Oxidizing reagent	Dark current of PMT
Photoionization (PID)	Vapor and gaseous Compounds	.002 to .02 μg/L

Table: Typical gas chromatography detectors and their detection limits.

Mass Spectrometry Detectors

Mass Spectrometer (MS) detectors are most powerful of all gas chromatography detectors. In a GC/MS system, the mass spectrometer scans the masses continuously throughout the separation. When the sample exits the chromatography column, it is passed through a transfer line into the inlet of the mass spectrometer . The sample is then ionized and fragmented, typically by an electron-impact ion source. During this process, the sample is bombarded by energetic electrons which ionize the molecule by causing them to lose an electron due to electrostatic repulsion. Further bombardment causes the ions to fragment. The ions are then passed into a mass analyzer where the ions are sorted according to their m/z value, or mass-to-charge ratio. Most ions are only singly charged.

The Chromatogram will point out the retention times and the mass spectrometer will use the peaks to determine what kind of molecules are exist in the mixture. The figure below represents a typical mass spectrum of water with the absorption peaks at the appropriate m/z ratios.

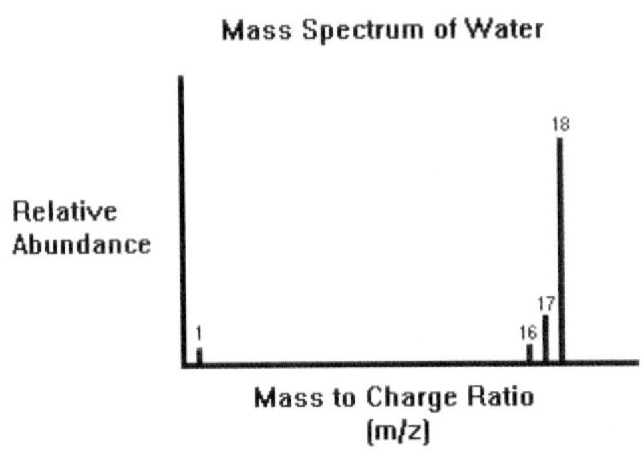

Figure: Mass Spectrum of Water

One of the most common types of mass analyzer in GC/MS is the quadrupole ion-trap analyzer, which allows gaseous anions or cations to be held for long periods of time by electric and magnetic fields. A simple quadrupole ion-trap consists of a hollow ring electrode with two grounded end-cap electrodes as seen in figure. Ions are allowed into the cavity through a grid in the upper end cap. A variable radio-frequency is applied to the ring electrode and ions with an appropriate m/z value orbit around the cavity. As the radio-frequency is increased linearly, ions of a stable m/z value are ejected by mass-selective ejection in order of mass. Ions that are too heavy or too light are destabilized and their charge is neutralized upon collision with the ring electrode wall. Emitted ions then strike an electron multiplier which converts the detected ions into an electrical signal. This electrical signal is then picked up by the computer through various programs. As an end result, a chromatogram is produced representing the m/z ratio versus the abundance of the sample.

GC/MS units are advantageous because they allow for the immediate determination of the mass of the analyte and can be used to identify the components of incomplete separations. They are rugged, easy to use and can analyze the sample almost as quickly as it is eluted. The disadvantages of mass spectrometry detectors are the tendency for samples to thermally degrade before detection and the end result of obliterating all the sample by fragmentation.

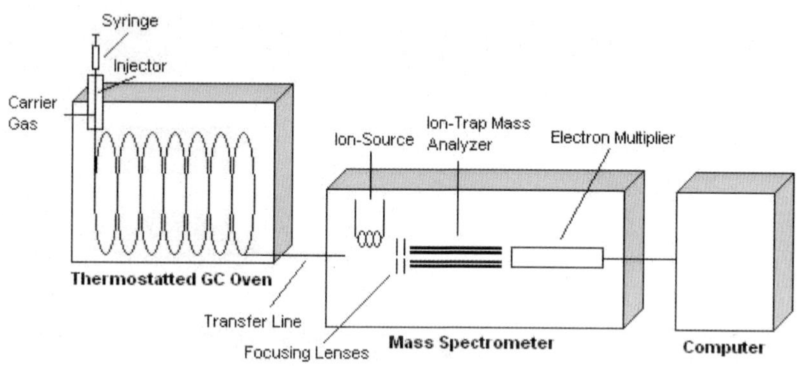

Figure: Schematic of the GC/MS system.

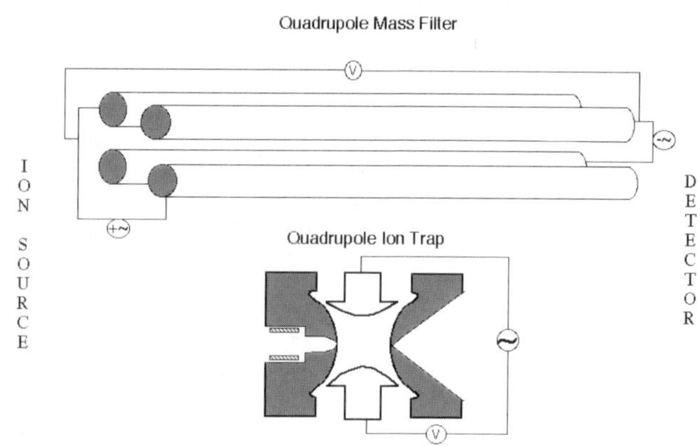

Figure: Arrangement of the poles in Quadrupole and Ion Trap Mass spectrometers

Flame Ionization Detectors

Flame ionization detectors (FID) are the most generally applicable and most widely used detectors. In a FID, the sample is directed at an air-hydrogen flame after exiting the column. At the high temperature of the air-hydrogen flame, the sample undergoes pyrolysis, or chemical decomposition through intense heating. Pyrolized hydrocarbons release ions and electrons that carry current. A high-impedance picoammeter measures this current to monitor the sample's elution.

It is advantageous to used FID because the detector is unaffected by flow rate, noncombustible gases and water. These properties allow FID high sensitivity and low noise. The unit is both reliable and relatively easy to use. However, this technique does require flammable gas and also destroys the sample.

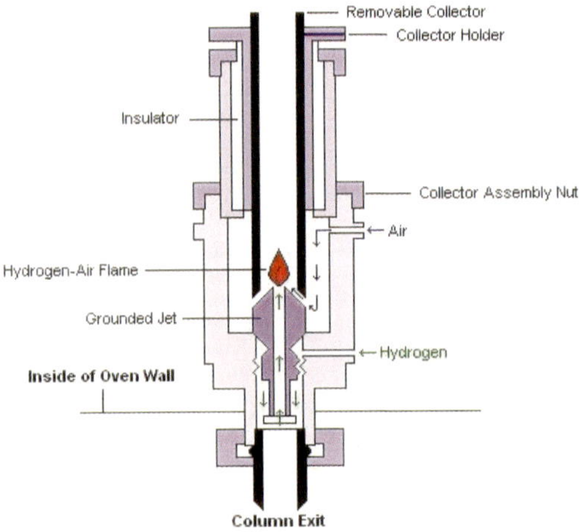

Figure: Schematic of a typical flame ionization detector.

Thermal Conductivity Detectors

Thermal conductivity detectors (TCD) were one the earliest detectors developed for use with gas chromatography. The TCD works by measuring the change in carrier gas thermal conductivity caused by the presence of the sample, which has a different thermal conductivity from that of the carrier gas. Their design is relatively simple, and consists of an electrically heated source that is maintained at constant power. The temperature of the source depends upon the thermal conductivities of the surrounding gases. The source is usually a thin wire made of platinum, gold or . The resistance within the wire depends upon temperature, which is dependent upon the thermal conductivity of the gas.

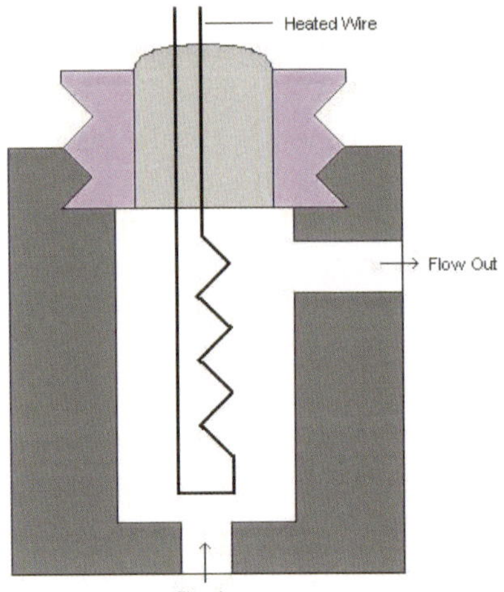

Figure: Schematic of thermal conductivity detection cell.

TCDs usually employ two detectors, one of which is used as the reference for the carrier gas and the other which monitors the thermal conductivity of the carrier gas and sample mixture. Carrier gases such as helium and hydrogen has very high thermal conductivities so the addition of even a small amount of sample is readily detected.

The advantages of TCDs are the ease and simplicity of use, the devices' broad application to inorganic and organic compounds, and the ability of the analyte to be collected after separation and detection. The greatest drawback of the TCD is the low sensitivity of the instrument in relation to other detection methods, in addition to flow rate and concentration dependency.

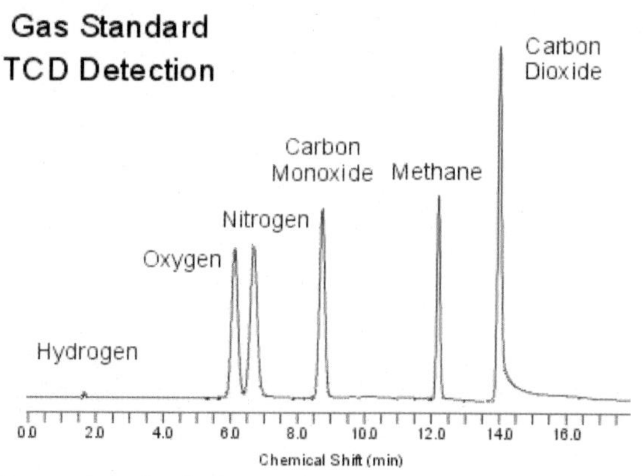

Figure: Standard Chromatogram of a Mixture of Gases

Chromatogram

Figure represents a standard chromatogram produced by a TCD detector. In a standard chromatogram regardless of the type detector, the x-axis is the time and the y-axis is the abundance or the absorbance. From these chromatograms, retention times and the peak heights are determined and used to further investigate the chemical properties or the abundance of the samples.

Electron-capture Detectors

Electron-capture detectors (ECD) are highly selective detectors commonly used for detecting environmental samples as the device selectively detects organic compounds with moieties such as halogens, peroxides, quinones and nitro groups and gives little to no response for all other compounds. Therefore, this method is best suited in applications where traces quantities of chemicals such as pesticides are to be detected and other chromatographic methods are unfeasible.

The simplest form of ECD involves gaseous electrons from a radioactive ? emitter in an electric field. As the analyte leaves the GC column, it is passed over this ? emitter, which typically consists of nickle-63 or tritium. The electrons from the ? emitter ionize the nitrogen carrier gas and cause it to release a burst of electrons. In the absence of organic compounds, a constant standing current is maintained between two electrodes. With the addition of organic compounds with electronegative functional groups, the current decreases significantly as the functional groups capture the electrons.

The advantages of ECDs are the high selectivity and sensitivity towards certain organic species with electronegative functional groups. However, the detector has a limited signal range and is potentially dangerous owing to its radioactivity. In addition, the signal-to-noise ratio is limited by radioactive decay and the presence of O_2 within the detector.

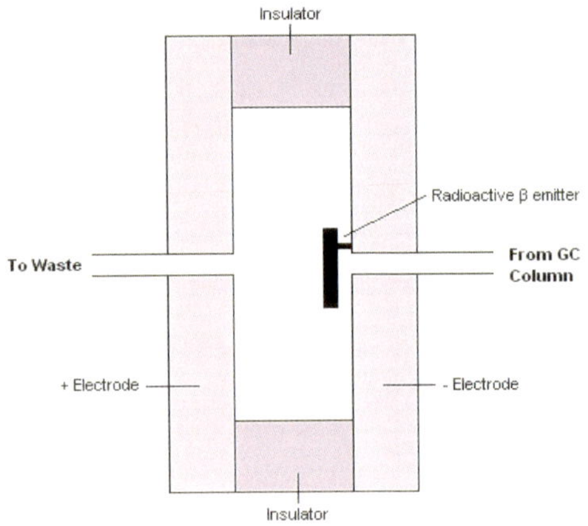

Figure: Schematic of an electron-capture detector.

Atomic Emission Detectors

Atomic emission detectors (AED), one of the newest addition to the gas chromatographer's arsenal, are element-selective detectors that utilize plasma, which is a partially ionized gas, to atomize all of the elements of a sample and excite their characteristic atomic emission spectra. AED is an extremely powerful alternative that has a wider applicability due to its based on the detection of atomic emissions. There are three ways of generating plasma: microwave-induced plasma (MIP), inductively coupled plasma (ICP) or direct current plasma (DCP). MIP is the most commonly employed form and is used with a positionable diode array to simultaneously monitor the atomic emission spectra of several elements.

The components of the Atomic emission detectors include:

- An interface for the incoming capillary GC column to induce plasma chamber,
- A microwave chamber,
- A cooling system,
- A diffration grating that associated optics, and
- A position adjustable photodiode array interfaced to a computer.

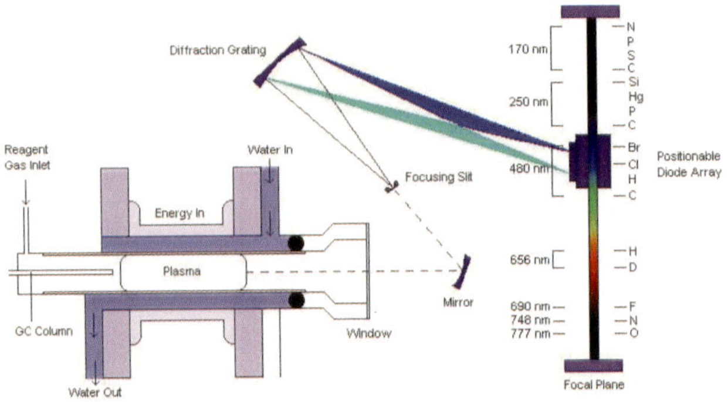

Figure: Schematic of atomic emission detector.

GC Chemiluminescence Detectors

Chemiluminescence spectroscopy (CS) is a process in which both qualitative and quantitative properties can be be determined using the optical emission from excited chemical species. It is very similar to AES, but the difference is that it utilizes the light emitted from the energized molecules rather than just excited molecules. Moreover, chemiluminescence can occur in either the solution or gas phase whereas AES is designed for gaseous phases. The light source for chemiluminescence comes from the reactions of the chemicals such that it produces light energy as a product. This light band is used instead of a separate source of light such as a light beam.

Like other methods, CS also has its limitations and the major limitation to the detection limits of CS concerns with the use of a photomultiplier tube (PMT). A PMT requires a dark current in it to detect the light emitted from the analyte.

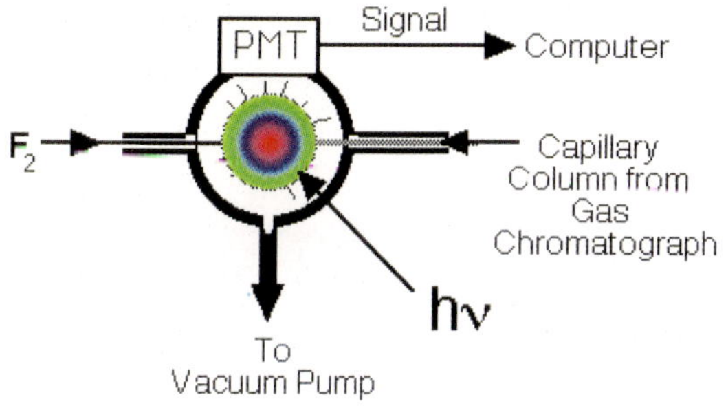

Figure: Schematic of a GC Chemiluminescence Detector

Photoionization Detectors

Another different kind of detector for GC is the photoionization detector which utilizes the properties of chemiluminescence spectroscopy. Photoionization detector (PID) is a portable vapor and gas detector that has selective determination of aromatic hydrocarbons, organo-heteroatom, inorganice species and other organic compounds. PID comprise of an ultrviolet lamp to emit photons that are absorbed by the compounds in an ionization chamber exiting from a GC column. Small fraction of the analyte molecules are actually ionized, nondestructive, allowing confirmation analytical results through other detectors. In addition, PIDs are available in portable hand-held models and in a number of lamp configurations. Results are almost immediate. PID is used commonly to detect VOCs in soil, sediment, air and water, which is often used to detect contaminants in ambient air and soil. The disavantage of PID is unable to detect certain hydrocarbon that has low molecular weight, such as methane and ethane.

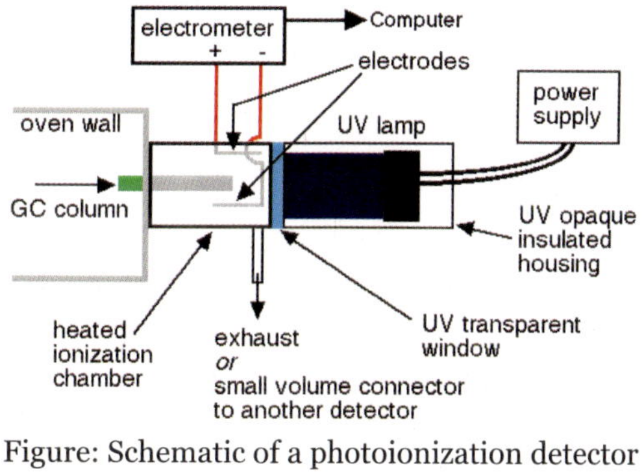

Figure: Schematic of a photoionization detector

Limitations

1. Not suitable for detecting semi-volatile compounds.
2. Only indicates if volatile organic compounds are presents.
3. High concentration so methane are required for higher performance.
4. Frequent calibration are required.
5. Units of parts per million range.
6. Enviromental distraction, especially water vapor.
7. Strong electrical fieldsRapid variation in temperature at the detector and naturally occurring compounds may affect instrumental signal.

Applications

Gas chromatography is a physical separation method in where volatile mixtures are separated. It can be used in many different fields such as pharmaceuticals, cosmetics and even environmental toxins. Since the samples have to be volatile, human breathe, blood, saliva and other secretions containing large amounts of organic volatiles can be easily analyzed using GC. Knowing the amount of which compound is in a given sample gives a huge advantage in studying the effects of human health and of the environment as well.

Air samples can be analyzed using GC. Most of the time, air quality control units use GC coupled with FID in order to determine the components of a given air sample. Although other detectors are useful as well, FID is the most appropriate because of its sensitivity and resolution and also because it can detect very small molecules as well.

GC/MS is also another useful method which can determine the components of a given mixture using the retention times and the abundance of the samples. This method be applied to many pharmaceutical applications such as identifying the amount of chemicals in drugs. Moreover, cosmetic manufacturers also use this method to effectively measure how much of each chemical is used for their products.

Liquid Chromatography

Liquid chromatography (LC) is an analytical chromatographic technique that is useful for separating ions or molecules that are dissolved in a solvent.

If the sample solution is in contact with a second solid or liquid phase, the different solutes will interact with the other phase to differing degrees due to differences in adsorption, ion-exchange, partitioning, or size.

These differences allow the mixture components to be separated from each other by using these differences to determine the transit time of the solutes through a column.

Liquid Chromatography Technique

In column liquid chromatography, as the liquid mobile phase passes through the column, components in the mobile phase interact to varying degrees with the solid stationary phase, also known as the chromatography media or resin. Molecules of interest in the mobile phase are separated based on their differing physicochemical interactions with the stationary and mobile phases.

These interactions can be based on molecular size (size exclusion chromatography), charge (ion exchange chromatography), hydrophobicity (hydrophobic interaction chromatography), specific binding interactions (affinity chromatography), or a combination of these (multimodal or mixed-mode chromatography).

The compostion of the mobile phase is typically changed during a separation run so as to alter the strengths of the interactions of the compounds of interest, that is, to change the phase partitioning of each compound between the stationary and mobile phases. Each compound then elutes from the column in a particular order depending on the relative strengths of its interaction with the resin and the mobile phase.

As the mobile phase continues to flow through the column, the column effluent, or eluate, is typically collected in fractions while monitoring the concentrations of the compounds eluted from the column over time to yield an elution curve, or chromatogram. The mode of detection varies with the analyte to be detected. For protein separations by column chromatography, protein concentration can be monitored manually using a dye-based protein assay such as the Bradford assay; however, such manual monitoring is labor intensive.

More commonly, a UV detector or spectrophotometer is attached to the chromatography system to continuously monitor protein elution from the column by measuring light absorption at 280 nm (A_{280}) by the amino acid tryptophan. The resulting chromatogram is then analyzed to quantitate proteins in the eluate. Each distinct peak represents a unique component resolved by the column, and the area under the curve corresponds to the amount of that compound eluted from the column. It is important to note that a single peak may contain more than one protein species; therefore, further analysis of the eluted fractions may be required, for example, by gel electrophoresis.

A typical chromatogram displays an initial broad peak of eluted protein that interacts weakly with the resin or, in some cases, not at all. In most cases, the column is washed with binding buffer until this first flowthrough peak (labeled "Binding Conditions" in figure) has completely eluted, and the A_{280} reading returns to baseline.

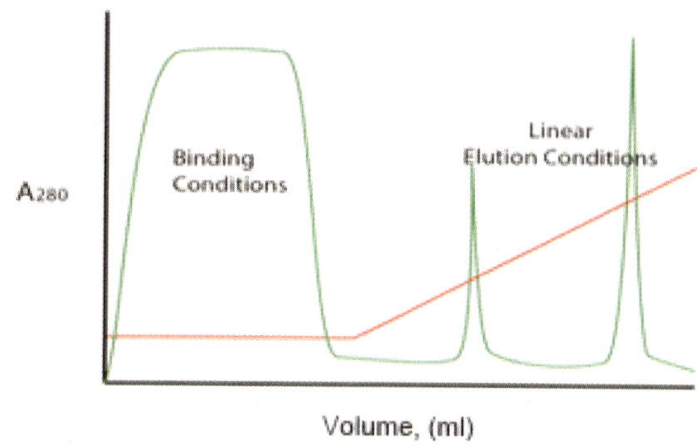

Figure: Example chromatogram showing a linear gradient elution.

Proteins that interact strongly with the resin are then eluted by changing the composition of the elution buffer. The specific composition of the elution buffer depends on the physicochemical properties of the molecules to be separated and the chromatography media used, often referred to as different media chemistries. Ion exchange chromatography, for example, which relies on interactions based on the net charges of the molecules, uses elution buffers of increasing ionic strength.

The red line in figure shows a linear gradient elution, but proteins can also be eluted using stepwise isocratic elution, where buffer conditions are changed in a stepwise fashion. By altering the elution gradient, flow rate, column length, and resin particle size, the protein separation ability (or resolution) of a given resin can be changed. Chromatography methods are optimized to yield sharp, tall peaks that are well separated.

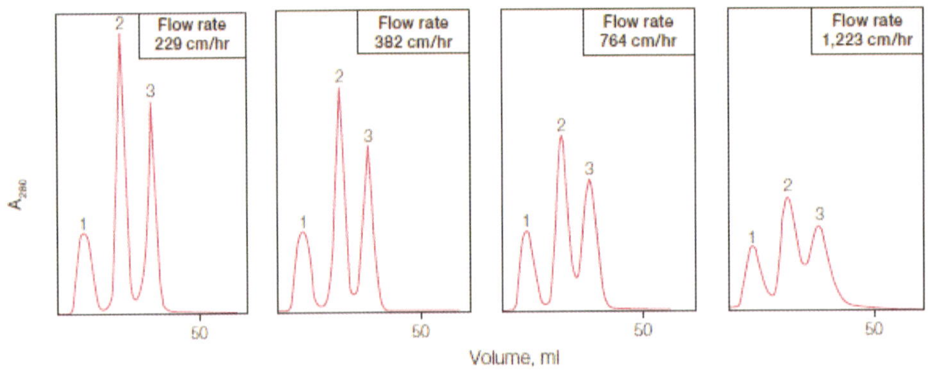

Figure: Optimization of resolution by changing flow rate. Optimal peak separation is achieved at a flow rate of 229 cm/hr.

Liquid Chromatography Workflow

Regardless of the interactions that are being exploited, liquid column chromatography is carried out in six steps:

- Column equilibration
- Sample loading
- Washing
- Elution
- Final column washing
- Column regeneration

Column Equilibration

Most liquid chromatography protocols begin with a resin equilibration step. A buffer that is compatible with the protein of interest and the resin of choice is passed over the column. A common practice is to equilibrate the column with 5–10 column volumes (CVs) of equilibration buffer.

For example, binding of proteins to hydrophobic interaction resins is most efficient at high ionic strength. Prior to sample application, the resin is therefore equilibrated in a buffer of high ionic strength.

The properties of the protein of interest are also considered in equilibration buffer selection, as buffer factors such as ionic strength are limited by protein stability; typically, one would avoid equilibration buffer conditions that would denature the protein of interest or prevent it from interacting with the stationary phase.

Sample Loading

After equilibration, the sample is loaded onto the column. The sample is generally loaded in a buffer with the same composition as the equilibration buffer to maximize protein interaction with the stationary phase.

Sample can be loaded manually or using a sample pump. Some types of chromatography limit the volume of sample that can be loaded onto the column. Another important sample loading consideration is that most resins have a finite capacity to bind protein; overloading a column by applying too much sample can adversely affect separation.

Column Washing

Once proteins have been immobilized on the stationary phase, proteins that interact only weakly or nonspecifically with the resin are removed by washing the column with several column volumes of wash buffer. This wash buffer can have the same composition as the equilibration buffer or contain components that disrupt weak specific interactions.

For example, immobilized-metal affinity chromatography (IMAC) elutes proteins bound to the resin with a high concentration of immidazole. A common practice is to use a wash buffer that includes an intermediate concentration of immidazole to eliminate contaminating proteins that are only weakly bound to the resin.

The column is washed until no protein is detected in the eluate. When using a chromatography system with a UV detector, the column is washed until the 280 nm absorption reading returns to the baseline.

Sample Elution

After all nonspecifically and weakly interacting proteins have been washed off of the resin, proteins that interact strongly with the resin are eluted from the column by changing the composition of the buffer that is passed over the resin.

In ion exchange chromatography, proteins are eluted with high–ionic-strength buffers or with a change in pH to disrupt the electrostatic interactions that immobilized the protein of interest. Proteins bound to a hydrophobic interaction resin, conversely, are eluted by lowering the ionic strength of the buffer. In affinity chromatography, proteins are commonly eluted from the column by the introduction of a competing ligand or by cleaving the affinity tag and may also be eluted using high-salt buffers or altering pH. Other elution protocols may involve mixing solvents of varying polarity to tune the solubility of each component in the mobile phase.

Elution conditions can either be changed in a linear gradient fashion or in a stepwise fashion. Often, a gradient elution protocol, in which the composition of the mobile phase changes linearly over time, is chosen to determine the elution profile and the elution buffer concentration at which the protein of interest is freed from the resin. Once this concentration has been determined, to save time, a stepwise isocratic elution protocol, in which the composition of the mobile phase is constant at each step, can be designed for future purifications.

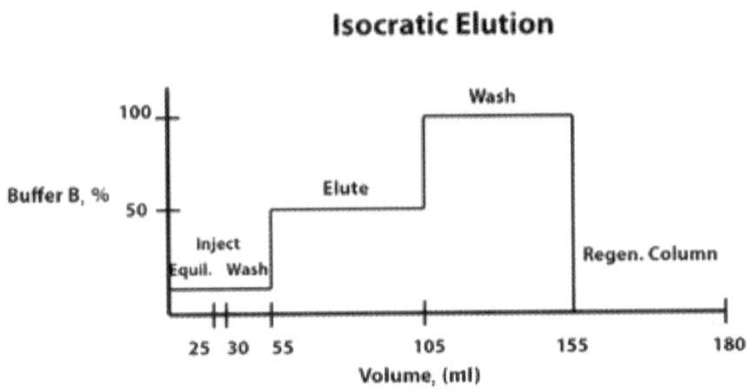

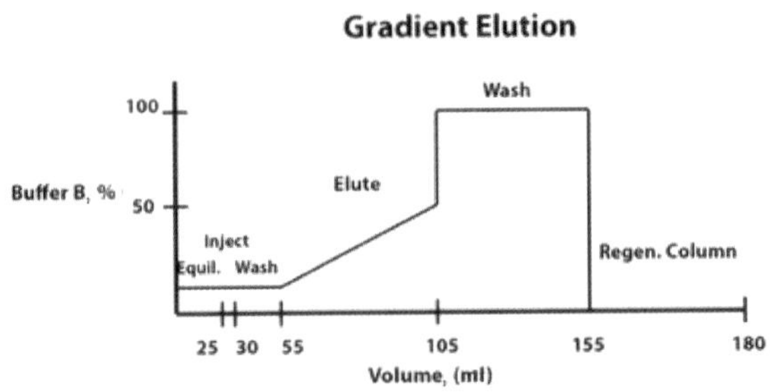

Size exclusion chromatography does not require buffer changes since it does not depend on specific interactions between the mobile phase and the stationary phase. There are no true wash and elution steps, as SEC relies solely on the fact that large molecules are retarded by porous beads, whereas small molecules pass through the resin with minimal resin interaction.

Final Column Washing

After the protein of interest has been eluted from the resin, any proteins that remain bound to the resin are eluted by increasing the strength of the elution buffer. This step permits columns to be reused for future separations.

Column Regeneration

After stripping the remaining compounds bound to the media, the column is then either saturated with equilibration buffer for subsequent reuse or filled with a storage buffer.

Fast Protein Liquid Chromatography

Fast protein liquid chromatography (FPLC) is, as the term implies, an effective (fast) liquid chromatography technique for the separation of protein molecules. The term also implies that the mobile phase is a liquid; the stationary phase is often a resin.

In this method, a pump is used to maintain the flow of the mobile phase at the desired rate. The composition of the eluent is varied by using different proportions of the fluids that make up the eluent. The conditions of separation are maintained in such a way that when the composition of the eluent changes there is a differential adsorption to the stationary phase of the components of the sample. Optimization of this technique leads to the separation of proteins and peptides in the sample.

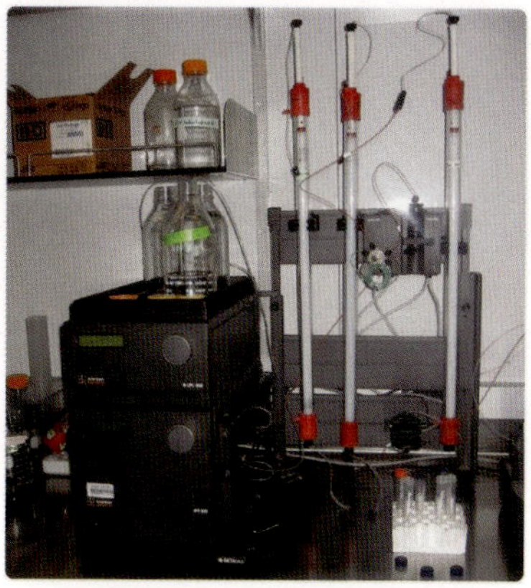

A laboratory set up of FPLC

Significance of FPLC

High pressure liquid chromatography (HPLC) is often the method of choice for separation of small molecules. Extraction and purification of biomolecules, however, is challenging for the following reasons:

- They function in the temperature range within which organisms and human beings survive
- They cannot tolerate high temperatures
- They lose their structure or functionality (or both) in the environment of organic solvents used in HPLC

Keeping these concepts in mind, Pharmacia developed the technique of fast protein liquid chromatography in 1982 to provide a method for the separation of biomolecules.

Features of FPLC

The goal of FPLC is purification of a biomolecule (biopolymer, protein, peptide), the size of which may be several thousand Daltons. The sample volume can range from about 100 ml to several liters. As is common with biomolecules, the eluent is a buffer, in turn often made up of two or more buffers. It is also important the eluent components used are biocompatible and do not react with the substance being purified (examples would be Teflon, titanium, glass).

The whole FLP chromatography apparatus is often used in a cold room (or at 4°C) so that structure of the protein being purified is not disturbed. As the pressures used are often in the range of a few bars, the method is sometimes referred to as medium-pressure liquid chromatography; and as this is used for purifying biomolecules, the technique also referred to as biopurification or biochromatography.

The stationary phase in FPLC is often a resin. Ion exchange, gel filtration, reversed phase, and affinity chromatography are the common modes used in the FPLC process, and there is also a wide choice of resins available for each of these modes. Sometimes, a combination of these modes is used. The first step in the isolation of a protein from a crude extract is often affinity chromatography. This may then be subject to ion-exchange chromatography to remove contaminants. When it is not possible to obtain a pure protein in two step a third step such as ion-change chromatography may also be used to obtain the final pure product.

The buffer used as the mobile phase and the type of resin used form a pair. Additionally and as already mentioned, the buffer itself may be a combination buffer, made up of a running buffer and an elution buffer. When the running buffer flows through a chromatographic column, the sample is applied by a sample pump. As the sample flows through the column, the protein of interest is selectively adsorbed by the resin; the other components in the sample simply flow out of the system. The buffer gradient is then changed by introducing the elution buffer with the help of buffer-blending valves. As more and more of the elution buffer flows into the chromatographic column, the protein of interest dissociates from the resin and flows out with the eluent buffer. The presence of the protein in the eluent is confirmed by a detector that measures the protein concentration.

Applications of FPLC

Body fluids such as plasma, urine, and cerebrospinal fluids contain a variety of proteins and peptides, and FPLC has proven to be a fast and reliable technique for a study of markers in these fluids. Lipoprotein profile, for example, can be obtained through FPLC. Profiling of proteins in cerebrospinal fluids has been helpful in detecting changes in the constitution of the fluid in disease states.

As polynucleotides can be also studied by FPLC, the technique has been used for the rapid purification of RNA and plasmid DNA.

FPLC protein profiling has also been used in the purification of animal venoms as well as in the diagnosis of beta-thalassemia. Its use in identifying microvariability within a single protein is also seen to have clinical importance.

FPLC has been used fairly extensively in analyzing the nitrogenous constituents of beer, which are believed to play a role in foam stability.

Purification of porphobilinogen deaminase from human erythrocytes, as well as purification of antibodies and vaccines for various purposes, has been facilitated by the use of FPLC.

Countercurrent Chromatography (CCC)

Counter-current chromatography (CCC), frequently referred to as high-speed counter-current chromatography, is a separation technique based on the distribution of analytes between two immiscible liquid phases, which is performed in a long and narrow tube. The main advantage of CCC is its high sample capacity paired with a low consumption of solvents. To give an example, analytical to semi-preparative CCC systems enable the separation of ~500 mg sample with ~100-500 mL solvent. Linking the separation power of chromatographic methods with the (semi-)preparative performance of CCC generates a strong tool suited for the isolation of natural products, for the purification of raw products and for the fractionation of complex samples, i.e. topics which are all relevant in lipid analysis.

Principle of CCC

Invented around 1970 , the principle of CCC is related to the liquid-liquid-extractions in a separation funnel, in which a compound is distributed between the two immiscible liquid phases (Figure a) . In CCC, one phase is kept stationary while the other, mobile phase is pumped through a long and hollow tube and serves to transport the analytes through the system. During this transport,

CCC takes advantage of thousands of such liquid-liquid-partitioning procedures, so providing the separation power expected from a real chromatographic technique.

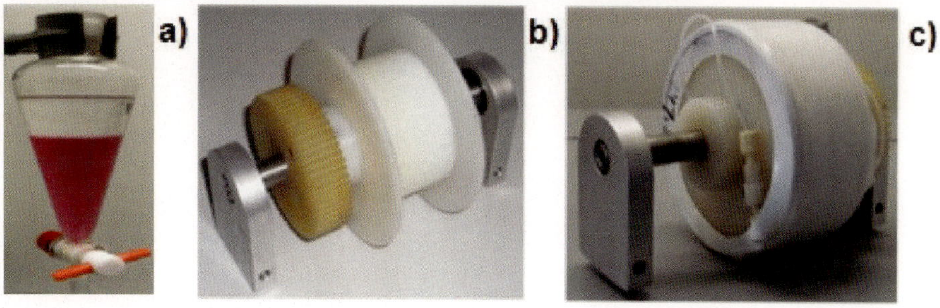

Figure: Photographs of (a) liquid-liquid extraction in a separation funnel with a biphasic solvent system in which the colorant is exclusively diluted in the lower phase, (b) an unwound CCC coil and (c) a CCC coil wound up with PTFE tubing. In CCC, thousands of "liquid-liquid extractions" on the way through the coil lead to the separation of compounds similarly soluble in both layers of the biphasic system.

The long and hollow tube is typically made of PTFE, which is wound multifold around a coil (or bobbin) (Figure b,c). Typically, two or three of such helical "multilayer coils" are serially connected and integrated into a centrifuge in such a way that they rotate in a planetary motion in a particular arrangement, which avoids any twisting of the tubing connections. During the separation, the coils are rotated at speeds of ~1000 rpm. During the CCC run, permanently changing gravitational forces are acting on the mobile and the stationary phase, resulting in frequent mixing and separation (= demixing) processes in the separation tube.

The introduction of the high-speed centrifuge system has led to the term "HSCCC". Nowadays this mode is more or less explicitly used so that there is no more an apparent need to distinguish high-speed CCC from CCC. As a consequence, "HSCCC" is more and more substituted with "CCC" in the scientific literature (similarly to the use of "GC" instead of "HRGC" for capillary or high resolution gas chromatography).

Figure: Photograph of the inside of a commercial CCC centrifuge with three serially connected multilayer coils labelled 1,2,3. During the CCC run the coils are rotated at ~1000 rpm.

Aside from the CCC centrifuge, modern benchtop CCC systems are constructed in the same way as HPLC systems. It is just replacing the HPLC column with the CCC centrifuge unit. Typically, a CCC system consists of a solvent reservoir, (HPLC) pump, injection system, detector and fraction collector. Currently, there are only a few constructors of commercial CCC systems, but these systems are suitable for the different problems that can be solved by CCC.

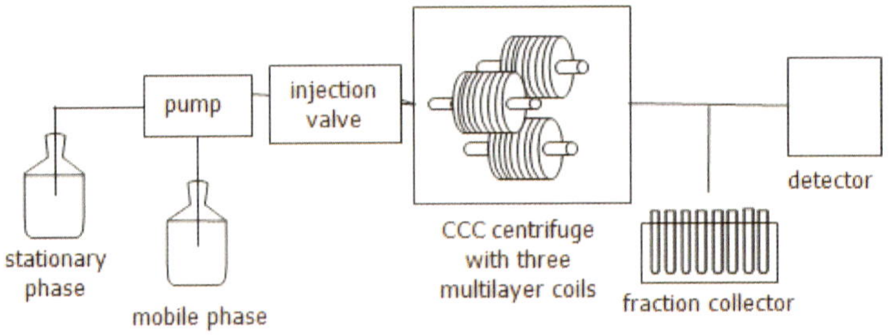

Figure: Schematic setup of a CCC system (modified from reference). The equipment is the same as in an HPLC setup except that the HPLC column is substituted with the CCC centrifuge

Factors Having an Impact on the CCC Separation

The solvent system consists of two immiscible phases, one of which is being used as the stationary phase and the other one as the mobile phase. The search for a proper solvent system for the target compounds is the most important preparatory work for a successful application of CCC. A suitable solvent system should meet the following requirements:

- The selected solvent components need to form two immiscible phases.
- The analytes should be stable and soluble in the system.
- The solvent mixture should form an acceptable volume ratio of the two phases, to minimize the solvent loss.

In a properly selected solvent system, the analytes should show a reasonable distribution coefficient $K_{U/L}$ (Equation):

$$K_{U/L} = \frac{\text{concentration of the analyte in the upper } (C_U)}{\text{concentration of the analyte in the lower } (C_L)}$$

$K_{U/L}$ is the concentration ratio C_U/C_L of a compound in the same volume of the two phases (i.e. the distribution coefficient). For a successful application of CCC, the KU/L value of the target compound should preferably be between 0.5 and 2.0. $K_{U/L}$ values should be determined in advance by shake-flask experiments using the following (or a similar) procedure:

- Place an aliquot of the analyte (dry) in a 6 mL derivation tube and add 2 mL from each phase of an equilibrated solvent system (solvent systems should be made ~24 h before the use).
- Close and shake the tube and wait for phase separation.
- Take an aliquot of both phases for determination of the concentration of the analytes, e.g. by GC or HPLC.

With the elution of the mobile phase, the analytes are the faster eluted the less they are miscible in the stationary phase. The mobile and stationary phase can substitute one another, which means that with the same solvent system CCC separations can be performed either in normal phase and reversed phase mode. In normal phase mode ("tail-to-head"), the upper (nonpolar) phase serves as the mobile phase while the lower (polar) phase is being used as the stationary phase. $K_{U/L}$ values from shake-flask experiments indicate the preferred mode of action for a successful separation with CCC. Compounds showing $K_{U/L}$ values <1 should be preferably analyzed in reversed-phase mode ("head-to-tail"). Vice versa, the normal-phase mode ("tail-to-head") is recommended for compounds with $K_{U/L}$ > 1 (Table). To give an example: in normal phase mode, long-chain saturated fatty acids are eluting first. In reversed phase mode ("head-to-tail"), the upper (nonpolar) phase serves as the stationary phase while the lower (polar) phase transports the analytes through the CCC system. In this mode, long-chain saturated fatty acids are eluting last (Table).

Table: Normal and reverse phase modes in CCC with selected characteristics.

Mode	CCC expression	Best suited for KU/L	Stationary phase	Mobile phase	Long chain saturated fatty acids
normal phase	"tail-to-head"	1 – 2*	Lower (polar)	Upper (non-polar)	elute before shorter ones
reversed phase	"head-to-tail"	0.5 – 1	Upper (non-polar)	Lower (polar)	elute after shorter ones

* switching to "tail-to-head" mode corresponds with (reciprocal) KL/U values of 1-0.5, i.e. the desired range of K values for CCC separations

Development and Selection of an Appropriate Solvent System for CCC Separations

The polarity of the analyte is a key-factor to be considered for the successful isolation of analytes when selecting a biphasic solvent system in CCC. Nonpolar analytes require a biphasic nonpolar solvent system because good solubility of the analyte in the solvent system is essential for large-scale fractionation in a single run. The rotation of the coil system usually increases the temperature by around 5° C. Therefore, solvent systems should be tested for temperature stability when the CCC system does not feature a tempering tool. Performance at different temperatures will also affect the phase solubility ratios, and in the worst case, the solvent system will not separate into two phases, so this parameter should be taken into account.

The settling time (i.e. the time required for phase separation after shaking) of the solvent system should be less than 20s because longer setting times decrease the retention of the analytes in the stationary phase. The larger the difference in the KU/L values of two compounds, the better they can be separated from each other. For lipid compounds, alkanes such as n-hexane, n-heptane and/or cyclohexane are the primary choice for the upper phase, whereas the lower phase may contain acetonitrile, methanol, ethanol, low amounts of water, and many others. Frequently, some 20 solvent mixtures have to be tested in such shake flask experiments.

Evaluation of the Solvent System and Pre-determination of the Elution Volume

In order to be independent of lab-to-lab and run-to-run variations, a standardization procedure has been suggested. Individual CCC fractions were related to the total volume of mobile phase (V_m), which has passed the coil at the end of each fraction. From this volume, the extruded volume of stationary phase is subtracted to give the corrected volume. Division of the corrected volume by the total coil volume allows predicting the elution of a substance from the CCC (Equation):

$$EV = \frac{V_m - V_e}{V_c} x100\%$$

with EV = corrected elution volume in percent; V_m = Volume of mobile phase; V_e = volume of extruded stationary phase; V_c = total coil volume.

With the help of equation, each CCC fraction can be expressed as a corresponding EV value. Accordingly, the elution of a compound (e.g. a fatty acid) from the CCC system can be described in relation to these values, especially for KU/L values near the recommended range of 0.5-1. For instance, the methyl ester of 14:0 showed an EV value range from 32.0 to 48.9% (mean EV 40.5%), which agreed well with the $K_{L/U}$ = 0.408. This measure facilitates the comparison of CCC fractionations with both different fraction volumes and different coil volumes (when using the same solvent system).

It should also be noted that solvents are not completely immiscible with each other. Even stable two-phase systems do not completely separate polar and nonpolar solvents. Low amounts of n-hexane are also soluble in aqueous solutions and the same applies the other way round (Table).

Table: Percentage composition of n-hexane, cyclohexane, tert-butyl methyl ether (TBME), methanol and water in the upper and lower phase of the solvent systems n-hexane/TBME/methanol/water (5/2/5/3)

	Upper phase	Lower phase
n-hexane	80.5	0.4
TBME	18.9	11.8
Methanol	0.4	57.3
Water	0.2	30.5

This and other issues lead to the fact that the stationary phase is not entirely kept in the system, but aliquots are transported with the mobile phase out of the system (i.e. a kind of column bleed or flooding of the stationary phase). In order to achieve a good chromatographic resolution a minimum of 65–70% of the stationary phase should remain in the column. The more of the stationary phase that remains in the system, the better the separations that can be achieved.

Challenges in the use of CCC in the Field of Lipids

Lipid compounds (fatty acids and components of the unsaponifiable matter) are characterized by being soluble in organic solvents and immiscible in water. For this purpose, most lipid compounds can be effectively enriched in a separation funnel in a two phase system consisting of n-hexane/water. Lipids will rather exclusively move to the organic n-hexane phase. That means, these compounds have partitioning coefficients between the n-hexane and aqueous phase of KU/L >100. Such situations do not require the use of CCC which is powerful at KU/L = 0.5 – 2 (Table). This is where the problem arises, i.e. the finding of two immiscible liquid phases that distribute lipid compounds in about the same ratio, and with high concentration.

Application of CCC for the Separation and Isolation of Fatty Acids

Fractionation and isolation of specific fatty acids from natural sources (e.g. oils) is an important task which can be managed by CCC. Liquid chromatographic separations of fatty acids are governed by the so-called equivalent chain length (ECL) rule, which has initially been developed for GC separations (Equation):

$$ECL=N-2n$$

where N is the number of carbon atoms n is the number of double bonds in the fatty acid.

The ECL rule compares the elution of different fatty acids and says that one double bond is equal to two carbons. One prominent example for fatty acids having the same ECL is 16:0 and 18:1n-9. Both major fatty acids have an ECL of 16, that means they are difficult to separate by liquid chromatographic methods. If this thought is pursued, polyunsaturated fatty acids (PUFAs) share the ECL with short-chain saturated fatty acids. For instance 22:6n-3 has an ECL of 10 (i.e. the same ECL as 10:0 and 12:1 isomers). However, 10:0 and 12:1 are rarely found (together with 22:6n-3) in samples. For this reason, CCC offers excellent opportunities for the isolation of PUFAs from marine oils. Accordingly, it was a rather simple task to isolate the methyl ester of 16:4n-1 (ECL = 8), i.e. a PUFA with the fourth double bond on the terminal carbon, from the fatty acid methyl esters obtained from fish oil. A simple two-phase mixture of n-hexane/methanol/water (~160/80/1) could be used for this purpose in tail-to-head mode. However, 22:6n-3 and 20:5n-3 (ECL = 10) could not be separated under these conditions. Stepwise modification of a simple two phase system enabled Bousquet and Le Goffic to separate the (free) PUFAs 22:6n-3, 20:5n-3, 18:4n-3, and 16:3n-3 (all ECL = 10) from each other. This was achieved by the use and stepwise increase of low amounts of water in n-heptane/methanol. Without water, the four PUFAs showed the same $K_{U/L}$ values of ~0.45 in n-heptane/methanol. By the addition of 10% water, the $K_{U/L}$ range increased from 0.55 to ~1.2 (22:6n-3). Under these conditions these PUFAs with the same ECL could all be separated from each other.

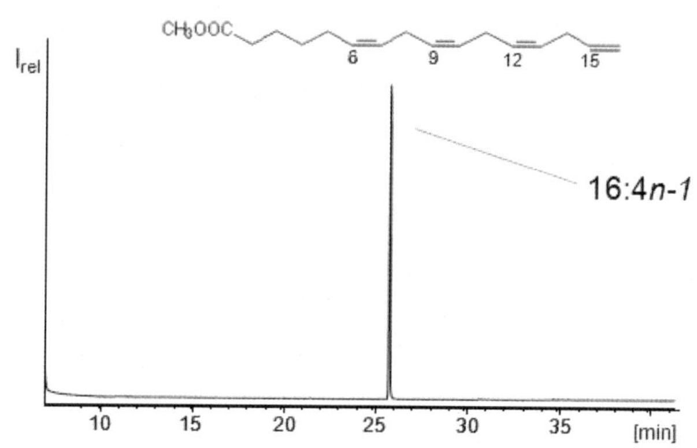

Figure: GC/MS total ion scan of the rare polyunsaturated fatty acid 6,9,12,15-hexadecatetraenoic acid (16:4*n*-1) methyl ester (with a terminal double bond) isolated by means of CCC from a fish oil.

Yet, in the field of lipids, the use of water in a two-phase solvent system reduces the solubility of fatty acids so that the injected sample amount has to be reduced. Typically, such drawbacks can be overcome by performing CCC in a two-step mode. In a first step, the whole PUFA fraction is enriched by pre-separation of the undesired compounds (e.g. saturated fatty acids when the goal is the isolation of PUFAs). This can be achieved by using only minute amounts of water. In a second step, the optimum conditions (here: n-heptane/methanol/water, 500/415/85) are used, and higher amounts of the PUFA fraction can be injected.

This strategy has also been used in form of a two-stage CCC method based on

1. an enrichment step followed by
2. an isolating step:

1. The Enrichment Step

The enrichment step is based on the CCC fractionation of the sample (extraction of for example 30 fractions). These are examined thoroughly by means of GC/MS. Usually GC/MS chromatograms of fatty acid methyl esters (FAMEs) are characterized by the major compound. Minor compounds are only noticeable when present at an abundance of ~1% relative to the dominating compound. By the CCC fractionation, minor fatty acids in the original sample can become major compounds in the corresponding fraction. As a consequence many more fatty acids than initially detected prior to the CCC run can be identified in the sample. Due to the good distribution of the compounds in the single fractions, the GC/MS analyses allow the identification of many compounds, which are not detectable without CCC fractionation. This strategy can be used to determine and identify trace compounds in samples. For instance, 430 different fatty acids could be detected in one butter sample (mainly in the filtrate obtained from urea complexation).

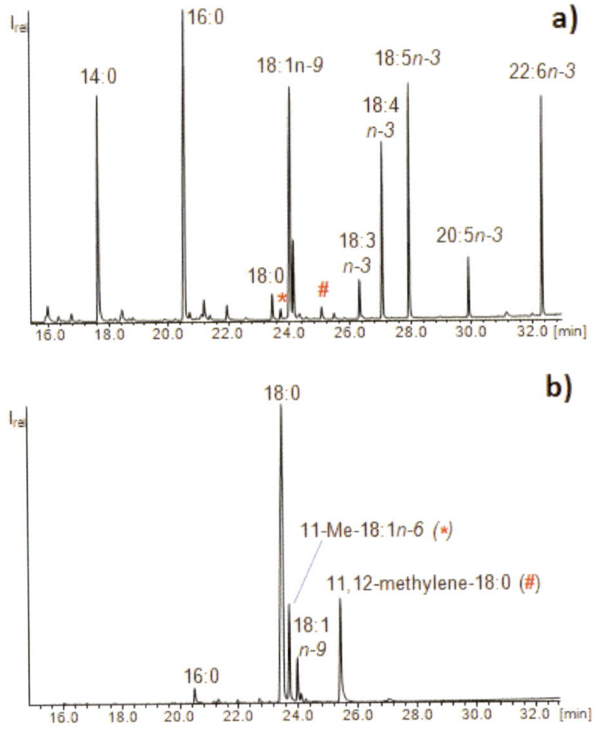

Figure: GC/MS chromatograms of (a) fatty acid methyl esters from a microalga and (b) a CCC fraction of the same sample illustrating the enrichment of the methyl esters of the minor fatty acids 11-Me-18:1n-6 (*) and 11,12-methylene-18:0 (#) ("enrichment step"). The full isolation of these rare fatty acids would require a second CCC ("isolation step").

Such attempts are best performed with FAMEs because the CCC fractions can be directly (or after dilution) measured by GC/MS. In such an enrichment mode, it makes little sense to optimize the separation in the case of complex mixtures of fatty acids because resolution of one pair will be accompanied with the co-elution of another pair.

The separation power of CCC lags behind HPLC. Fatty acids usually elute as comparably broad peaks which (in dependence of both the fractionation volume and the amount of the compound) distributed over 2 to 5 fractions. As a consequence, traces of 16:0 already eluted into the fraction of 18:0, and so did 18:1n-9 and other related fatty acids. These separation problems have then to be solved in the second CCC run, i.e. the isolation step.

2. The Isolation Step

As for other chromatographic systems, the retention (elution) time of a compound from the CCC system is largely independent of the injected amount. However, peaks of highly concentrated compounds are broader and this reduces resolution and separation power. For the isolation of minor compounds, a second CCC run (even with the same solvent system) can help to solve separation problems as illustrated in figure.

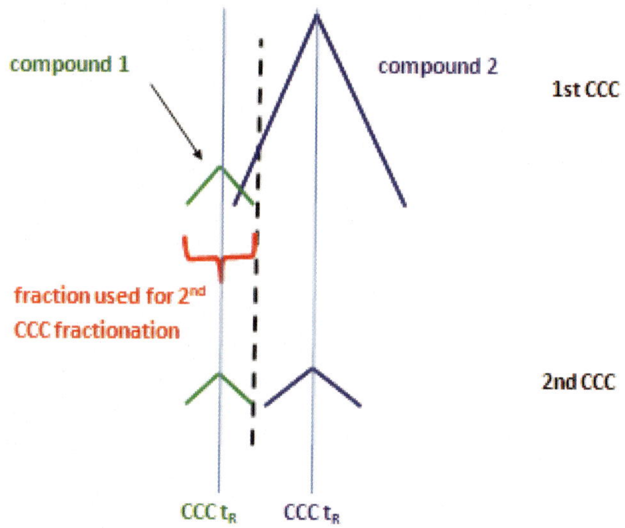

Figure: Simplified scheme of the repeated CCC fractionation using the same solvent system. In the 1st CCC (top, "enrichment step") major fatty acids and minor fatty acids can be enriched in individual fractions. When fractions from the 1st CCC containing compound 1 are analyzed in the 2nd CCC (bottom, "isolation step), they can be fully separated from compound 2.

In the upper plot, a minor compound 1 is overlapping with a major compound 2 and compound 1 cannot be isolated in neat form in one run. However, fractions containing compound 1 will contain lesser amounts of compound 2 than in the first run. When these fractions are re-injected, compound 1 can be separated from compound 2. This strategy will only be successful when both compounds elute differently and have significantly different abundances.

From the fractions of the enrichment step, those containing interesting compounds can be selected. Therefore, CCC has become an important tool in the field of lipid separations.

CCC separation of Compounds of the Unsaponifiable Matter of Lipids

Carotenoids

Crude plant extracts with carotenoids usually contain only a few compounds, with carotenoids being particularly nonpolar. For this reason, further enrichment steps can be avoided, and these compounds can be isolated in high purity with one CCC run. Carotenoids were successfully fractionated by using n-heptane/chloroform/acetonitrile (10:3:7) or n-hexane/ethanol/water (4/3/1 or 6/5/1.3) for the isolation of lutein. Due to their huge difference in polarity, free xanthophylls and esterified ones are eluting well-separated by CCC.

Sterols

CCC has also been used for the enrichment of β-sitosterol from a phytosterol standard (97% purity) with n-heptane-ethyl acetate-acetonitrile (5:1:5, v/v/v) as the solvent system. The solvent system n-hexane-ethyl acetate-acetonitrile (5:1:5, v/v/v) was used to isolate α-spinasterol (96.7 purity) and β-sitosterol (97.2% purity) from the root of Chinese herb Adenophora tetraphlla. The two-phase solvent system consisting of n-hexane-ethyl acetate-butanol-methanol-water (3.5:0.3:0.5:2.5:0.3, v/v) was used to isolate ergosterol (36.5 mg, purity 92.0%) and stigmasterol (43.6 mg; purity 95.5%) from 100 g of the Chinese herb A. roxburghii. Higher purities (99% or more) usually require two or even three CCC runs.

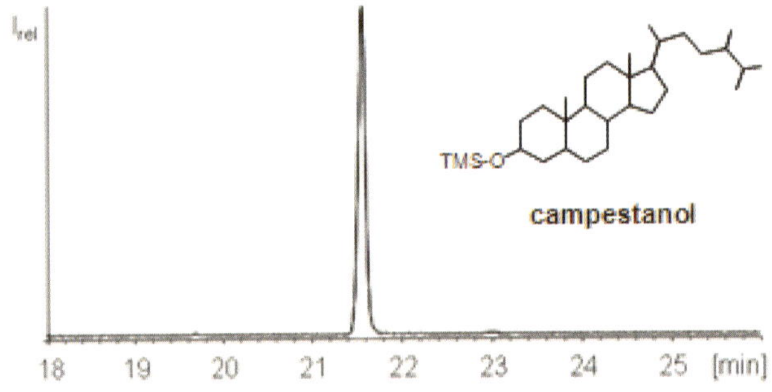

Figure: GC/MS chromatogram of the trimethylsilyl ether of campestanol isolated from vegetable oil by means of CCC.

Droplet Countercurrent Chromatography (DCC)

The technique of countercurrent chromatography (CCC) has seen a rapid expansion following the introduction of new methods such as droplet countercurrent chromatography (DCCC), rotation locular countercurrent chromatography (RLCC), and coil planet centrifugation. These methods have the advantage of being more rapid and less solvent consuming than traditional CCC. Furthermore, the advent of commercially available, compact apparatus has led to a widespread acceptance of these new liquid-liquid techniques as standard laboratory procedures for the separation of natural products. The detection of compounds that are eluted from a DCCC can be performed by three methods:

- UV detection for suitable UV-active substances,
- Monitoring of the fractions that are collected by TLC, and
- Weighing of fractions after evaporation of solvent.

The majority of DCCC separations involve polar compounds, especially glycosides, which are often difficult to purify. Chloroform-methanol-water systems of varying compositions remain the most widely used, in view of the good formation of droplets and the convenient viscosity of this combination. The most notable developments in the application of DCCC have occurred in the field of polyphenols, in particular in the separation of tannins. DCCC has also been applied in the separation of natural products such as alkaloids, triterpene glycosides, steroid glycosides, basic steroid saponins, and glycosides of flavonoids. Rotation locular countercurrent chromatog-raphy (RLCC) relies on the percolation of one layer of a two-phase solvent system through compartments (loculi) that contain the second layer. During passage of the mobile phase, the loculi (connected into tubes) are constantly rotated, to increase contact between the two phases. Basically, RLCC has the same advantages as DCCC. As in DCCC, the apparatus can be run in either ascending or descending solvent modes but the formation of droplets is not a necessary condition of RLCC. Consequently, a broader range of solvent system is possible, and a system containing ethyl acetate (often incompatible with DCCC) has been used, for example, in the separation of flavonoids— an important application of this method has been the separation of enantiomers of (±)-norephedrine on an instrument consisting of 16 columns and each column containing 37 loculi. The stationary phase was sodium hex-afluorophosphate solution at pH 4, and the mobile phase was(R, R)-di-nor-5-yl tartrate in 1,2-dichloroethane. Presumably, the enantiomers of (t)-norephedrine form different diastereotopic complexes with the tartrate ester, and these complexes are then partitioned differently between the two solvent phases. Separations by RLCC of a range of natural products, including flavones, xanthone glycosides, and antitumor antibiotics have been reported. RLCC provides a useful complementary method to DCCC in instances in which suitable solvent systems are not available.

Techniques Based on Separation

Size-exclusion Chromatography

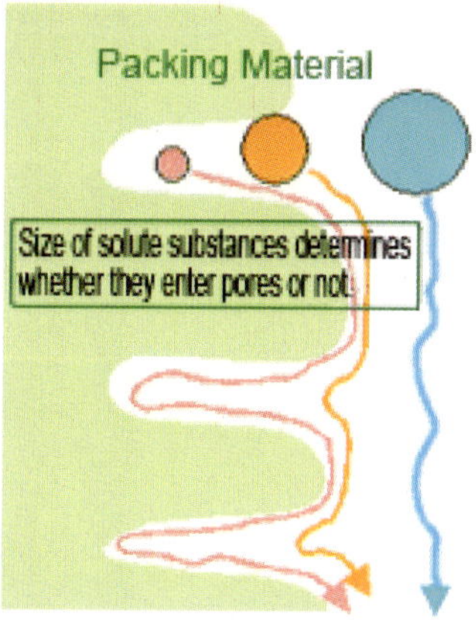

Principle of Size Exclusion Mode

Size exclusion chromatography is one of the HPLC separation modes. The column used is filled with material containing many pores. When dissolved molecules of various sizes flow into the column, smaller dissolved molecules flow more slowly through the column because they penetrate deep into the pores, whereas large dissolved molecules flow quickly through the column because they do not enter the pores. Consequently, larger molecules elute from the column sooner and smaller molecules later, which effectively sorts the molecules by size. This is the separation principle of size exclusion chromatography.

There are two basic types of size exclusion chromatography. One is gel permeation chromatography (GPC), which uses a hydrophobic column packing material and a non-aqueous mobile phase (organic solvent) to measure the molecular weight distribution of synthetic polymers. The other is gel filtration chromatography (GFC), which uses a hydrophilic packing material and an aqueous mobile phase to separate, fractionate, or measure the molecular weight distribution of molecules soluble in water, such as polysaccharides and proteins.

Information Obtained from Measuring Molecular Weight Distributions

Methods used to measure the molecular weight of polymers include the light scattering method, osmotic pressure method, and viscosity method. These allow obtaining the average molecular weight values. However, many polymers consist of molecules with not just one weight, but rather a collection of molecules with many different weights. Even for synthetic polymers with the same average molecular weight, the distribution of their molecular weights can differ, which can result in differences in their properties, as shown in the table below.

In this regard, size exclusion chromatography enables obtaining information about how much a sample contains of respective molecular weights. Therefore, it is widely used for quality control, such as to identify differences in properties of polymers.

Synthetic polymer properties that correlate to various average molecular weights and measurement methods.

	Correlating Polymer Properties	Measurement Method (absolute value)
Number Average Molecular Weight (Mn)	Tensile strength, impact strength, and hardness	Osmotic pressure and vapor pressure methods
Weight Average Molecular Weight (Mw)	Brittleness	Light scattering and ultracentrifuge methods
Z-Average Molecular Weight (Mz)	Deflection and rigidity	

Types of average molecular weight include the number average molecular weight (Mn), the weight average molecular weight (Mw), z-average molecular weight (Mz), and viscosity average molecular weight (Mv).

The number average molecular weight (Mn) is calculated by dividing the total polymer weight by the total number of polymer molecules, using equation.

$$\text{Mn}=\frac{\text{W}}{\sum \text{Ni}}=\frac{\sum(\text{Mi Ni})}{\sum \text{Ni}}=\frac{\left(\sum \text{Hi}\right)}{\left(\sum \text{Hi/Mi}\right)}$$

The weight average molecular weight (Mw) is calculated using equation, which emphasizes the contribution of polymers with larger molecular weights.

$$Mw = \frac{\sum(Wi\,Mi)}{W} = \frac{(\sum Hi\,Mi)}{\sum(Hi)}$$

W: Total weight of polymers

Wi: Weight of the polymer

Mi: Molecular weight of the elution time.

Ni: Number of molecules with molecular weight Mi

Hi: Height of the elution time

These are the relative average molecular weight values determined using size exclusion chromatography. To determine absolute values, the osmotic pressure and vapor pressure methods are used to determine Mn and light scattering and ultracentrifuge methods are used to determine Mw.

The peak molecular weight in size exclusion chromatography changes between Mn and Mw. The poly dispersity index (Mw/Mn) is used as the parameter that determines whether molecular weight distributions are narrow or broad. If the polydispersity value is close to 1.0, the distribution is narrow. Larger values mean a broader distribution.

Calibration Curve

The following equation generally describes the relationship between the elution volume (Ve) and molecular weight (M) of polymers.

log M = b - c Ve (where b and c are constants)

A graph of this relationship is called a calibration curve. Calibration curves are prepared by using molecular weight markers with known weights to find the elution volume and plotting the logarithm of the molecular weight on the vertical axis versus the elution volume on the horizontal axis (retention time). Molecules larger than the pore size pass straight through (are excluded). This is called the exclusion limit. Conversely, molecules below a certain size completely penetrate the pores and tend to elute almost in the same position. This is called the permeation limit. The calibration curve is valid in the range between this exclusion limit and permeation limit. Within this range, the larger the molecule, the sooner it elutes and the smaller the molecule the later it elutes.

Calibration curves with a steeply sloped portion are obtained using packing materials with large pores intended to measure larger molecules. Curves with more gradual slopes are obtained using packing materials with small pores, intended for measuring smaller molecules.

The lower the slope, the larger the difference in elution volume between small differences in molecular weight. Therefore, if two columns are available that can measure the target sample molecules; the one with the less sloped calibration curve will provide more accurate molecular weights.

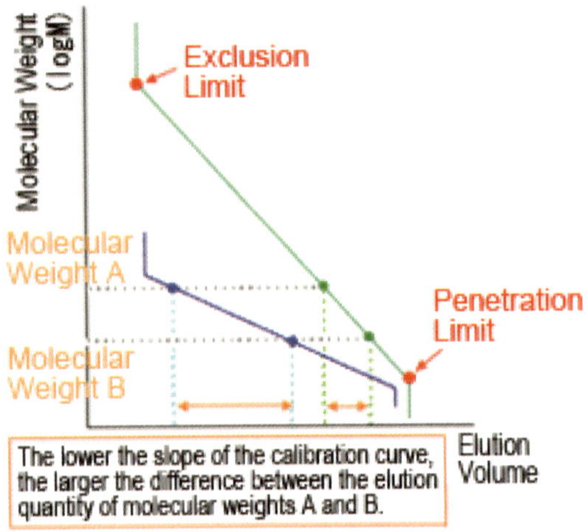

Slope of Calibration Curve and Differences in Elution Volume

Choosing a Column

Size exclusion chromatography columns are commercially available with a group (series) of different packing material pore sizes (different exclusion limits). Compare the calibration curves and select a column that is best suited to the range of molecule weights to be measured.

If the sample is distributed over a wide range of molecular weights, choose several columns from the same product series that are able to measure given sample range, then use them connected in series.

Alternatively, in some cases, a column packed with a mixture of materials with different pore sizes (mixed gel) is used. Such columns are used if the molecular weight range of target components in a sample is unknown or extremely broad.

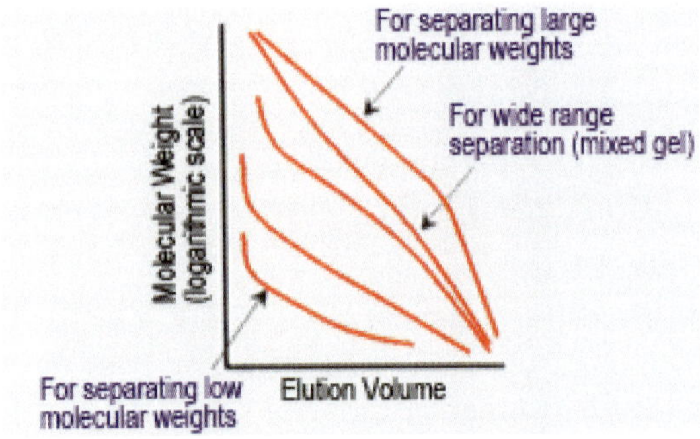

Calibration Curves for Various Columns

If samples contain molecules larger than the packing material pores, they are excluded and can cause a peak to appear near the exclusion limit. If this occurs, a column with larger pores should be selected.

Many size exclusion chromatography columns have limitations on the eluents that can be used. Using an eluent that is not specifically permitted could irreversibly expand the packing material or otherwise cause physical deterioration. Therefore, investigate which solvents readily dissolve the sample in advance and then choose a column compatible with one of those solvents.

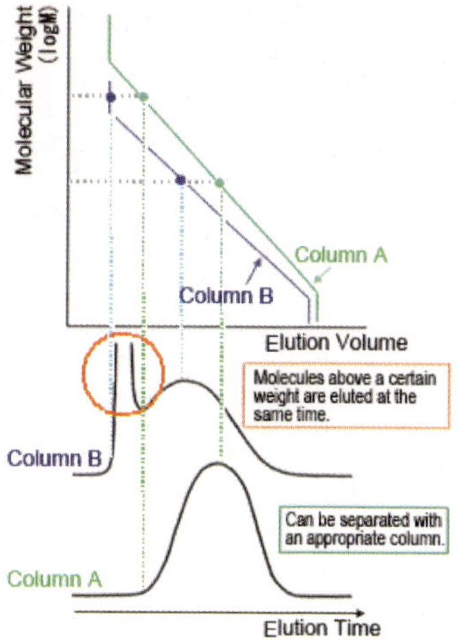

Selecting a Column (Select a column that allows using the sloped portion of the calibration curve.)

Expanded Bed Adsorption Chromatographic Separation

The development of molecular biology techniques has enabled researchers to generate large quantities of biologically important molecules from alien sources—transgenic bacteria, plants, and animals. Although this ability has revolutionized the production and delivery of pharmaceutical and therapeutic products, one problem has remained: All of the protein in the world won't do you any good if you can't purify it.

Anyone who has purified protein from bacteria will recall the hassles of lysing bacterial pellets. Regardless of whether you are performing enzyme- or detergent-based lysis, or physical cell disruption, there is still the problem ofseveral rounds of lysis and solution clarification. Even at this stage, many chromatographic resins are prone to clogging, and sample feedlots may require further rounds of clarification by ultracentrifugation or ultrafiltration. And with each stage of purification, not only is valuable time lost, but also valuable sample, and yields can drop dramatically in these first few steps. What chromatographers need is a method that combines sample preparation with the first stage of chromatography. Welcome to expanded-bed adsorption (EBA) chromatography.

Technology

EBA uses apparatus that is familiar to most users of standard liquid chromatography. The column has a flow adapter that is positioned to suit the specific step of resin preparation or protein purification. And a series of pumps and valves, connected through the adapter and bottom of the column, control the flow rate and direction of the buffer and sample loading. Thus, it is feasible to perform preliminary EBA trials with a little ingenuity and standard chromatographic equipment.

In the more traditional packed-bed methods, where the resin is confined between the bottom of the column and the flow adapter, clogging occurs when particulate matter and cell debris cannot flow around the closely packed resin beads. In contrast, EBA columns are fed from below, and the adapter is held away from the packed resin level, giving the resin room to expand and thus creating spaces between the beads.

The Mechanics of EBA

When the resin is packed in the column, the beads sit close together and leave little room for large aggregates and clumps to maneuver. Under typical chromatographic conditions, where the flow comes from the top of the column, this would lead to a buildup of particulate matter on the resin surface, eventually clogging the column.

As buffer is injected from below, the resin becomes fluidized and the beads form a stable concentration gradient when their sedimentation velocity equals the upward liquid flow velocity. The goal is to achieve a flow rate that exceeds the terminal velocity of the feedlot particulates but not that of the resin beads.To accentuate this difference, many standard resin beads have been modified to include inert quartz or metal alloy cores, depending on the application and source.

Unlike traditional resins, in which the beads tend to be uniform in size,the beads of EBA resins are variable, typically ranging from 50 to 400 μm. Thus, the larger particles populate the lower portion of the fluidized bed while the smaller particles populate the upper portions. If the beads are too small, expansion will occur at velocities comparable to the escape velocity of the particulate contaminants, lowering purification efficiency. Likewise, if the beads are too large, fluidization requires higher flow rates and protein binding is impaired because of improper diffusion among the beads.

As the sample feedlot is injected, the particulates and cell debris move freely around the resin beads and eventually leave through the top of the column. As with any chromatographic step, the resin then undergoes strenuous washing to limit nonspecific interactions between the particulates and resin. Meanwhile, the compounds of interest interact with the beads and are retained on the column. The column is then allowed to pack, the flow is reversed, and the compound is eluted from the beads as in traditional methods.

It is this last step that separates EBA from both batch-mode and fluidized-bed adsorption chromatography. In the latter methods, the compounds of interest can be eluted from the resin only in a batch- or stepwise manner, so there is little or no resolution of eluted molecules.By contrast, a packed column allows the use of elution gradients and improves elution peak resolution.

"The advantage is a higher recovery," says Guenter Jagschies, vice president of industrial separations for Amersham Biosciences. "That's where the money comes out for the customer." Jagschies says that EBA users often see a 25% increase in their recovery. "That is very important," he continues, "because we are talking about the first step in the downstream process, and you can never have more product after any later step. You always lose something."

Process Applications

As scientists scale up biopharmaceutical production, however, they are quickly reaching a point at which they will have two options. They must either install new fermentation capacity or increase the productivity of their cell lines. "This is an issue in the monoclonal antibody business," says Jagschies, "where in 3 or 4 years' time, we are not going to have enough production capacity in terms of fermenters to meet the demands from the biopharmaceutical industry."

According to Jagschies, recent advances in increasing the monoclonal antibody (mAb) productivity of Chinese hamster ovary cells from a few hundred micrograms per liter to 1 g/L have been paralleled by increasing problems with cell death and the contamination of mAb solutions with cellular proteases. This decreases the final yield, purity, and homogeneity of the mAbs. “People have found that by using EBA very early in the process,” adds Jagschies, “they can capture these proteases and can reduce these kinds of problems. They actually got purer products as compared to [using] packed-bed columns.”

Beyond mAb purification, Kaoru Kobayashi of Yoshitomi Pharmaceutical Industries recently reported on the purification of pharmaceutical-grade recombinant human serum albumin (rHSA) that had been expressed in the methylotrophic yeast Pichia pastoris. HSA is used in the treatment of severe hypoalbuminemia and traumatic shock, but it suffers from the fact that it is purified from whole blood and may thus contain bloodborne contaminants such as HIV. Expressing rHSA in a nonhuman system, however, eliminates these contaminants. Similarly, Jörg Thömmes and colleagues at the Heinrich-Heine University and Research Centre Jülich in Jülich (Germany) and Mucos Pharma GmbH and Co. in Geretsried (Germany) used EBA as part of an integrated fermentation and purification process for the isolation of recombinant human chymotrypsinogen B expressed in P. pastoris.

Nonprotein Applications

EBA also may have a large impact in the preparation of pharmaceutical-quality DNA for nonviral vaccines and gene therapeutics. Owen Thomas and colleagues at the Centre for Process Biotechnology in the Technical University of Denmark worked closely with developers at UpFront Chromatography to explore the use of EBA in the separation of plasmid DNA from the shear-sensitive solids that result from the lysis of microbial cells. The solids include flocculant material from the cell membranes and organelles as well as host nucleic acids.

The Danish researchers developed a series of polyethylene imine- and diethylaminoethyl-linked agarose resins that bind DNA more tightly than commercial anion exchange resins, although at low ionic strength they have experienced some problems of individual DNA strands binding to more than one resin bead, effectively cross-linking the beads.

Further, the researchers developed a plasmid-specific affinity resin by linking avidin to the resin bead and biotin to an oligonucleotide, the sequence of which matches a portion of the target plasmid. By first forming a triple helix between the biotinylated oligonucleotide and the plasmid, the researchers can draw the plasmid from solution through the binding of biotin by avidin.

Nonprocess Applications

Perhaps the most interesting use of EBA is not in the isolation of potential therapeutics but in diagnostics and epidemiology. Isabelle Accoceberry and her colleagues in Bordeaux (France) recently reported on the use of immunoaffinity EBA in the isolation of parasite spores from human stool samples. The researchers created a mAb that bound a protein on the surface of the spores of Enterocytozoon bieneusi, an opportunistic protistan parasite that causes diarrhea, malabsorption, and weight loss. Conjugating the mAbs to a Streamline rProtein A adsorbent, the French group isolated spores from numerous human stool samples, detecting the spores by an indirect immunofluorescence antibody test.

“The ability to obtain a pure population of whole mature spores,” wrote Accoceberry, “free of all fecal contaminants, should facilitate biological, biochemical, and immunological studies of the infective stage of E. bieneusi.”

Supercritical Fluid Chromatography

A supercritical fluid is the phase of a material at critical temperature and critical pressure of the material. Critical temperature is the temperature at which a gas cannot become liquid as long as there is no extra pressure; and, critical pressure is the minimum amount of pressure to liquefy a gas at its critical temperature. Supercritical fluids combine useful properties of gas and liquid phases, as it can behave like both a gas and a liquid in terms of different aspects. A supercritical fluid provides a gas-like characteristic when it fills a container and it takes the shape of the container. The motion of the molecules are quite similar to gas molecules. On the other hand, a supercritical fluid behaves like a liquid because its density property is near liquid and, thus, a supercritical fluid shows a similarity to the dissolving effect of a liquid.

The characteristic properties of a supercritical fluid are density, diffusivity and viscosity. Supercritical values for these features take place between liquids and gases. Table demonstrates numerical values of properties for gas, supercritical fluid and liquid.

	Gas	Supercritical fluid	Liquid
Density (g/cm^3)	$0.6 \times 10^{-3} – 2.0 \times 10^{-3}$	0.2 - 0.5	$0.2 \times 10^{-5} - 2.0 \times 10^{-5}$
Diffusivity (cm^2/s)	0.1 - 0.4	$10^{-3} - 10^{-4}$	$0.2 \times 10^{-2} - 3.0 \times 10^{-2}$
Viscosity (cm/s)	$1 \times 10^{-4} - 3 \times 10^{-4}$	$1 \times 10^{-4} - 3 \times 10^{-4}$	0.6 - 2.0

Supercritical fluid properties compared to liquids and gases

Just like supercritical fluids combine the benefits of liquids and gases, SFC bring the advantages and strong aspects of HPLC and GC together. SFC can be more advantageous than HPLC and GC when compounds which decompose at high temperatures with GC and do not have functional groups to be detected by HPLC detection systems are analyzed.

There are three major qualities for column chromatographies:

- Selectivity.
- Efficiency.
- Sensitivity.

Generally, HPLC has better selectivity that SFC owing to changeable mobile phases (especially during a particular experimental run) and a wide range of stationary phases. Although SFC does not have the selectivity of HPLC, it has good quality in terms of sensitivity and efficiency. SFC enables change of some properties during the chromatographic process. This tuning ability allows the optimization of the analysis. Also, SFC has a broader range of detectors than HPLC. SFC surpasses GC for the analysis of easily decomposable substances; these materials can be used with SFC due to its ability to work with lower temperatures than GC.

Instrumentation for SFC

As it can be seen in figure, SFC has a similar setup to an HPLC instrument. They use similar stationary phases with similar column types. However, there are some differences. Temperature is critical for supercritical fluids, so there should be a heat control tool in the system similar to that of

GC. Also, there should be a pressure control mechanism, a restrictor, because pressure is another essential parameter in order for supercritical fluid materials to be kept at the required level. A microprocessor mechanism is placed in the instrument for SFC. This unit collects data for pressure, oven temperature, and detector performance to control the related pieces of the instrument.

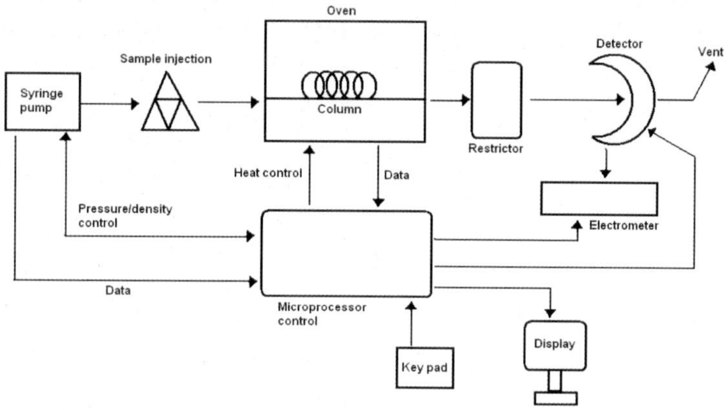

Scheme of a supercritical fluid chromatography instrument.

Stationary Phases

SFC columns are similar to HPLC columns in terms of coating materials. Open-tubular columns and packed columns are the two most common types used in SFC. Open-tubular ones are preferred and they have similarities to HPLC fused-silica columns. This type of column contains an internal coating of a cross-linked siloxane material as a stationary phase. The thickness of the coating can be 0.05-1.0 μm. The length of the column can range from of 10 to 20 m.

Mobile Phases

There is a wide variety of materials used as mobile phase in SFC. The mobile phase can be selected from the solvent groups of inorganic solvents, hydrocarbons, alcohols, ethers, halides; or can be acetone, acetonitrile, pyridine, etc. The most common supercritical fluid which is used in SFC is carbon dioxide because its critical temperature and pressure are easy to reach. Additionally, carbon dioxide is low-cost, easy to obtain, inert towards UV, non-poisonous and a good solvent for non-polar molecules. Other than carbon dioxide, ethane, n-butane, N_2O, dichlorodifluoromethane, diethyl ether, ammonia, tetrahydrofuran can be used. Table shows select solvents and their T_c and P_c values.

Solvent	Critical Temperature (°C)	Critical Pressure (bar)
Carbon dioxide (CO_2)	31.1	72
Nitrous oxide (N_2O)	36.5	70.6
Ammonia (NH_3)	132.5	109.8
Ethane (C_2H_6)	32.3	47.6
n-Butane (C_4H_{10})	152	70.6
Diethyl ether (Et_2O)	193.6	63.8
Tetrahydrofuran (THF, C_4H_8O)	267	50.5
Dichlorodifluoromethane (CCl_2F_2)	111.7	109.8

Detectors

One of the biggest advantage of SFC over HPLC is the range of detectors. Flame ionization detector (FID), which is normally present in GC setup, can also be applied to SFC. Such a detector can contribute to the quality of analyses of SFC since FID is a highly sensitive detector. SFC can also be coupled with a mass spectrometer, an UV-visible spectrometer, or an IR spectrometer more easily than can be done with an HPLC. Some other detectors which are used with HPLC can be attached to SFC such as fluorescence emission spectrometer or thermionic detectors.

Advantages of Working with SFC

The physical properties of supercritical fluids between liquids and gases enables the SFC technique to combine with the best aspects of HPLC and GC, as lower viscosity of supercritical fluids makes SFC a faster method than HPLC. Lower viscosity leads to high flow speed for the mobile phase.

Thanks to the critical pressure of supercritical fluids, some fragile materials that are sensitive to high temperature can be analyzed through SFC. These materials can be compounds which decompose at high temperatures or materials which have low vapor pressure/volatility such as polymers and large biological molecules. High pressure conditions provide a chance to work with lower temperature than normally needed. Hence, the temperature-sensitive components can be analyzed via SFC. In addition, the diffusion of the components flowing through a supercritical fluid is higher than observed in HPLC due to the higher diffusivity of supercritical fluids over traditional liquids mobile phases. This results in better distribution into the mobile phase and better separation.

Applications of SFC

The applications of SFC range from food to environmental to pharmaceutical industries. In this manner, pesticides, herbicides, polymers, explosives and fossil fuels are all classes of compounds that can be analyzed. SFC can be used to analyze a wide variety of drug compounds such as antibiotics, prostaglandins, steroids, taxol, vitamins, barbiturates, non-steroidal anti-inflammatory agents, etc. Chiral separations can be performed for many pharmaceutical compounds. SFC is dominantly used for non-polar compounds because of the low efficiency of carbon dioxide, which is the most common supercritical fluid mobile phase, for dissolving polar solutes. SFC is used in the petroleum industry for the determination of total aromatic content analysis as well as other hydrocarbon separations.

Centrifugal Partition Chromatography

Centrifugal Partition Chromatography (CPC) is one of the various techniques constituent of Countercurrent Chromatography (CCC). Centrifugal Partition Chromatography is also known as Hydrostatic Countercurrent Chromatography. This refers to the rotary movement of the column which rotates around one single rotation axis, contrary to Hydrodynamic Countercurrent Chromatography systems that have a planetary rotation movement around two rotating axes.

Working Principle

Centrifugal Partition Chromatography is a liquid-liquid chromatography technique. The kromaton fast centrifugal partitioning chromatograph thus functions based on the principles of liquid/liquid partitioning chromatography: two immiscible liquid phases are mixed together to form a two-phase system, and are then separated multiple times. The individual solutes are isolated based on the different partitioning coefficients of each compound in this two-phase system.

One of the liquid phases of the two-phase system is used as a stationary liquid phase: it is fed into the column (the rotor) while the latter is spinning at moderate rotational speed. The stationary phase is retained inside the rotor by the centrifugal force generated.

The second phase of the two-phase system is used as the mobile phase containing the solutes to be extracted. It is fed under pressure into the rotor and pumped through the stationary phase.

Both phases are mixed together. It is at that time that the exchange of molecules between the two phases occurs. The separation of the solutes is achieved as a function of the specific partitioning coefficient (Kd) of each solute between the mobile and stationary phases. The mobile phase then decants at each cell outlet thus entering the next cell.

The eluted fractions of the mobile and stationary phases are collected over a period of several minutes to several hours. These fractions, or eluates, will contain the individual purified solutes.

Application Fields

Centrifugal Partition Chromatography is a preparative chromatography technique. It is an alternative to classic purification techniques such as preparative HPLC (High Performance Liquid Chromatography) or Flash Chromatography.

Centrifugal Partition Chromatography can be used for:

- Final purification (purity > 99.5%) of molecules derived from natural extracts, synthetic mixtures or biological matrices.
- Fractionization or pre-fractionization of highly complex crude extracts.
- Liquid-liquid extraction using the newly developed Centrifugal Partition Extractors (CPE) prototypes.

APOLAR POLAR

Simple lipids - Sterols - Inorganic elements - Polyphenols - Polysaccharides-Proteins
Phospholipids - Various inorganic acids - Anthocyanins - Peptides
Glycolipids - Alkaloids - Flavonoids - α amino acids

Benefits

Because there is no packed solid phase required, Centrifugal Partition Chromatography and the FCPC range offers many benefits:

- Cost effective
  - No solid packing material that is costly to buy, replace or reprocess.
  - Low solvent consumption (10 to 25% less).
  - One single rotor adapted to all types of separations.
- A soft technique
  - No irreversible adsorption or sample loss -> 100% recovery.
  - No denaturation of fragile molecules due to interaction with silica -> adapted to fragile molecules.
- Efficient
  - High purification levels up to 99.9%.
- Faster
  - 3 to 5 times faster than other current solutions.
- Flexible
  - Adapted to all types of products, natural or synthetic.
  - Well-adapted to a very wide range of polarities.
- Ideal for easy scale up
  - One single separation/purification technique adapted to samples ranging from mg to several kg.

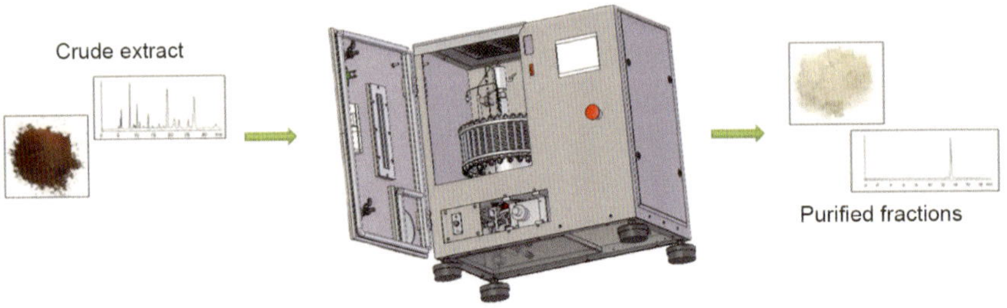

Electrochromatography

Electrochromatography is a separation method of mixture of chemical compounds which uses simultanously occurring interphase reactions and electric field influence. Theelectroosmotically driven flow is employed in the discussed systems. Sometimes a pressure-driven flow is ,also used.

Various types of interactions, mentioned above provide the new separation possibilities. However, a selection of separation conditions cannot be only limited to generally known methods as chromatography and electrophoresis. Mutual relations between interactions in both kinds of systems, mentioned above, should be also considered. For example, selecting the chemical composition of both phases - the electrical properties (e.g. conductivity) should be taken into account at the same time. Therefore, there is still basic question to resolve: what is actual relation between the two kinds of influences. This is the fundamental problem.

Thin-layer Electrochromatography

For investigation of the electrophoretic mobility, in thin-layer electrochromatography (electrophoresiswith chromatographic phenomena) the following relationship has been used:

$$u = \frac{x - x_i}{tE}\left(\frac{l'}{l}\right)^2 = \frac{(x - x_i)(1 + T)}{Et}$$

where x denotes apparent (macroscopic) length of migration, under the influence of electric field and solution movement measured along the medium surface, x_i – apparent length of electroosmotically driven flow indicator (non-charged compound e.g. glucose), E -electric field strength, i-apparent (macroscopic) length of the pore (capillary), measured along the medium layer, l' - actual length of the tortuos pore (capillary) of the medium layer, and $T(l'/l)-1$ is the path length parameter or structure coefficient for the layer.

But in thin-layer electrochromatography, when "chromatognlphic phenomena" (adsorption, partition, etc.) influence the rate of migration 'of compounds moving under the action of electric field the coefficient R_D analogous to the chromatographic coefficient R_F, should be taken into account. Therefore we obtain for the length of migration:

$$x = \frac{x_s}{R_{D,s}} \quad x_i = \frac{x_i}{R_{D,i}}$$

where x_s x' denote apparent length of migration of the particles influenced by "chromatographic phenoména".

Because the above mentioned phenomena retard the movement of the substances subject to electric field strength and/or solution flow action - the $R_{D,i,s}$ coefficients should be lower than unity ($R_{D,i,s}$ < 1.0). Table gives, for some cases, the experimentally obtained values of coefficient R_D.

Plots in figure represent electroosmotic solution flow velocities as a function of thin-layer porous media compositions.

Kowalczyk and Czapliliski proposed the use ofa hydraulic flow (electrolyte cells on various levels) as additional to electroosmotic one for separation of substances by means of thin-layer electrophoresis. The increased flow of the liquid phase, obtained in that way can be useful for separation optimization also for electrochro- matography, instead of pressure-driven flow. Ten years later also Pretorius and coworkers have been using thin-layer electrochromatography and discussed the advantages of this technique over conventional chromatography and electrophoretic methods ofseparation.

Table: Val ues ofcoefticient R_D for thin-layer electrochromatography for electcoosmotically driven chroillatography for glucose, analogous to R_F in TLC, which is the retardation coefficient for substance molecules, subject to interphase phenomena, during the movement in electric field.

Solution :	Veronal buffer, pH 8.6	0.05 mol l^{-1} hydrochloric acid							
Stationary phases:	Filter Paper Whatman 1	Glass fiber	Filter paper whatman 1	TiO_2	starch	kaolin	Al_2O_3	Glass fiber	SiO_2
Substances:	R_D-values								
Glucose	0.95	1.0	0.97	0.97	0.97	1.0	1.0	1.0	0.9
Ox albumin	0.91	1.0	-	-	-	-	-	-	-
Fe^{3+}	-	-	1.0	1.0	0.0	0.37	1.0	0.0	0.0
Cu^{2+}	-	-	1.0	1.0	0.61	0.91	1.0	1.0	0.0
Co^{2+}	-	-	1.0	1.0	-	-	1.0	1.0	0.0

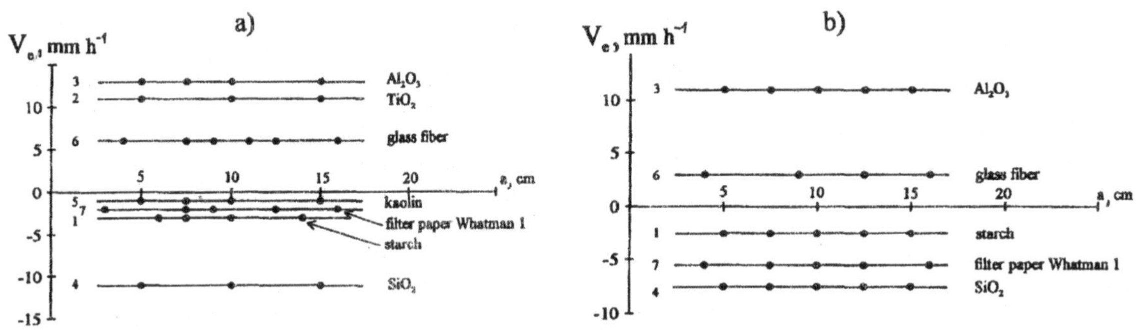

Figure: Diagrams of electroosmotic flow in different layers: solution flow velocities as a functions along the medium layers. Voltage: 7.5 V cm^{-1}, time: 60 min. a) 0.05 mol l^{-1} hydrochloric acid, b) veronal buffer, pH =8.6.

Chromatography Special Techniques

Hydrophobic Interaction Chromatography

Hydrophobic interaction chromatography (HIC) is a powerful technique used for the purification of proteins in analytical and preparatory applications.

HIC separates and purifies protein molecules on the basis of their hydrophobicity – it is more popular than other chromatography techniques for the separation of proteins as it employs a less denaturing environment compared to them. Therefore the biological activity of the proteins is kept intact.

HIC can also be effectively used to remove impurities or product aggregate species in aqueous solutions as it exploits the difference in hydrophobic properties of the aggregates and the target molecules. It is often used in combination with techniques such as ion exchange or gel filtration chromatography.

The Principle of Hydrophobic Interaction Chromatography

In HIC, the sample protein molecules are introduced into the column containing a high-salt buffer. The salt promotes interaction between the hydrophilic and hydrophobic regions of the protein and the medium by reducing the solvation of sample molecules and exposing their hydrophobic regions.

The amount of salt needed to promote binding is inversely proportional to the hydrophobicity of the molecules. Hence the sample molecules can be eluted out in the order of increasing hydrophobicity using a decreasing salt gradient. Bound protein molecules can be effectively desorbed by washing with a dilute buffer or water.

The Steps in Hydrophobic Interaction Chromatography

- HIC media are composed of alkyl or aryl ligands coupled to an inert, porous matrix which are then packed into a chromatography column as a packed bed arrangement.
- A moderately high salt buffer is used to fill the pores and space between particles in the matrix. Typical salts used are 1–2M ammonium sulfate or 3M sodium chloride which are selected to promote the key interaction between the protein sample and the medium whilst minimizing that of other less hydrophobic proteins (impurities).
- The column is washed to remove non-bound proteins.
- The salt concentration is gradually lowered to start eluting proteins. The manipulation of salt gradients allows differential elution of the proteins - Proteins having the lowest hydrophobicity are eluted first.
- A final wash using salt-free buffer helps remove tightly-bound proteins. Additives in the buffer can further promote desorption of the bound proteins. Additives may include water-miscible alcohols, chaotropic ("salting-in") salt solutions and detergents.
- Rarely, harsher conditions such as 0.5–1.0M sodium hydroxide, 70% ethanol, or 30% isopropanol are required to remove all bound proteins.

Crucial Factors that Affect Hydrophobic Interactions

- Ligand - The type of ligand used determines the adsorption behavior of proteins e.g. straight chain alkyl ligands is hydrophobic while aryl ligands show both aromatic and hydrophobic interactions.
- Matrix - Hydrophilic carbohydrates such as cross-linked agarose and synthetic copolymer materials are the most commonly used supports. Different supports will vary in their selectivity even with the same ligand.
- Degree of substitution - The binding capacity of a protein is directly proportional to the degree of substitution of the ligand. However, high levels of ligand substitution increase the strength of the interaction and make it difficult to elute the proteins.
- Temperature – There is a direct and positive correlation between temperate and the affinity of hydrophobic interactions. High temperature also influences the structure and solubility of proteins. Temperature is seldom used to modulate elution of molecules in HIC.

- pH – The mobile phases used in HIC mostly have a neutral pH range of 5 – 7. The effect of pH on protein-medium interactions varies from protein to protein. Generally, the hydrophobic interaction between media and protein decreases with increase in pH as the protein charge tends to increase. Whilst pH can have an effect on the degree of protein binding, it is not considered significant enough to use pH gradients for the elution of solute molecules.
- Salt concentration – Salt addition to the buffer and sample promotes ligand-protein binding but there is a risk of protein precipitation at high salt concentrations. Sodium, ammonium, or potassium sulfates are known to produce higher precipitation effects though they are very effective in promoting interaction between the ligand and the protein.

Two-dimensional Chromatography

Two-dimensional chromatography is a two-dimensional separation method where the separation processes are both chromatography. initial separation proceeds in a primary column and 'cuts' of the eluent from this column are directed into one or more secondary columns.

Two Dimensional Liquid Chromatography (2DLC)

The term "two dimensional liquid chromatography (2DLC)" refers to the technique in which two independent liquid phase separation systems are applied to the sample. 2DLC can be done by transferring either only the interesting portion of the first dimension effluent onto the second dimension, this is referred to as "heartcutting" chromatography, or by sequentially transferring the entirety of the first dimension effluent, in many small aliquots, onto the second dimension; this is known as "comprehensive" 2D chromatography. From an analytical perspective it is very fruitful to think of the second column and its associated detector as comprising a very chemically selective detector for the separation on the first column. The simple block diagram of a generalized 2DLC system in figure shows all of the components contained by the dashed box effectively act as a chemically selective detector for peaks as they elute from the first dimension column.

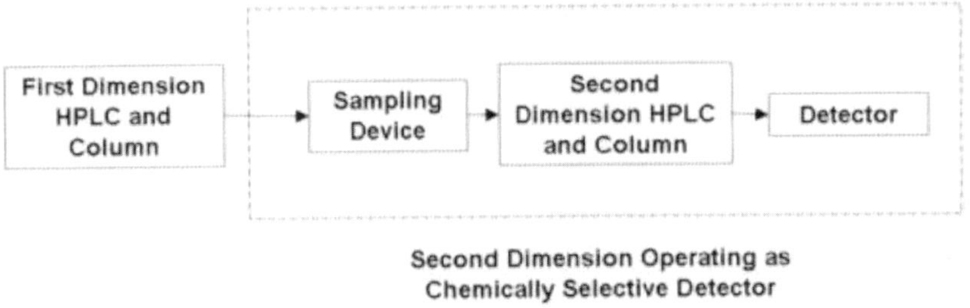

Figure: Block diagram of instrumentation for 2DLC. The dashed box indicates that the second dimension of the system effectively acts as a chemically selective detector for the peaks that elute from the first dimension column.

Karger, Snyder, and Horvath, and later Giddings and Guiochon pointed out the value of a true comprehensive two-dimensional separation lies in the idea that under ideal circumstances the overall peak capacity (n_c,2D) should be equal, not to the sum, but rather to the product of the individual peak capacities of the first and second dimension separations (${}^1n_c$ and ${}^2n_c$, respectively):

$$n_{c,2D} = nc1 \times nc2$$

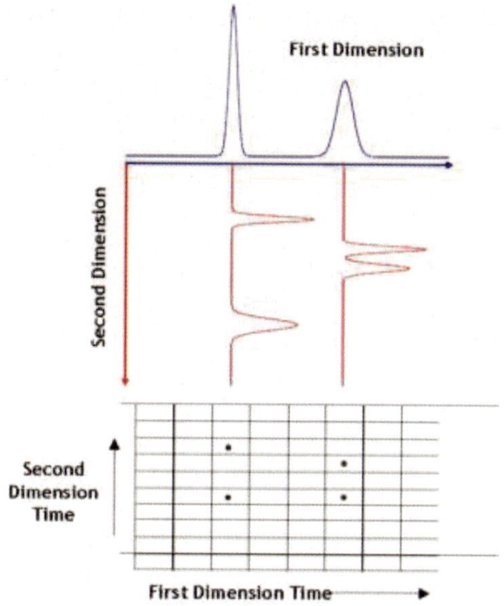

Figure: Illustration of the multiplicative relationship between the peak capacities of the independent first and second dimensions in comprehensive two-dimensional separations.

The history of the development of the theory and practice of multi-dimensional liquid chromatography was carefully reviewed by Schure . Work on "online" comprehensive 2DLC was initiated by Erni and Frei in 1978. In 1990, Jorgenson and coworkers utilized comprehensive 2DLC for protein separations; the full separation took six hours. By the early part of this century various forms of 2D liquid phase separation methods had become a mainstay of proteomics research. In 2006, Stoll and Carr used supra-ambient temperatures (110° C) and high linear velocities (> 3 cm/s) on the second dimension column to speed up second dimension gradient elution separations to about 20s, and thereby greatly shortening the overall comprehensive 2DLC separation to about 30 min. A schematic of the instrumentation used for this work is shown in figure. This opened a new era in the fast, routine use of 2DLC. A comparison of the separating ability demonstrated in Jorgenson's initial work of 1990 and this lab's work in 2006 is given in Fig. While Jorgenson's first work achieved a peak capacity production rate of 0.36/min, the latest work has a peak capacity productivity of 33/min, a nearly 100-fold improvement.

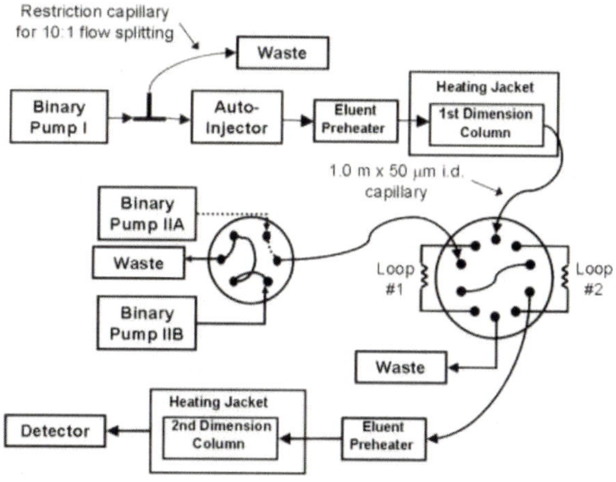

Figure: Schematic of instrumentation used for comprehensive, fast 2DLC using high temperature and high velocity to achieve high speed. From ref.

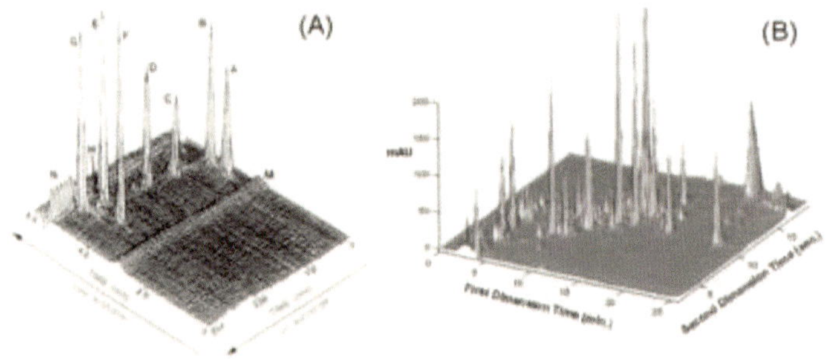

Figure: Comparison of (A) Jorgenson et al.'s 2D- and (B) Carr et al.'s 2D-separation chromatograms. Conditions: (A) protein sample, 5 µL/min, 0% to 100% buffer B from 20 to 260 mins; buffer A, 0.2 M NaH_2PO_4, pH 5; buffer B, 0.2 M NaH_2PO_4/0.25 M Na_2SO_4, pH 5. Detection at 215 nm. (B) corn seedling extract, 50 mm × 2.1 mm i.d. Discovery HS-F5; 0.10 mL/min; 5 to 70% B from 0 to 23 mins; buffer A, 20 mM NaH_2PO_4, 20 mM $NaClO_4$, pH 5.7; solvent B, aceto-nitrile; 40 °C; Second Dimension, 50 mm × 2.1 mm i.d. ZirChrom-CARB; 3.00 mL/min; 0 to 70% B in 17.4 s; buffer A, 20 mM $HClO_4$ in water; solvent B, acetonitrile; 110 °C; detection at 220 nm.

Because of its high resolving power, 2DLC has received a great deal more attention during the past few years, especially by those dealing with complex samples. Our group has successfully developed fast 2DLC based on a single pair of two judiciously selected RPLC phases for the separation of a number of complex mixtures, including corn seedling extracts, coffee, red wine and human urine. Samples derived from proteomic and metabolomic research represent two of the more complex but common types of samples which require multi-dimensional liquid chromatography for the successful resolution of a large number of constituents. Proteomics is the large-scale study of proteins as expressed in biological systems. Metabolomics is the study of the low molecular weight compounds present in organisms which participate in the metabolism required for growth, maintenance and normal functioning. Typically proteomic and metabolomic samples contain thousands of constituents whose concentrations are widely different. Figures illustrate some possibilities of what can be learned when a single 1DLC peak was subjected to a second very different separation mechanism. Figure shows the limited resolving power of a 1D-separation, wherein one peak in the first dimension produced nine peaks in the second dimension. Figure shows the serious dynamic range problem wherein the peaks of low abundance species are obscured in the 1D-separation by a single dominant peak of much higher concentration in the 1D-separation. In both cases, the separation requires high resolving power, that is, in chromatographic terms both problems require high peak capacities.

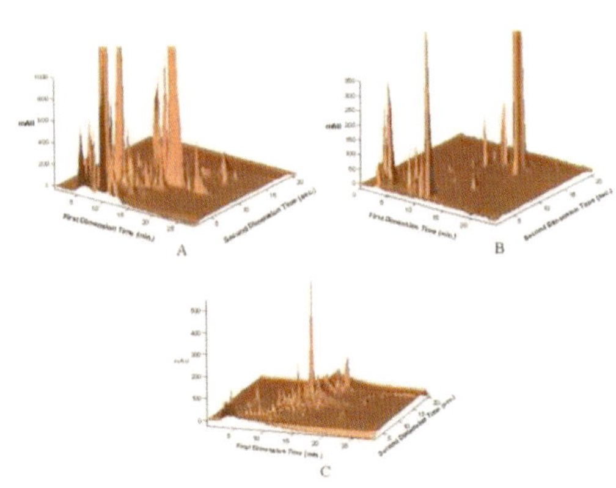

Figure: 2DLC chromatograms of (A) urine; (B) coffee; (C) red wine. Conditions: Second Dimension – Flow rate, 3.00 mL/min.; Gradient elution from 0-85 %B from 0-21 seconds, where A is 10 mM perchloric acid in water, and B is acetonitrile; injection volume, 34 µL, Temperature, 110 °C, 33 mm × 2.1 mm i.d. ZirChrom-CARB (8% carbon (w/w)). First Dimension-, Urine - Flow rate, 0.10 mL/min.; Gradient elution from 0-50 %B from 0-30 minutes, where A is 20 mM sodium dihydrogen phosphate, 20 mM sodium perchlorate, pH 5.7, and B is acetonitrile ; injection volume, 20 µL, Temperature, 35 °C, 200 mm × 2.1 mm i.d. Discovery HS-F5; Coffee - Flow rate, 0.10 mL/min.; Gradient elution from 0-40 %B from 0-30 minutes, where A is 20 mM sodium dihydrogen phosphate, 20 mM sodium perchlorate, pH 5.7, and B is acetonitrile ; injection volume, 20 µL, Temperature, 35 °C, 100 mm × 2.1 mm i.d. Discovery HS-F5; Red Wine - Flow rate, 0.10 mL/min.; Gradient elution from 0-50 %B from 0-23 minutes, where A is 20 mM sodium dihydrogen phosphate, 20 mM sodium perchlorate, 0.2 mM EDTA, pH 5.7, and B is acetonitrile ; injection volume, 20 µL, Temperature, 35 °C, 100 mm × 2.1 mm i.d. Discovery HS-F5.

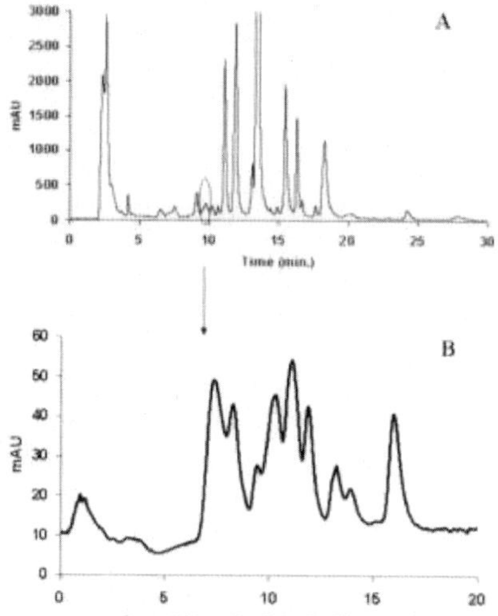

Figure: First (A) and second (B) dimension chromatograms of corn seedling separation, where the second dimension chromatogram is obtained from 9.80-10.15 min in the first dimension

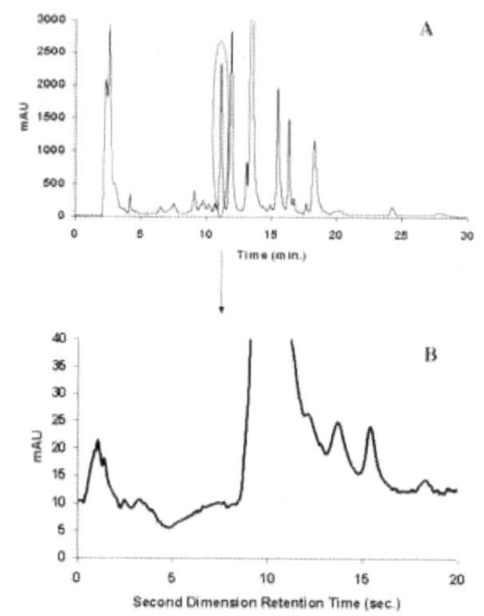

Figure: First and second dimension chromatograms of corn seedling separation in Fig, where the second dimension chromatogram is obtained from 11.55-11.90 min in the first dimension.

Scope of 2DLC as an Analytical Tool

The capabilities of different variations of the liquid chromatography technique range quite widely from dealing with very simple samples containing a few unique chemical constituents, to very complex samples containing literally thousands of constituents. The ability to deal with increasingly complex samples invariably comes at the cost of increased analysis times. Figure shows the historical view of the role of 2DLC relative to other modes of LC including isocratic and gradient elution 1DLC; indeed, until very recently 2DLC has been a niche technique accessible to only a handful of research laboratories. Figure also indicates a number of ways that fundamental research to improve the performance of LC separations has impacted the scope of different modes of LC. The general trend has been to improve existing modes of LC to handle increasingly complex samples in less and less analysis time.

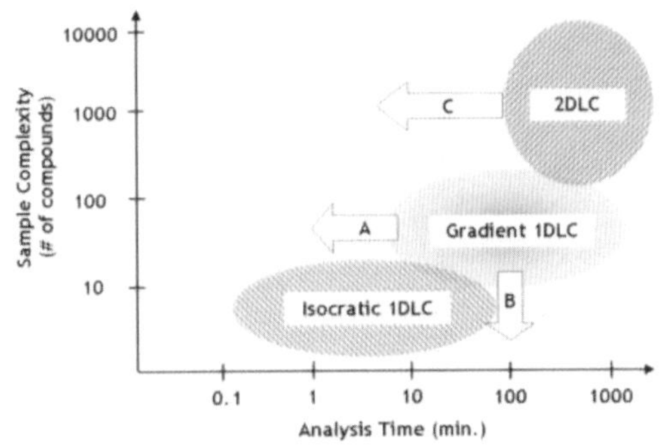

Figure: Domain of various LC modes as a function of sample complexity and analysis time. The elipses indicate the historical domains of the different modes; the arrows indicate changes in the perceived limits of these domains as a result of recent fundamental research in these areas.

A few areas of improvement are worth mentioning here. Schellinger and coworkers studied the factors limiting the speed of gradient elution 1DLC and found that the historical view that column sets the re-equilibration time and thus the speed of gradient elution 1DLC is misguided, and that in most cases the design of instruments used for these separations is the biggest limiting factor. This finding has resulted in a significant reduction in the limiting analysis time of gradient elution 1DLC, as indicated by arrow 'A' in figure; this trend will likely continue to even shorter times.

The improvement in the speed of gradient elution 1DLC discussed above has also led to a systematic comparison of the performance of gradient vs. isocratic elution 1DLC for the analysis of rather simple samples. It was found that gradient elution 1DLC can be very effective in the analysis of mixtures containing less than ten chemical constituents; historically this type of mixture has been thought to be best separated using isocratic elution. This change in mindset is represented by arrow 'B' in figure.

The advances in the speed and performance of gradient elution 1DLC discussed above have contributed considerably to the improvement of the speed of 2DLC when gradient elution is performed in both dimensions, as in our own work. This improvement in the speed of 2DLC without sacrificing resolving power is represented by arrow 'C' in figure.

Two Dimensional Gas Chromatography (2DGC)

Two-dimensional gas chromatography (2DGC) has a much longer history than 2DLC and provides the basis for many insights relevant to 2DLC. 2DGC has been successful in handling complex mixtures of volatile compounds, such as pesticides, petroleum derived materials, and environmental samples. There are several requirements which much be met to achieve a successful separation. First, as for almost any hyphenated technique, the second dimension has to be fast. This requirement for much faster separation of the second dimension than the first was realized in 2DGC a relatively long time ago. Fortunately, high speed gas chromatography has been well-known for many years and is relatively easier to implement in GC than in LC from both theoretical and practical perspectives. Phillips's cryogenic modulation technique enables fast cooling and heating during sample transfer from the first to the second dimension. The way in which each second dimension separation is carried out in 2DGC does not require thermal re-equilibration between successive runs. This is an inherent advantage in terms of the speed of 2DGC over 2DLC in those cases where gradient elution is used in each second dimension LC separation; this comprises a majority of 2DLC separations using RPLC in the second dimension. Unlike liquid chromatography, the flow velocity in gas chromatography is not limited by pressure; furthermore diffusion and hence inter-phase mass transfer are much faster in GC than in LC. It is not difficult to generate peaks which are significantly less than one second wide in GC; peak widths of only tens of milliseconds have been achieved with fast temperature gradients and short columns.

Second, the requirement of retention "orthogonality" between the first and second dimension separations and maximal usage of the "separation space" to achieve the highest possible "conditional peak capacity" are easily accomplished in 2DGC. It must be realized that dispersion (London) interactions are the dominant type of interaction in all forms of gas-liquid partition chromatography and as a consequence of this fact log k' vs. log k' plots for any two GC columns are inherently rather strongly correlated for two columns as different as a permethylsilicone and Carbowax, even for a highly variegated set of probe solutes. Phillips et al. showed that to create orthogonality in GC, some factor that controls selectivity of the second dimension column must be readily tunable. For gas chromatography, temperature tuning is easily achieved and serves as an effective method for creating an "orthogonal" separation. As shown in Fig., excellent orthogonality and coverage can be achieved in 2DGC. The fact that a high degree of orthogonality can be produced in GC suggests that 2DLC might be feasible with two reversed-phase columns. Typically 2DLC has been done by combining very different modes of LC such as ion exchange and reversed-phase or size exclusion and reversed-phase. The third requirement involves fast detection and data handling. More details on these basic requirements are given below.

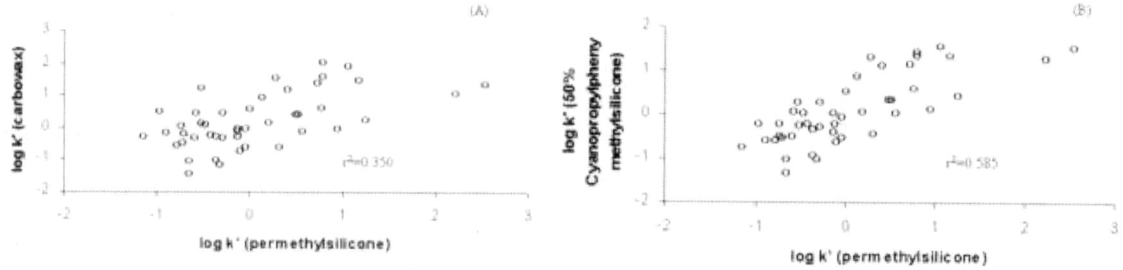

Figure: (A) Plot of log k' vs. log k' of carbowax and permethylsilicone phases, (B) Plot of log k' vs. log k' of 50% cyanopropylphenyl methylsilicone and permethylsilicone phases. Both show a significant lack of orthogonality.

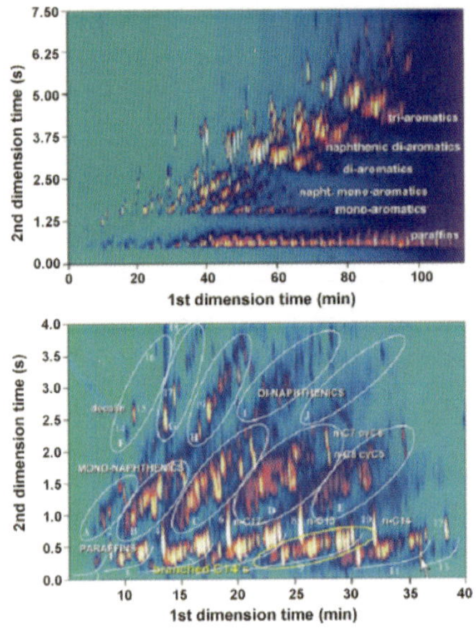

Figure: 2DGC chromatogram of a light cycle oil using a (25 m × 0.25 mm DB-1) × (1.5 m × 0.1 mm OV-1701) column combination (top), and a non-aromatic hydrocarbon solvent using a (10 m × 0.25 mm CP Sil-2 CB) × (2.5 m × 0.1 mm BPX-50) column combination (bottom). 1 through 13: alkanes; 1: branched C_{10}s; 2: n-C_{10}, 3: branched C_{11}s; 4: n-C_{11}, 5: branched C_{12}s; 6: n-C_{12}; 7: branched C_{13}s; 8: n-C_{13}; 9: branched C_{14}s; 10: n-C_{14}; 11: branched C_{15}s; 12: n-C_{15}; 13: branched C_{16}s; 14: unknown; 15: trans-decalin; 16: cis-decalin;17: trans-methyl-decalins; 18: cis-methyl-decalins; A through E: mono-naphthenes C_{10} through C_{14}; F through J: di-naphthenes C_{10} through C_{13}.

Simulated Moving-bed Chromatography

Compared to traditional batch chromatography, Simulated Moving Bed (SMB) chromatography leads to greater yields of purified substances while consuming less eluent and packing material.

Unlike batch chromatography, in a SMB process not only the liquid phase is in motion but the solid phase also moves in counter-current mode to the liquid. At the same time, the individual areas are operated with different volume currents. The concept that is demonstrated in figure refers to the classic four zone version SMB configuration.

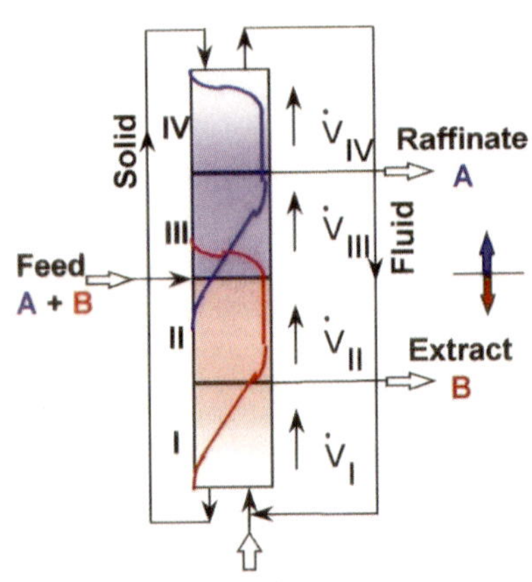

SMB process scheme.

Separation Zones

the solid phase is regenerated. The feed (component A + B) is fed in the middle of the illustrated "SMB column". The liquid phase (desorbent or eluent) travels from the bottom to the top and transports component A (blue).

Component A is continuously led out of the system in the raffinate as a pure substance. The solid phase (solid) passes from the top to the bottom and transports component B (red). Component B is pure when it is led out of the system as an extract. The distribution of both components over the whole column system is effortlessly recognized by the concentration profile.

The movement of the solid phase is recognized by connecting a specific number of columns in series in a closed circuit and by switching the columns counter-clockwise in defined time intervals one position upstream. The build-up of the process is schematically displayed in figure. Four pumps are combined into the system.

Two pumps are employed for the add-on of feed and fresh eluent and two pumps are employed for the maintenance of the internal flow. Generally 8, 12, or 16 columns are employed based on the need. The columns in general are equally distributed in the individual zones (for example, 2/2/2/2). However, based on the separation problem, they can be asymmetrically distributed (for example, 1/3/3/1).

In addition to the four zones, concepts with three, five or eight zones can also be recognized. The latter two can be drawn for a three component separation. Complex mixtures can also be separated into two valuable substance fractions (pseudo-components).

In standard chromatographic separations with just one column, a small part of the column takes part in the separation, while the predominant part does not make a contribution to the productivity. In a SMB process, the whole column bed is efficiently used. A stationary state is developed when all volume currents are optimized. This means that the middle concentrations of the purified extract and raffinate currents no longer change. The system can be continuously operated under these conditions.

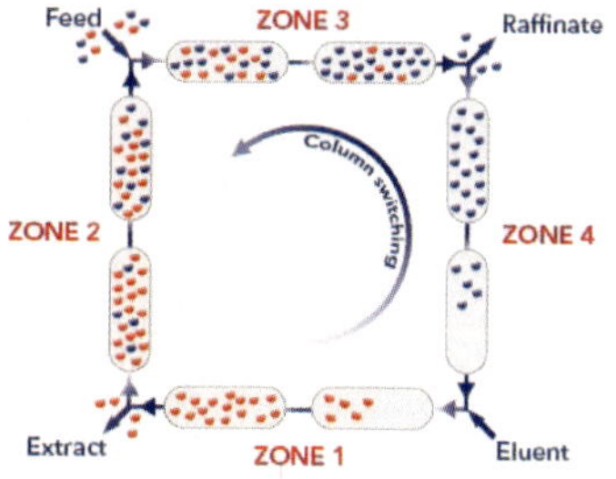

Classical 4 zone SMB configuration.

Benefits

- Higher recovery and purity.
- Lower solvent consumption.
- Higher productivity – up to 20-fold vs. batch systems.

- Larger stationary phase particles, lower pressures.
- Scalable from milligrams to tons of purified product.
- Continuous process.

Applications

- Bulk separations.
- End product purifications.
- Impurity analysis.
- Chiral separation.
- Process development.

Periodic Counter-current Chromatography

Periodic counter-current chromatography (PCC) is a method for running affinity chromatography in a quasi-continuous manner. Today, the process is mainly employed for the purification of antibodies in the biopharmaceutical industry as well as in research and development. When purifying antibodies, Protein A is used as affinity matrix. However, periodic counter-current processes can be applied to any affinity type chromatography.

Basic Principle

In conventional affinity chromatography, a single chromatography column is loaded with feed material up to the point before target material (product) cannot be retained by the affinity material anymore. The resin with the adsorbed product on it is then washed to remove impurities. Finally, the pure product is eluted with a different buffer. Notably, if too much feed material is loaded onto the column, the product can break through and product is consequently lost. Therefore, it is very important to only partially load the column to maximize the yield.

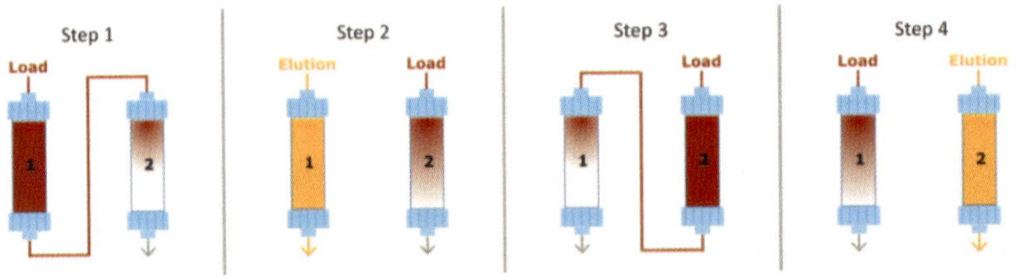

Process diagram for the 2-column periodic counter-current process "

Periodic counter-current chromatography puts this problem aside by utilizing more than one column. PCC processes can be run with any number of columns, starting from two.The following paragraph will explain a two-column version of PCC, but other protocols with more columns rely on the same principles. A diagram depicting the individual process steps is shown on the right.

- In Step 1, the so-called sequential loading phase, columns 1 and 2 are interconnected. Column 1 is fully loaded with sample (red) while its breakthrough is captured on column 2.

- In Step 2, column 1 is washed, eluted, cleaned and re-equilibrated while loading separately continues on column 2.
- In Step 3, after regeneration of column 1, the columns are again inter-connected and column 2 is fully loaded while its breakthrough is captured on column 1.
- Finally, in Step 4 column 2 is washed, eluted, cleaned and re-equilibrated while loading continues independently on column 1. This cyclic process is repeated in a continuous way.

Several variations of periodic counter-current chromatography with more than two columns exist. In these cases, additional columns are either placed within the feed stream during loading, having the same effect as using longer columns. Alternatively, additional columns can be kept in an unoccupied stand-by mode during loading. This mode offers additional assurance that the main process is not influenced by washing and cleaning protocols, albeit in practice this is rarely required. On the other hand, the underutilized columns reduce the theoretical maximum productivity for such processes. Generally, the advantages and disadvantages of different multi-column protocols are the subject of debate. However, without a doubt, compared to single column batch processes, periodic counter-current processes provide significantly increased productivity.

Dynamic Process Control

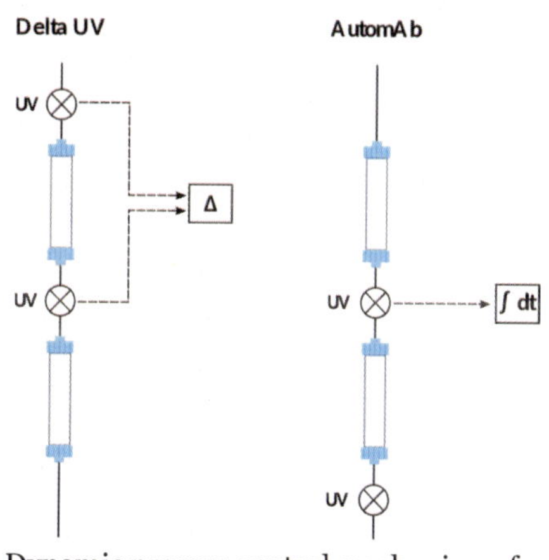

Dynamic process control mechanisms for periodic counter-current chromatography.

On the time scale of continuous chromatography runs, it is fairly common to observe changes in important process parameters, such as column health, buffer quality, feed titer (concentration) or feed composition. Such changes result in an altered maximum column capacity, relative to the amount of loaded feed material. In order to achieve a steady quality and yield for each process cycle, the timing of the individual process steps therefore has to be adjusted. Manual changes are in principle conceivable, but rather impractical. More commonly, dynamic process control algorithms monitor the process parameters and apply changes as needed automatically.

There are two different operating modes for dynamic process controllers in use today. The first one, called DeltaUV, monitors the difference between two signals from detectors situated before and after the first column. During initial loading, there is a large difference between the two signals, but it is diminishing as the impurities make their way through the column. Once the column is fully saturated with impurities and only additional product is being held back, the difference between the signals reaches a constant value. As long as the product is completely being captured on the column, the difference between the signals will remain constant. As soon as some of the product breaks through the column (compare above), the difference diminishes. Thus, the timing and amount of product breakthrough can be determined. The second possibility, called AutomAb, requires only the signal of a single detector situated behind the first column. During initial loading, the signal increases, as more and more impurities make their way through the column. When the column is saturated with impurities and as long as the product is completely being captured on the column, the signal then remains constant. As soon as some of the product breaks through the column (compare above), the signal increases again. Thus, the timing and amount of product breakthrough can again be determined.

Both iterations work equally well in theory. In practice, the requirement for two synced signals and the exposure of one detector to unpurified feed material, makes the DetaUV approach less reliable than AutomAb.

Pyrolysis Gas Chromatography

Pyrolysis Gas Chromatography/Mass Spectrometry (Pyrolysis-GC-MS) is an analytical tool used to characterize a wide variety of polymers and composite materials that cannot be analyzed using traditional GC-MS. The fundamental concept of Pyrolysis-GC-MS is not complicated but highly sensitive. A sample is introduced to a heated environment and is broken down into smaller stable fragments/components (aka pyrolyzates) for analysis. The pyrolyzates are then chromatographically separated in the same manner as traditional GC-MS. Similar to traditional GC-MS, the fragments are separated utilizing a compound's intrinsic affinity for a "stationary phase" (solid support with specialized coating). Once the pyrolyzates are separated, the data is analyzed by highly skilled scientists who can deconvolute and interpret the data and subsequently provide useful information and insight on the original material.

In order to conduct a thorough investigation, the Pyrolysis temperature program must be set up in a very specific way. When a sample is introduced into the Pyrolysis unit, it is instantly heated from ambient to the desired temperature in an inert atmosphere. This rapid heating provides reproducible results. If additional tests are needed, the test sample is cooled in an inert atmosphere to room temperature and the Pyrolysis unit is reprogrammed. Several different experiments can be performed and they are explained below.

Types of Pyrolysis-GC-MS Analysis

- A Single Shot Pyrolysis-GC-MS: The single shot is run at a single temperature, normally >500° C (dependent on the material being examined). The temperature of the sample is rapidly increased at a temperature rate <20msec to break bonds. This is done in order to characterize the original sample.
- A Double Shot Pyrolysis: This is normally run at both low (80-350° C) and high temperatures (500-800° C). The lower temperature analysis is known as a thermal desorption step. This step can be used to examine low molecular weight components (monomers, oligomers, etc.), additives and residual solvents. The higher temperatures allow for the total fragmentation of the higher molecular weight items such as the crosslinked material or the entire polymer backbone.

- Evolved Gas Analysis (EGA): This is run by dropping the sample into the Pyrolysis unit set at a low temperature. The furnace is then increased to a high temperature at a set ramp rate. The components that are off-gassed/evolved from the sample are then examined. This testing is done in order to identify the temperature range for the desired component. The Pyrolysis-GC-MS can then be set up the desired temperature range for a more in-depth study of the identified compound. As a side note, Thermal Gravimetric Analysis (TGA) can also be used to aid in the determination of the temperature analysis profile for Pyrolysis-GC-MS.
- Low Temperature (normally considered <350° C) Thermal Desorption: This testing would be done for products where additives, residual solvents or other off-gassed components are desired but the polymer identification is not needed.
- Reactive Pyrolysis-GC-MS: There are certain classes of compounds, for example polyesters, that have complex Pyrolysis mixtures (i.e. many degradation products, etc.). These complex mixtures make the interpretation of the data difficult, if not impossible. When a derivatization technique is utilized, a chemical reaction occurs while the sample is in the Pyrolysis chamber, allowing for a much easier interpretation of the derivatized products.
- Heart Cut-EGA GC-MS (HC-EGA-GC-MS): This method is useful when looking for specific components in a highly complex matrix or when the full composition of a complex system is desired. Heart-Cut Pyrolysis involves using the EGA technique to isolate the desired elution times (i.e. zones). The instrument is then programmed to isolate each specified zone identified by EGA. Each zone is then individually analyzed. This technique allows for the examination of regions that might contain multiple components under one peak but are not obvious during the regular Pyrolysis GC-MS analysis. While extremely useful, this technique can add to the complexity and amount of time for the analysis.
- Quantitative Pyrolysis-GC-MS: This would be done on a component that the amount present could not be determined using another technique. An example component would be a low ppm additive (such as an Irganox, Irgafos, etc.), FAME components in rubbers, phthalates plus many other compounds.
- Cryotrapping capabilities are available for the Pyrolysis GC-MS. Prior to introduction into the GC-MS, the pyrolyzates are cryotrapped using liquid nitrogen. This is done to narrow the chromatographic band and improve the detection limit.

Pyrolysis-GC-MS virtually eliminates the need for the various sample preparation techniques required by traditional GC-MS (extraction, dilutions, concentrating the sample, various chemical reactions, heating/cooling, etc. are unnecessary). The samples can be analyzed in an "as received" state unless they fall into the category of needing reactive Pyrolysis-GC-MS.

Chiral Chromatography

Chiral column chromatography makes use of the concept of chirality. Chirality relates to study of the three-dimensional structure of molecules. The term comes from the Greek root chir-, hand, and means "handedness." Chiral molecules show a left- and right-handed nature. They exist as mirror image forms that cannot be made to overlap byrotation. Molecules showing chirality rotate the plane of polarized light. The degree of rotation is called specific rotation or optical rotation.

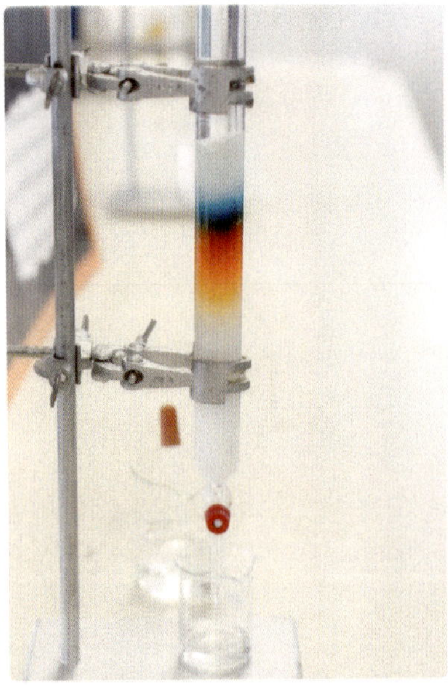

The two mirror image forms of a chiral compound are called enantiomers. Those that rotate the plane of polarized light to the right are called (+) or dextrorotatory forms; those that rotate to the left are called (-) or laevo rotatory. Racemic mixtures are those that have both enantiomers in equal proportion. They do not rotate polarized light, as the enantiomeric effects cancel out each other.

The Need for Chiral Separation

A majority of bioorganic molecules (amino acids, proteins, sugars, nucleic acids) are chiral, and in nature they exist in only one of two possible enantiomeric forms. The absorption, metabolism, and excretion of enantiomers have been found to be different in animal and human studies. The implication of this is that living organisms show preferential positive responses to one of a pair of enantiomers in drugs and agrochemicals.

The human body is so selective in its responses to chiral compounds that one isomer may produce a desired effect, while the other may be ineffective, and sometimes even toxic, so it is necessary to separate enantiomers in the pharmaceutical and agrochemical industries. Other reasons for separation include lower dosing requirements, non-interference by the undesirable enantiomer, and cost and time savings.

Separation of Chiral Compounds by Column Chromatography

Enantiomer separation from numerous compound groups, including nonsteroidal anti-inflammatory drugs (NSAIDs), agricultural compounds, natural products, β-blockers, and pharmaceutical drugs, have been achieved through chiral column chromatography.

There are two approaches to the separation of chiral compounds by column chromatography. One is to use a chiral additive to the mobile phase, and the other is to use a chiral stationary phase (CSP). Either way, the enantiomers are introduced into an asymmetric environment, and this gives rise to two diasteriomeric compounds with different physicochemical profiles, which in turn is used for physical separation of the individual enantiomer either by high-pressure liquid chromatography (HPLC) or gas chromatography (GC). Of the two possibilities, use of the chiral stationary phase is more common.

Chiral Stationary Phase

The chiral stationary phase is prepared by binding of a chiral selector with a support material, usually silica. Transient diasteriomeric complexes with different stabilities are formed between the enantiomers and the bound chiral selector, which is then used for their separation.

To date over a hundred chiral stationary phases have been developed and made commercially available. However, none of these can be considered universal for all the groups of racemic compounds. Choosing the right CSP for the separation of a specific pair of enantiomers is often difficult.

One method is to choose the CSP based on predictive rules developed for empirical structures. This can be complemented with knowledge of the chemistry of the racemic mixture and the CSP. Reactions of the mobile phase with the solutes and the CSP may add to the complications, so interactions between all three components must be considered. The key is knowing the possible chiral recognition mechanisms involved in the entire process.

Currently available CSPs have been categorized into 5 types according to their chiral recognition mechanisms:

- Type I: solute–CSP complexes formed by attractions (attractive interactions, hydrogen bonding, dipole stacking, pi–pi interactions).
- Type II: solute–CSP complexes formed by attractive interactions, but inclusion complexes also play important roles.
- Type III: solute–CSP complexes formed as inclusion complexes in chiral cavities.
- Type IV: solute–CSP complexes formed as metal diasteriomeric complexes (chiral ligand exchange chromatography).
- Type V: CSP is a protein, and the solute–CSP complexes formed in these cases are a result of polar and hydrophobic interactions.

Chiral Mobile Phase Additives in Column Chromatography

As CSPs are relatively expensive, chiral mobile phase additives (CMPA) are an attractive alternative. In this technique, the additives dissolved in the mobile phase form transient complexes with the solutes. The forces that guide the separation include differences in the formation kinetics and relative stabilities, as well differences in their partitioning between the mobile and the stationary phases.

However, other factors such as concentration of the additives, pH, and temperature of the mobile phase will have to be carefully considered for successful separation of enantiomers. Common CMPAs include camphorsulphonic acid, ligand exchangers, macrocyclic antibiotics, and cyclodextrins.

Multicolumn Countercurrent Solvent Gradient Purification

The increasing production volumes of biomolecules, especially therapeutic proteins, and a rising cost pressure from the market, has engendered a strong interest in the chromatographic purification step in the downstream processing of biomolecules. This is because the chromatography step is often irreplaceable and also a major cost driver in downstream downstream.

The generic purification problem to be solved, particularly in the area of chromatographic bioseparations, is a separation of the feedstock into three fractions: the early eluting, weakly adsorbing impurities; the desired product(s), and the late eluting, strongly adsorbing impurities. A literature overview of chromatographic three-fraction separation processes can be found elsewhere. Conventionally, the generic three-fraction purification problem is solved using batch column chromatography, often incorporating a linear variation of the eluting composition with respect to the elution time (also called solvent gradient chromatography). The column effluent is collected in several consecutive, small portions and some of these portions ideally contain the desired product with an average purity being higher than the required one, thereby making up the product fraction. To increase the yield of solvent gradient chromatography, product-rich portions collected from the column effluent, which do not fulfill the purity specifications, can be recycled to the feed point.

Besides batch solvent gradient chromatography and its derivative including a recycle, the multicolumn countercurrent solvent gradient purification (MCSGP) process is a third suitable process developed recently to perform three-fraction separations of biomolecules. This process involves three columns and continuously purifies a complex feed mixture into three fractions, thereby incorporating solvent gradients.

Conventional multicolumn processes such as simulated-moving-bed (SMB) chromatography cannot be used for the purification of biomolecules, because an SMB can only perform a binary separation and solvent gradients cannot be implemented properly. SMB chromatography is therefore more suitable for a binary separation of small molecules.

A numerical comparison between the conventional batch chromatography process and the MCSGP process with respect to yield and process productivity is also performed for a polypeptide purification. The Pareto curves obtained from the optimization procedure are compared and the performance superiority of the MCSGP process is explained.

Principle of the MCSGP Process

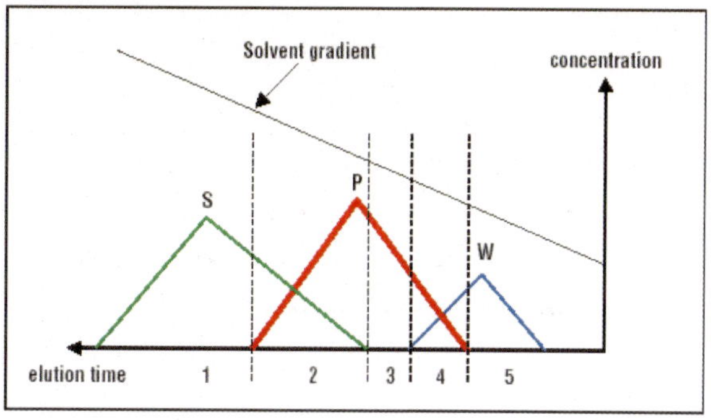

Figure: Simplified chromatogram of generic chromatographic purification problem (time axis from right to left); S: strongly adsorbing, late eluting impurities; P: product; W: weakly adsorbing, early eluting impurities.

The generic problem in the chromatographic purification of biomolecules can be simplified to the chromatogram as shown in figure.

To facilitate the following explanation, the time axis of the chromatogram has been reversed (i.e., it goes from right to left). The generic chromatogram can be cut into five fractions as indicated by the numbers on the time axis:

1 = strongly adsorbing impurities

2 = product contaminated by strongly adsorbing impurities

3 = product

4 = product contaminated with weakly adsorbing impurities

5 = weakly adsorbing impurities.

The aim of an ideal purification process would be to collect fraction 3, to drain fractions 1 and 5, and to recycle fractions 2 and 4. Transferring these tasks to a multicolumn chromatographic process results in a 6-column flowsheet (Figure), which shows the basic principle of the MCSGP process.

Figure shows six columns. Each column is connected to a gradient pump. The outlet of columns 2 and 4 is recycled into columns 4 and 6, respectively. Fractions 1 to 5 as shown in the chromatogram in figure are eluted from a column with the same number. For example, the strongly adsorbing impurities are drained from column 1. In column 2, the product fraction contaminated with strongly adsorbing impurities, P+S (i.e., fraction 2) is eluted, mixed with the outlet of the gradient pump, and recycled to column 4. In column 3, the product is collected. The outlet of column 4, P+W, is recycled to column 6. In column 5, the raw feed mixture is fed to the column inlet and the weakly adsorbing impurities are eluted. In order to run this process continuously, after a certain time all columns are switched one position to the left (e.g., column 4 becomes column 3), and column 1, being void of any solutes at the end of the switch time, can be switched to the position of column 6.3–4 This column switching results in a countercurrent movement of liquid and solid flow. This countercurrent operation improves the efficiency of the process.

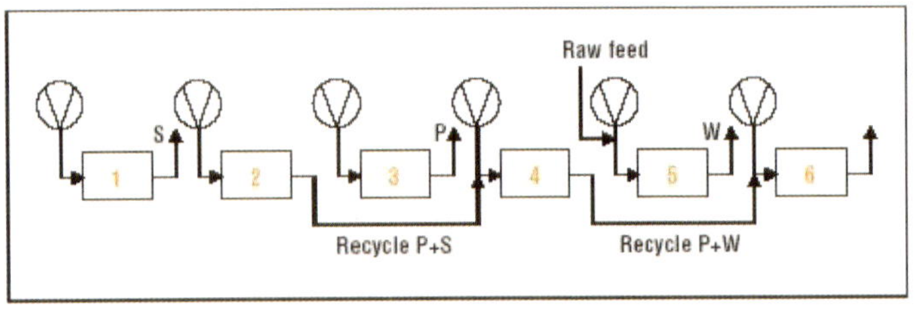

Figure: Basic principle of the MCSGP process

To reduce equipment costs, the MCSGP process used with just three columns. This reduction from six to three columns can be achieved by performing the draining or collecting of S, P, and W and the recycling of P+S and P+W, not in parallel but in two sequential steps. This results in two separate flowsheets, which are applied to the three columns in an alternating fashion.

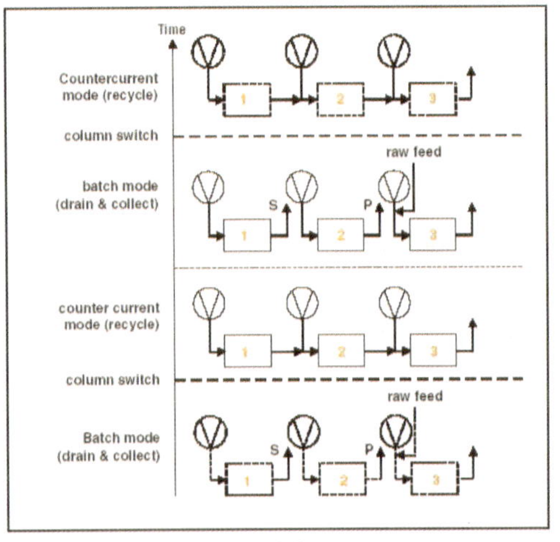

Figure: Flowsheet of the MCSGP process

After the columns have been switched one position to the left (where column 1 is switched to the position of column 3), the columns are operated in the countercurrent mode (i.e., recycling of P+S and P+W into the subsequent column takes place). After a certain time, t_{CC} , the columns are connected in a different manner and then operated for a time, t_B , in the batch mode (i.e., while the draining of S and W and the collecting of P take place). Then the columns are switched again and operate in the countercurrent mode. The column switching plus countercurrent and batch operation mode makes one cycle of time (t_{CC} + t_B) and this cycle is repeated.

References

- Partition-chromatography, analytical-chemistry: tutorvista.com, Retrieved 14 April 2018
- Normal-phase-chromatography: studyread.com, Retrieved 21 May 2018
- Ion-exchange-chromatography: biochemden.com, Retrieved 18 July 2018
- Column-chromatography: chemdictionary.org, Retrieved 21 March 2018
- Column-chromatography, science-education-10217: jove.com, Retrieved 19 April 2018
- Tlc-thin-layer-chromatography-Principle-Procedure: owlcation.com, Retrieved 17 June 2018
- Continuous-processing-multicolumn-countercurrent-solvent-gradient-purification-process: biopharminternational.com, Retrieved 25 March 2018

4

Extraction

Extraction is a technique that is used for the separation of a substance from a matrix. This chapter discusses in detail the most significant extraction techniques, such as electromembrane extraction, accelerated solvent extraction, leaching, raffinate, magnetic separation, etc.

After a reaction is completed, the solution often times does not only contain the desired product, but also undesired byproducts of the reaction, unreacted starting material(s) and the catalyst (if it was used). These compounds have to be removed in the process of isolating the pure product. A standard method used for this task is an extraction or often also referred to as washing. Strictly speaking, the two operations are targeting different parts in the mixture: while the extraction removes the target compound from an impure matrix, the washing removes impurities from the target compound i.e., water by extraction with saturated sodium chloride solution. Washing is also used as a step in the recrystallization procedure to remove the impurity containing mother liquor adhering to the crystal surface.

Many liquid-liquid extractions are based on acid-base chemistry. The liquids involved have to be immiscible in order to form two layers upon contact. Since most of the extractions are performed using aqueous solutions (i.e., 5 % NaOH, 5 % HCl), the miscibility of the solvent with water is a crucial point as well as the compatibility of the reagent with the compounds and the solvent of the solution to be extracted. Solvents like dichloromethane (=methylene chloride in older literature), chloroform, diethyl ether, or ethyl ester will form two layers in contact with aqueous solutions if they are used in sufficient quantities. Ethanol, methanol, tetrahydrofuran (THF) and acetone are usually not suitable for extraction because they are completely miscible with most aqueous solutions. However, in some cases it is possible to accomplish a phase separation by the addition of large amounts of a salt ("salting out"). Commonly used solvents like ethyl acetate (8.1 %), diethyl ether (6.9 %), dichloromethane (1.3 %) and chloroform (0.8 %) dissolved up to 10 % in water. Water also dissolves in organic solvents: ethyl acetate (3 %), diethyl ether (1.4 %), dichloromethane (0.25 %) and chloroform (0.056 %). Oxygen containing solvents are usually more soluble in water (and vice versa) because of their ability to act as hydrogen bond donor and hydrogen bond acceptor. The higher water solubility lowers the solubility of weakly polar or non-polar compounds in these solvents i.e., wet Jacobsen ligand in ethyl acetate. Other solvents such as alcohols increase the solubility of water in organic layers significantly because they are miscible with both phases and act as a mediator. This often leads to the formation of emulsions.

The most important point to keep in mind throughout the entire extraction process is which layer contains the product. For an organic compound, it is relatively safe to assume that it will dissolve better in the organic layer than in most aqueous solutions unless it has been converted to an ionic specie, which makes it more water-soluble. If a carboxylic acid (i.e., benzoic acid) was deprotonated using a base or an amine (i.e., lidocaine) was protonated using an acid, it would become more water-soluble because the resulting specie carries a charge. Chlorinated solvents (i.e., dichloromethane, chloroform) exhibit a higher density than water, while ethers, hydrocarbons and many esters possess a lower density than water, thus form the top layer. One rule that should always be followed when performing a work-up process:

Never dispose of any layer away until you are absolutely sure (=100 %) that you will never need it again. The only time that you can really be sure about it is if you isolated the final product in a reasonable yield, and it has been identified as the correct compound by melting point, infrared spectrum, etc. Keep in mind that it is always easier to recover the product from a different layer in a beaker than from the waste container or the sink. In this context it would be wise to label all layers properly in order to be able to identify them correctly later if necessary.

In order to separate compounds from each other, they are often chemically modified to make them more ionic i.e., convert a carboxylic acid into a carboxylate by adding a base. Standard solutions that are used for extraction are: 5 % hydrochloric acid, 5 % sodium hydroxide solution, saturated sodium bicarbonate solution (~6 %) and water. All of these solutions help to modify the (organic) compound and make it more water-soluble and therefore remove it from the organic layer. More concentrated solutions are rarely used for extraction because of the increased evolution of heat during the extraction, and potential side reactions with the solvent.

What Do We use When to Extract

a. Removal of a carboxylic acid or mineral acid

In order to remove an acidic compound from a mixture, a base like NaOH or $NaHCO_3$ is used. The carboxylic (or mineral) acid and the base react to form a sodium salt, which is usually exhibits a higher solubility in aqueous solutions due to its negative charge and higher polarity (as indicated by a more negative log K_{ow} value i.e., CH3COOH: -0.17, Na^+CH_3COO-: -3.72).

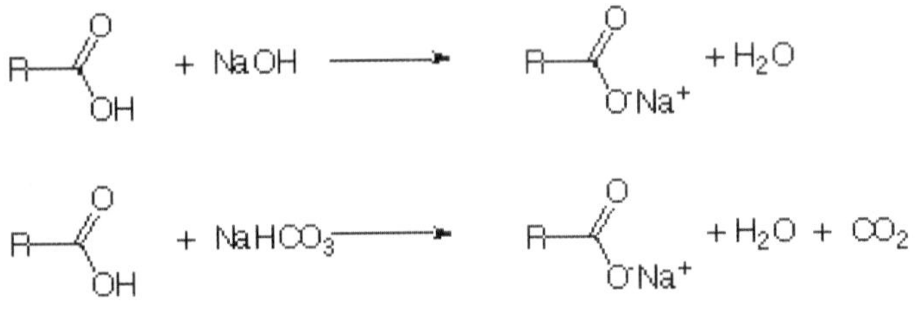

Which of the two reagents should be used depends on the other compounds present in the mixture. Sodium hydroxide is usually easier to handle because it does not evolve carbon dioxide as a byproduct. In addition, the concentration can be increased significantly if is needed. However, if compounds were present that are sensitive towards strong bases or nucleophiles (i.e., esters, ketones, aldehydes, etc.), sodium bicarbonate should be used. It does not react with these compounds because it is a weaker base and a weak nucleophile (due to its resonance stabilization). Note that the formation of carbon dioxide as a byproduct causes a pressure build-up in the separatory funnel, the centrifuge tube or the conical vial. Thus, additional precautions (i.e., frequent venting) have to be taken to prevent any accidents resulting from the pressure build up in the extraction vessel. The target compound can subsequently be recovered by adding a mineral acid to the basic extract i.e., benzoic acid in the Grignard experiment in Chem30CL.

b. Removal of a phenol

Most phenols are weak acids (pK_a= ~10) and do not react with sodium bicarbonate, which is a weak base itself ($pK_a(H_2CO_3)$=6.37, 10.3). However, they do react with a strong base like NaOH. This difference in acidity can be exploited to separate carboxylic acids and phenols from each other in an organic layer. While many phenols dissolve poorly in water (8.3 g/100 mL at 20 °C, log K_{ow}=1.46), phenolates dissolve very well in aqueous solutions. After the extraction, the phenol can be recovered by adding a mineral acid to the basic extract.

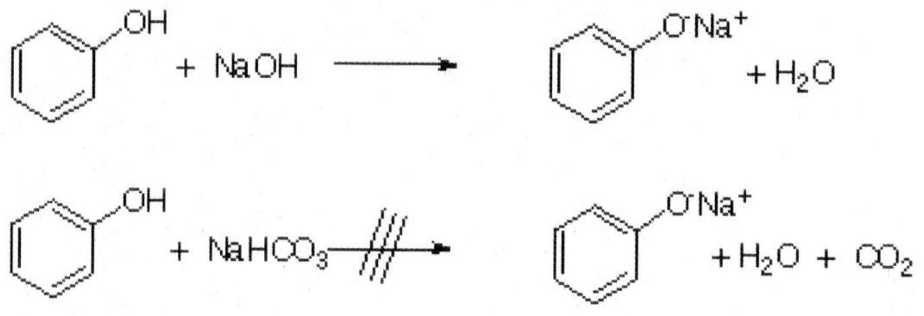

c. Removal of an amine

Depending on the chain length, amines might or might not be soluble in water i.e., propylamine is miscible with water (log K_{ow}=0.48), triethylamine displays a limited solubility at room temperature (17 g/100 mL, log K_{ow}=1.44), while tributylamine hardly dissolves at all (0.37 g/100 mL, log K_{ow}=4.60). Amines are basic and can be converted to ammonium salts using mineral acids i.e., hydrochloric acid. The resulting salts dissolve in water. However, the solubility of the ammonium salts decreases as the number and size of R-groups increases. Ammonium salts from primary amines are much more soluble in water than salts from tertiary amines due the increased ability to form hydrogen bonds [(H_3NEt)Cl: 280 g/100 g H_2O, (H_2NEt_2)Cl: 232 g/100 g H_2O, ($HNEt_3$)Cl: 137 g/100 g H_2O (all at 25° C)].

$$R{-}NH_2 + HCl \longrightarrow R{-}NH_3^+ + Cl^-$$

After separation of the organic and the aqueous layer, the amine can be recovered by addition of a strong base like NaOH or KOH to the acidic extract i.e., lidocaine synthesis. Note that amides are usually not basic enough to undergo the same protonation (pK_a of conjugate acid: ~ -0.5).

d. Isolation of a neutral species

Most neutral compounds cannot be converted into salts without changing their chemical nature. Many of these neutral compounds tend to react in undesired ways i.e., esters undergo hydrolysis upon contact with strong bases or strong acids. One has to keep this in mind as well when other compounds are removed. For instance, epoxides hydrolyze to form diols catalyzed by acids and bases. Ketones and aldehydes undergo condensation reactions catalyzed by both, acids and bases. Esters also hydrolyze to form carboxylic acids (or their salts) and the corresponding alcohol. In order to separate these compounds from each other, chromatographic techniques are often used, where the compounds are separated based on their different polarities.

e. General Separation Scheme

Based on the discussion above the following overall separation scheme can be outlined. Which sequence is the most efficient highly depends on the target molecule. There is obviously no reason to go through the entire procedure if the compound sought after can be isolated in the first step already. Note that many of these steps are interchangeable in simple separation problems.

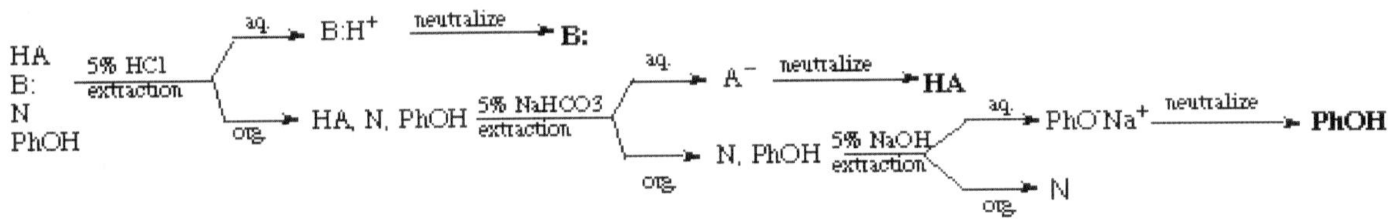

For instance, if the target compound was the base in the system, the extraction with HCl should be performed first. Whatever remains in the organic layer is not of interest anymore afterwards, unless one of the other compounds has to be isolated from this layer as well. If the target compound was an acid, the extraction with NaOH should be performed first. This strategy saves steps, resources and time, and most of all, greatly reduces waste.

Practical Aspects of an Extraction

An extraction can be carried out in macro-scale or in micro-scale. In macro-scale, usually a separatory funnel is used. Micro-scale extractions can be performed in a conical vial or a centrifuge tube depending on the quantities. Below are several problems that have been frequently encountered by students in the lab:

a. Which layer is the aqueous layer?

Most solutions are relatively diluted (~5 %) and their density is not much different from that of water (i.e., 5 % HCl: 1.02 g/cm³, 5 % NaOH: 1.055 g/cm³). Thus, the density of a solid i.e., sodium hydroxide (2.1 g/cm³ in the solid) does not provide the information sought. The density is determined by the major component of a layer, which is usually the solvent. About 5 % of a solute does not change the density of the solution much. However, this can change if very concentrated solutions are used! Thus, diethyl ether and ethyl acetate, which are both less dense than the dilute solutions that are usually used for extraction, form the top layer, while dichloromethane and chloroform form the bottom layer (currently both of them are not used in Chem30BL or Chem30CL due to safety concerns).

b. Why are three layers observed sometimes?

It is not uncommon that a small amount of one layer ends up on top of the other. Mixing with a stirring rod or gentle shaking usually takes care of this problem. Small amounts (compared to the overall volume of the layer) should be discarded here.

c. Why do the layers not separate?

This would usually happen if the mixture was shaken too vigorously. Subsequently, an emulsion is formed instead of two distinct layers. In such an event, the mixture can be stirred slowly with a glass rod to bring the small droplets together a little faster, which ultimately leads to the formation of a new layer. In some cases, a careful draining of the existing lower layer can also be helpful because it pushed the bubbles together in the smaller part of the extraction vessel. In cases, where the phases have similar polarity or density, the addition of more solvent can assist the separation. Sometimes, the addition of a salt (or salt solution) can also lead to a better phase separation ("salting out"). In many cases, centrifugation or gravity filtration works as well. When it is known, through experience, that some mixtures may form emulsions, vigorous shaking should be avoided. Instead, gently rocking the separatory funnel back and forth for 2-3 minutes will accomplish sufficient degree of mixing while minimizing the formation of emulsions.

d. How do we know that we are done extracting?

Strictly speaking, hardly ever all of the solute will be extracted since there is finite distribution coefficient for the compound. As a general rule, multiple extractions with small quantities of solvent or solution are more efficient than one extraction using the same amount of solvent. The amount of material left behind after two or three extractions is usually very small (less than 5 % in most cases) and does not justify the effort and resources (solvent and time to perform the extractions and to remove the solvent later on). Excessive washing will also lower the yield of the product, if the desired compound dissolves noticeably in the other phase.

e. Why does the extraction container (vial, centrifuge tube, separatory funnel) make funny noises?

This phenomenon will often be observed if sodium bicarbonate is used for the extraction in order to neutralize or remove acidic compounds. The reaction affords carbon dioxide (CO_2), which is a gas at ambient temperature. Pressure builds up that pushes some of the gas and the liquid out. The container should be vented immediately before the pressure build-up can cause an explosion, an ejection of the stopper on the top or excessive spillage upon opening. A similar observation will be made if a low boiling solvent is used for extraction. The shaking of the mixture increases the surface area, and therefore the apparent vapor pressure of the solvent. In addition, many extraction processes are exothermic because they involve an acid-base reaction.

f. The centrifuge tube leaks

Often times the cap is either the wrong cap in the first place or it is not properly placed on the top. If $NaHCO_3$ is used for extraction, the centrifuge tube has to be vented more frequently.

g. The separatory funnel leaks

Before using the separatory funnel, the user should check if the stopcock plug and the stopcock fit together well. In addition, the stopper on the top has to fit into the joint on the top to prevent leakage there.

h. Why is a centrifuge tube, a conical vial or a separatory funnel used for the extraction and not a beaker or test tube?

The conical shape of these pieces of equipment makes it easier to collect the solution on the bottom using a Pasteur pipette because of the smaller interface. The task of getting a clean phase separation will be more difficult if the liquids are spread out over a large, flat or curved surface.

i. Which layer should be removed, top or bottom layer?

The bottom layer is always removed first independently if this is the one of interest or not because it is much easier to do. If a centrifuge tube or conical vial was used, the bottom layer should be drawn using a Pasteur pipette. From this point of view, a solvent with higher density than water would be preferential, especially when very small quantities are used. This will allow to minimize the number of transfer steps required.

j. How much solvent/solution is used for the extraction?

This highly depends on the quantity of a compound that has to be removed. For most washing processes, 10-20 % of the volume of the solution to be washed will do an adequate job. If a large amount of a compound has to be transferred or neutralized, more concentrated solutions and larger quantities might be needed. Multiple extractions with smaller quantities are preferred over one extraction with the same quantity of solution/solvent. If solutions with higher concentrations are used, extra caution is advised because neutralization reactions are exothermic. This can pose a serious problem when using low boiling solvents i.e., diethyl ether, dichloromethane, etc. because a pressure build-up will be observed in the extraction container.

Types of Extraction

Liquid-liquid Extraction

Liquid-liquid extraction is an important separation technology for a wide range of applications in the chemical process industries (CPI). Unlike distillation, which is based on boiling point differences, extraction separates components based on their relative solubilities in two immiscible liquids. Extraction is typically chosen over distillation for separation applications that would not be cost-effective, or even possible, with distillation.

In a liquid-liquid extraction unit, a liquid stream (carrier) containing the component(s) to be recovered (solute) is fed into an extractor, where it contacts a solvent. The two liquids must be immiscible or only slightly miscible; this allows them to form a dispersion, with one liquid dispersed as droplets in the other.

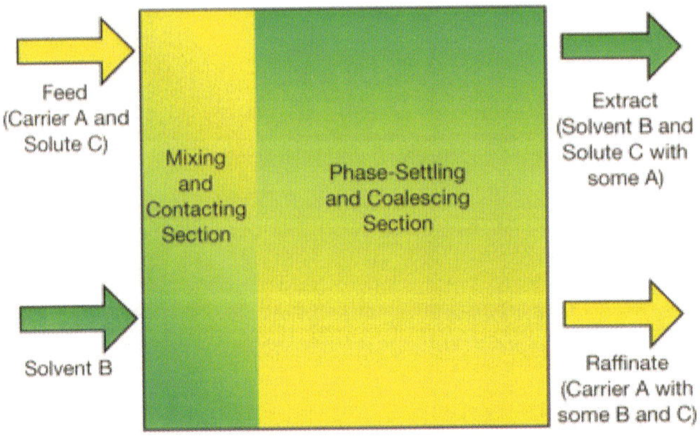

Figure: illustrates the general concept of liquid-liquid extraction.

Mass transfer occurs between the droplets (dispersed phase) and the surrounding liquid (continuous phase). In order for the two liquids to be subsequently separated, they must have different densities. The droplets then accumulate above or below the continuous phase, depending on the liquids' relative densities. The boundary between the continuous phase and the droplet dispersion is referred to as the interface, and can be at the top or bottom of the extraction column.

Extraction is typically carried out in continuous, staged units, which can be operated in either of two modes: with co-current mixing or with counter-current mixing. The co-current mixing mode is generally limited to one theoretical stage per extraction unit, whereas counter-current mixing is amenable to multiple stages per unit. For this reason, counter-current mixing is usually preferred over co-current mixing.

Counter-current extractors can be arranged in one of two ways, the choice of which depends on the density of the solvent relative to that of the solute carrier. If the solvent is less dense than the carrier liquid, the solvent is fed into the bottom of the column, the solute is carried upward to the top of the extractor, and the carrier liquid is removed from the bottom of the unit. If the solvent is more dense than the carrier liquid, the solvent is fed into the top of the column, the solute is carried downward to the bottom of the extractor, and the carrier stream is removed from the top.

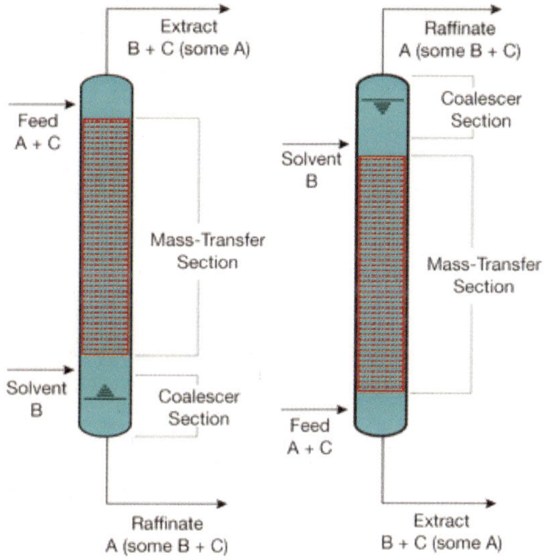

Figure. Counter-current extraction units can be set up in one of two ways. Left: When the solvent is lighter than the carrier liquid, the solvent is introduced at the bottom of the column and the solute is carried up toward the top of the extractor. Right: When the solvent is heavier than the carrier, the solvent is introduced at the top of the column and the solute is carried downward by the solvent toward the bottom of the column. The red dots represent the region in which the solute is transferred to the solvent.

Selecting a Solvent

Solvents used in liquid-liquid extraction are chosen to achieve maximum transfer of the solute from the carrier into the solvent. They must not be completely miscible with the carrier liquid and should have a high affinity for the solute molecules.

An ideal solvent for liquid-liquid extraction will typically have the following properties:

- High solubility for the solute and low solubility for the carrier liquid
- Density difference vs. the carrier liquid greater than 150 kg/m^3
- Mid-level interfacial tension (5–30 dyne/cm)
- High resistance to thermal degradation
- Nonreactive with other chemicals involved in the extraction process
- High boiling point (for ease of material handling)
- Low viscosity (for ease of handling)
- Nontoxic, nonflammable, and not corrosive to process equipment (for ease of handling)
- Low cost.

Engineers usually choose several solvents for process development. Often, the solvent candidates violate one or more of the criteria for an ideal solvent, so a comparison study is required to identify the best solvent for a particular application.

Dispersed-phase Droplet Size

An extraction column can be designed with either liquid as the dispersed phase: The solvent can be dispersed as droplets in the carrier liquid, or the carrier liquid can be dispersed as droplets in the solvent.

In packed columns, it is generally desirable to disperse the inlet stream with the higher volumetric flow rate into the continuous phase in order to maximize the interfacial area for mass transfer. In agitated columns, the liquid phase with the lower volumetric flow rate is typically dispersed, so that both phases (dispersed and continuous) have similar residence times. It is generally desirable to disperse the liquid containing the solute in the extracting solvent, so that the direction of mass transfer is out of the dispersed droplets and into the continuous phase.

The dispersion in the extraction unit is unstable, so the droplets will eventually separate into a bulk phase. Droplet size is a critical parameter in determining the rate at which that occurs. The maximum stable size of droplets in the contacting section of static (non agitated) extractors depends on the internal energy and physical properties of the fluids, and can be determined through statistical measurements taken during extraction tests. Droplets in mechanically agitated extractors will be smaller than those in static extractors. The maximum stable droplet diameter (D_M) for static extractors can be calculated by:

$$D_M = c\sqrt{\frac{\Pi}{\Delta \tilde{n}\ g}}$$

where c is the proportionality constant, P is the interfacial tension, $\Delta\rho$ is the difference between the two fluids' densities, and g is the gravitational constant.

The direction of mass transfer can influence the behavior of liquid-liquid dispersions, thereby affecting the proportionality constant in equation. The Marangoni effect explains how mass transfer of solute from one phase to the other creates interfacial tension gradients on the surface of the dispersed phase. Transfer of solute from the droplet outward into the continuous phase tends to create larger droplets and higher interfacial area of contact between the two phases, and is represented by a larger proportionality constant. Transfer of solute from the continuous phase into the droplets tends to form smaller droplets and the proportionality constant is smaller. The mass-transfer rate across the droplet surface can also influence droplet behavior, with high flux rates causing the droplets to shatter. Figure illustrates these effects.

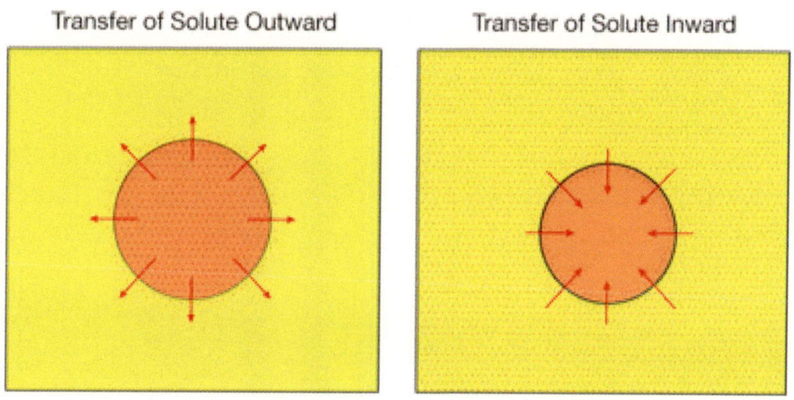

Figure: The solute can move either outward from the droplets to the continuous solvent phase, or inward from the continuous phase into the dispersed solvent. Left: The transfer of solute outward causes the droplets to grow and can cause them to shatter into smaller droplets. Right: The transfer of solute from the solvent into the droplets can create stable droplet surfaces.

Mechanically agitated extractors add energy to the system to influence the droplet size and thereby generate sufficient interfacial area of contact between the two phases. The high shear rate produced by turbulent agitation breaks the droplets into smaller droplets and distorts their shape. Equations have been developed to calculate the maximum drop size for agitated liquid-liquid extraction.

Axial Mixing

Axial (longitudinal) mixing causes the extractor to deviate from ideal plug-flow operation. As a result of axial mixing, the dispersed phase and the continuous phase will exhibit a range of residence times, which tends to reduce the concentration driving force for mass transfer in the column. Axial mixing can be a particular problem in large-diameter extraction columns, because redistributors have less profound effects in larger columns than they do in small-diameter extractors.

Several mechanisms and conditions can lead to axial mixing, including:

- Vessel walls or other fixed components can exert frictional drag on the liquids and cause turbulent mixing.
- Nonuniform droplet size distribution in the contacting section of the extractor can cause the extractor to deviate from plug-flow operation.
- Chaotic eddy patterns can form when fluids.

Solid Phase Extraction

Solid-phase extraction (SPE) is a sample preparation technique routinely used in analytical laboratories for the extraction of analytes from a complex matrix.

This sample preparation technique enables the extraction, cleanup and concentration of analytes prior to their quantification. Solid phase extraction prevents most problems encountered with liquid-liquid extraction and improves quantitative recovery yields. This technique is rapid (most extraction in less than 30 min), easy to perform and can be automated. In addition, low amount of solvents is handled and this technique is fully adapted for the pre-treatment of complex matrices such as urine, blood, food samples, water etc.

SPE Procedure

Solid phase extraction uses the difference of affinity between an analyte and interferents, present in a liquid matrix, for a solid phase (sorbent). This affinity allows the separation of the target analyte from the interferents.

A typical solid phase extraction involves four steps:

1. First, the cartridge is equilibrated or conditioned with a solvent to wet the sorbent.
2. Then the loading solution containing the analyte is percolated through the solid phase. Ideally, the analyte and some impurities are retained on the sorbent.
3. The sorbent is then washed to remove impurities.
4. The analyte is collected during this elution step.

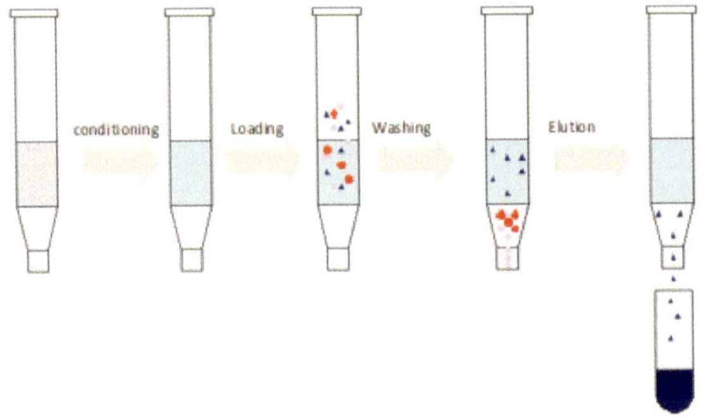

Importance of the Phase for A High Selectivity

A broad range of phases exists for SPE. Most of them are based on ionic, hydrophobic or polar interactions. Their selectivity is limited and the target analyte often co-elutes with interfering compounds.

Molecularly Imprinted Polymers are the most highly selective phases available today. This technology enables to reduce the presence of interferents during the elution step and obtain higher recoveries in chromatography analysis even at low levels.

Acid-base Extraction

An acid-base extraction is a type of liquid-liquid extraction. It typically involves different solubility levels in water and an organic solvent. The organic solvent may be any carbon-based liqiuid that does not dissolve very well in water; common ones are ether, ethyl acetate, or dichloromethane.

Acid-base extraction is typically used to separate organic compounds from each other based on their acid-base properties. The method rests on the assumption that most organic compounds are more soluble in organic solvents than they are in water. However, if the organic compound is rendered ionic, it becomes more soluble in water than in the organic solvent. These compounds can easily be made into ions either by adding a proton (an H^+ ion), making the compound into a positive ion, or by removing a proton, making the compound into a negative ion.

Suppose you have a mixture of two compounds. There is a neutral one, which doesn't react with any acids or bases. There is also a basic one, which reacts with acids by picking up a proton. In this case, a proton might be added via reaction with a strong mineral acid (represented by HX in the figure). Suppose an aqueous solution of mineral acid, such as HCl, were shaken vigorously with an ethereal solution of an organic base and an organic neutral. The proton would be transferred to a basic compound, but not to a neutral one. The basic compound would become ionic, and more water-soluble.

Note that in the figure, the ether is represented in yellow, whereas the water is shown in blue. The water is on the bottom in this case because water has a higher density than ether, so it will sink to the bottom (along with anything dissolved in it). Some organic solvents do have a higher density than water, so the aqueous solution would float to the top in those case.

As a result, the ethereal solution would contain only the neutral compound, not the basic one. The neutral compound could be isolated simply by evaporating the ether.

However, as a practical matter, the ether would have to be dried first. What's the difference between evaporating and drying? Have you ever been to the beach or taken a shower? Drying refers to the removal of water. This step is necessary because ether tends to dissolve a lot of water in it. Once the ether has been evaporated, there would be some neutral compound, but it would be mixed with water.

Water removal is most easily done by adding a drying agent, such as magnesium sulfate or sodium sulfate. The water sticks to these solids, which are then filtered off.

Now the neutral compound is alone in the ether. Evaporation of the ether gives the pure, neutral compound.

However, the basic compound is stuck in the water, and it isn't the same compound anymore. It's an ion, now. If we want the original compound in a pure form, we need to take that proton away. That can be done by adding a mineral base, such as sodium hydroxide.

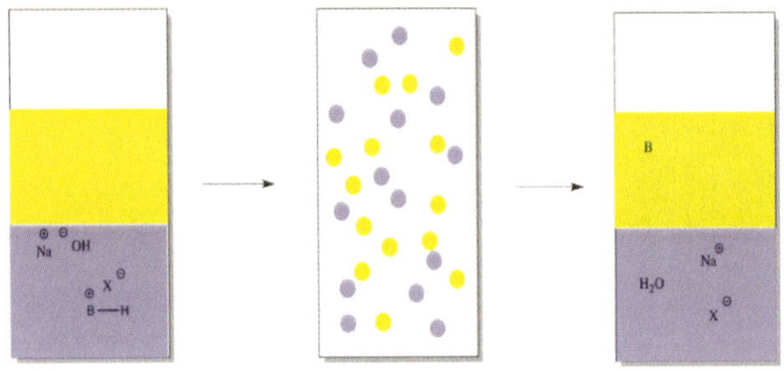

The mineral base will remove the proton, leaving the original organic compound. The organic compound is uncharged and not as soluble in water anymore. It will go back into the ether layer.

Conversely, we might have a mixture of an acidic organic compound and a neutral compound to start out with. In that case, we would add a mineral base in the first place, to take a proton away from the acidic compound. The mineral base might be something like sodium hydroxide or sodium bicarbonate. In the figure, it is just represented as Na^+ B^-.

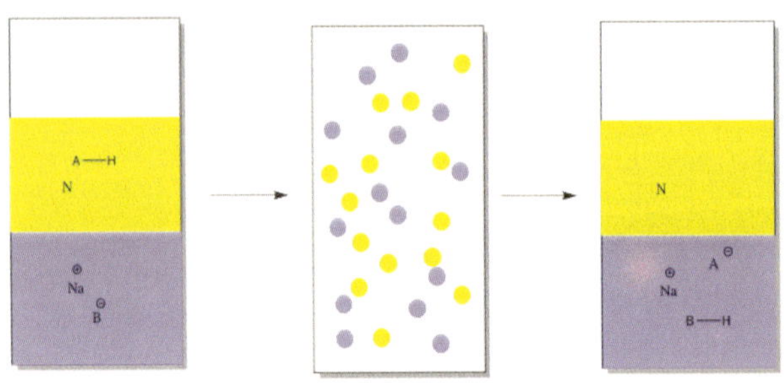

The acidic compound becomes ionic and water-soluble when it loses a proton. That leaves the neutral compound alone.

To get that acidic compound back, we would add a mineral acid such as hydrochloric acid in order to restore the missing proton.

Just as in the other case, the ether layer containing a pure compound could be separated, dried and evaporated in order to provide the pure compound.

In an acid-base extraction a chemical reaction is performed in the separatory funnel in order to change the polarity and therefore partitioning of a compound in the aqueous and organic layers. A common method is to perform an acid-base reaction, which can convert some compounds from neutral to ionic forms (or vice versa).

For example, imagine that a mixture of benzoic acid and cyclohexane is dissolved in an organic solvent like ethyl acetate in a separatory funnel. To separate the components, a water wash may be attempted to remove benzoic acid, but benzoic acid is not particularly water-soluble due to its nonpolar aromatic ring, and only small amounts would be extracted into the aqueous layer.

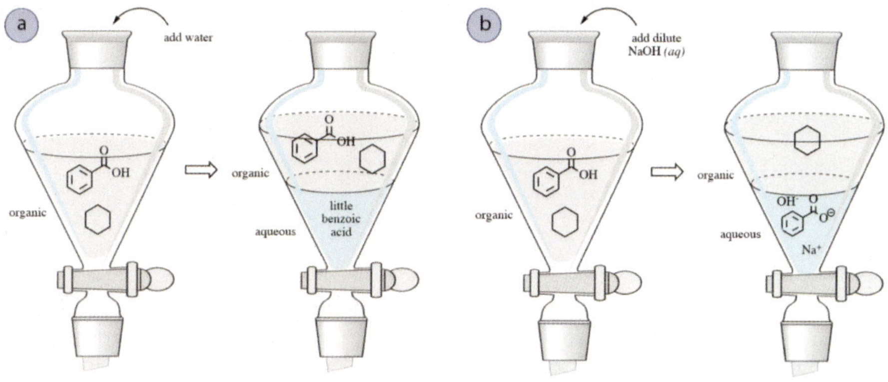

Figure: Washing a mixture of benzoic acid and cyclohexane with: a) water, b) aqueous NaOH

Separation of a mixture of benzoic acid and cyclohexane is however possible using a wash with a base such as NaOH NaOH. Due to its acidic nature, benzoic acid can undergo a reaction with NaOH NaOH as follows, resulting in the carboxylate salt sodium benzoate.

$$PhCO_2H(aq) \quad + \quad NaOH(aq) \quad \rightarrow \quad H_2O(l) \quad + \quad PhCO_2Na(aq) \qquad \left(\text{or } PhCO_2^- \ Na^+\right)$$

Benzoic acid $\qquad$ Sodium benzoate

The solubility properties of carboxylic acids are substantially different than their corresponding carboxylate salts. Sodium salicylate is roughly 350 times more soluble in water than salicylic acid due to its ionic character, and it is rather insoluble in organic solvents such as diethyl ether.

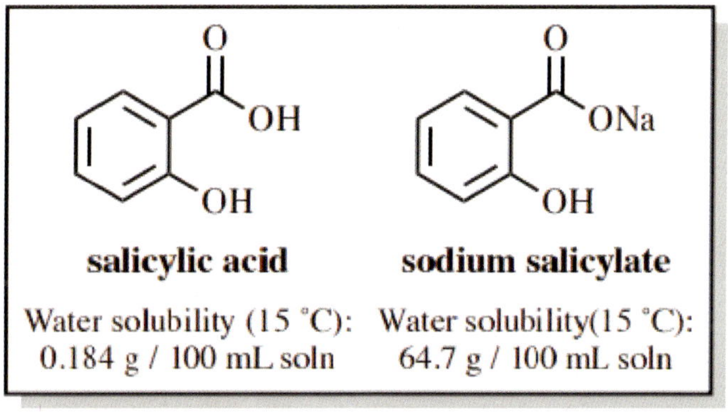

Figure: Aqueous solubility data for salicylic acid and sodium salicylate.

Therefore, a wash with NaOH NaOH would convert benzoic acid into its ionic carboxylate form, which would then be more soluble in the aqueous layer, allowing for the sodium benzoate to be extracted into the aqueous layer. Cyclohexane would remain in the organic layer as it has no affinity for the aqueous phase, nor can react with NaOH NaOH in any way. In this manner, a mixture of benzoic acid and cyclohexane can be separated. The aqueous layer may be later acidified with HCl(aq) HCl(aq) if desired to convert the benzoic acid back to its neutral form.

Sodium Bicarbonate Washes

An acid-base extraction can be used to extract carboxylic acids from the organic layer into the aqueous layer. NaOH can be used to convert a carboxylic acid into its more water-soluble ionic carboxylate form. However, if the mixture contains a desired compound that can react with NaOH NaOH, a milder base such as sodium bicarbonate should be used. A similar reaction occurs:

$$\underset{\text{Benzoic acid}}{PhCO_2H(aq)} + NaHCO_3(aq) \rightarrow \underset{\text{Sodium benzoate}}{PhCO_2Na(aq)} + H_2CO_3(aq) \rightleftharpoons H_2(l) + CO_2(g)$$

One difference in using the base $NaHCO_3$ instead of NaOH is that the byproduct carbonic acid (H_2CO_3) can decompose to water and carbon dioxide gas. When shaking an acidic solution with sodium bicarbonate in a separatory funnel, care should be taken to swirl gently and vent more frequently to release pressure from the gas.

An example of a reaction that often uses sodium bicarbonate wash in the work-up is a Fischer Esterification reaction. To demonstrate, benzoic acid was refluxed in ethanol along with concentrated sulfuric acid in order to form ethyl benzoate. A TLC plate of the reaction mixture at 1 hour of reflux showed residual unreacted carboxylic acid, which is not uncommon due to the energetics of the reaction.

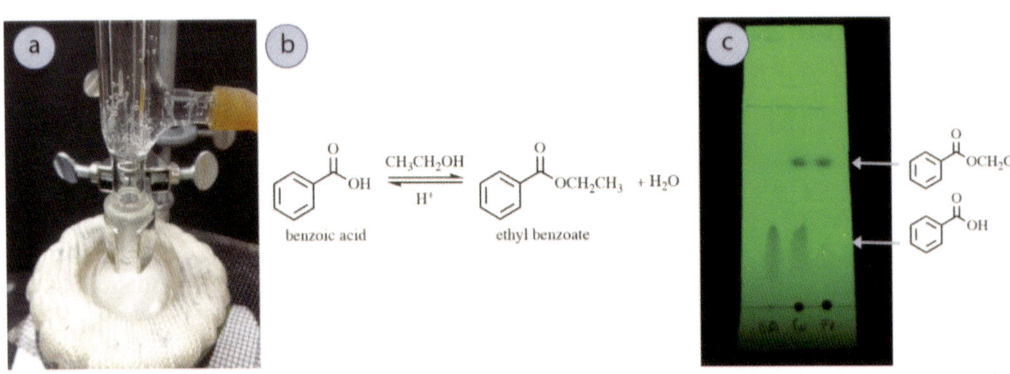

Figure : a) Refluxing reagents, b) Reaction scheme, c) TLC after 1 hour reflux, where the first lane (BA) is benzoic acid, the second lane (Co) is the co-spot and the third lane (Pr) is the reaction mixture (ran with 1:1 hexanes:ethyl acetate and visualized with UV light).

The residual carboxylic acid can be removed from the desired ester product using an acid-base extraction in a separatory funnel. A wash with sodium bicarbonate converts benzoic acid into its more water-soluble sodium benzoate form, extracting it into the aqueous layer. Additionally, the sodium bicarbonate neutralizes the catalytic acid in this reaction.

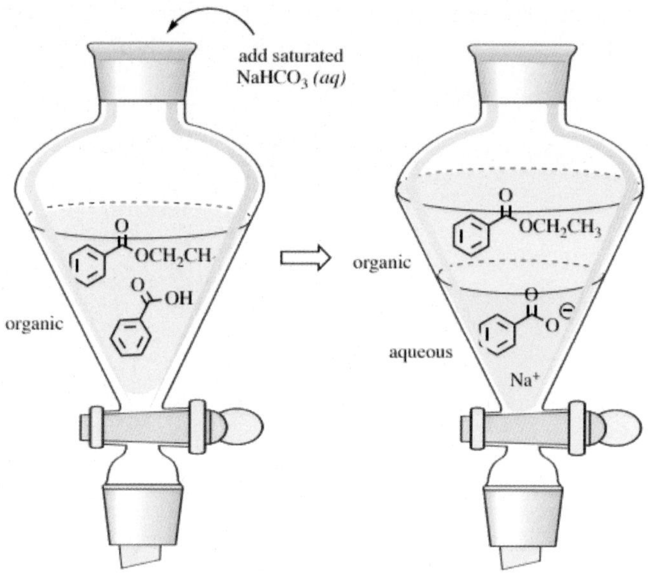

Figure: Saturated sodium bicarbonate wash to remove residual carboxylic acid from the reaction mixture of a Fischer esterification. The ester is then isolated in the organic layer.

Sodium bicarbonate is preferable to NaOH NaOH in this process, as it is a much weaker base; washing with NaOH NaOH could cause hydrolysis of the ester product.

Mixtures of Acids and Bases

The acid-base properties of compounds can be utilized to selectively extract certain compounds from mixtures. This strategy can be extended to other examples.

Extracting Bases

Basic compounds such as amines can be extracted from organic solutions by shaking them with acidic solutions to convert them into more water-soluble salts. In this way, they can be extracted from an organic layer into an aqueous layer.

$$\underset{\text{Basic amine}}{PhNH_2(aq)} + HCl(aq) \rightarrow \underset{\text{Ammonium salt}}{PhNH_3Cl(aq)} \quad \left(\textit{or } PhNH_3^+Cl^-\right)$$

Extracting Carboxylic Acids vs. Phenols

Carboxylic acids can be extracted from an organic layer into an aqueous layer by shaking them with basic solutions, which converts them into their more water-soluble salts.

$$\underset{\text{Carboxylic acid}}{PhCO_2\,H(aq)} + NaOH(aq) \rightarrow H_2O(l) + \underset{\text{Carboxylate salt}}{PhCO_2Na(aq)} \quad \left(\text{or } PhCO_2^-Na^+\right)$$

A similar reaction occurs with phenols (PhOH), and they too can be extracted into an aqueous NaOH layer.

However, phenols are considerably less acidic than carboxylic acids, and are not acidic enough to react completely with NaHCO3, a weaker base. Therefore, a solution of bicarbonate can be used to separate mixtures of phenols and carboxylic acids.

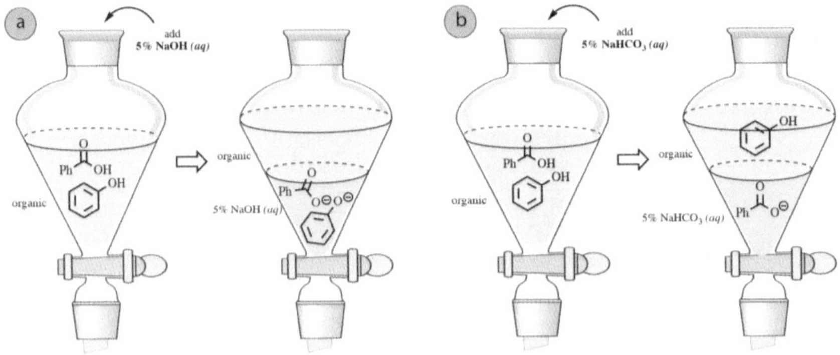

Figure: a) Extraction of both carboxylic acids and phenols into 5%NaOH(aq) 5%NaOH(aq), b) Extraction of only carboxylic acids into 5%NaHCO 3 (aq) 5%$NaHCO_3$(aq).

Extracting Acid, Base and Neutral Compounds

The acid-base properties allow for a mixture containing acidic (e.g. RCO_2H), basic (e.g. RNH_2), and neutral components to be purified through a series of extractions, as summarized in figure (which uses an organic solvent less dense than water).

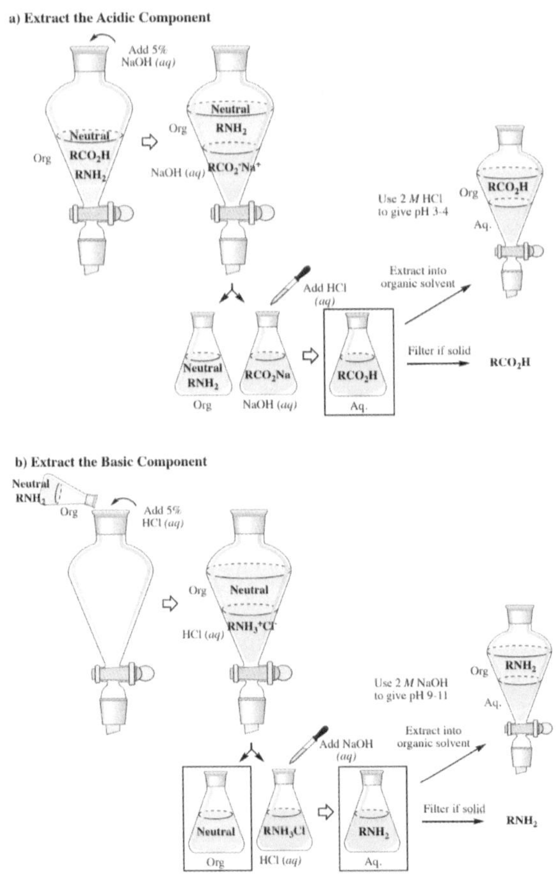

Figure: Flowchart for separating a mixture containing acidic, basic, and neutral components.

It is assumed that readers conducting this type of experiment are familiar with performing single and multiple extractions. In this topic are described differences between general extraction procedures and the process as summarized in figure.

1. Isolating the Acidic component
   a. When the acidic component is in the aqueous layer in an Erlenmeyer flask, it can be converted back to the neutral component through addition of 2MHCl(aq) 2MHCl(aq) until the solution gives a pH of 3-4 (as determined by pH paper). If large quantities of acid are present such that acidification would require too great a volume of 2MHCl(aq) 2MHCl(aq), concentrated HCl(aq) HCl(aq) may be instead added dropwise. Lower concentrations of HCl(aq) HCl(aq) are less hazardous, but increasing the volume of the aqueous layer by a large amount would affect the efficiency of subsequent extractions and filtering steps.
   b. After acidification, two routs may be taken, depending on if the acidic component is solid or liquid.
      i. If a solid forms upon acidification of the ionic salt, it can be collected through suction filtration. This method should only be used if large quantities of large-sized crystals are seen. If fine crystals form (which are quite common), they will clog the filter paper and interfere with adequate drainage. If only a small amount of solid is seen compared to the theoretical quantity, it is likely the compound is quite water-soluble, and filtration would lead to low recovery.
      ii. If no solid forms upon acidification (or if fine crystals or low quantity of solid forms), extract the acidic component back into an organic solvent (×3 ×3). As a general rule of thumb, use one-third as much solvent for the extractions as the original layer (e.g. if using 100mL 100mL aqueous solution, extract with 33mL 33mL organic solvent each time). Be sure to first cool the aqueous solution in an ice bath before extraction if the acidification created noticeable heat. Follow up with a brine wash (×1 ×1) if using diethyl ether or ethyl acetate, dry with a drying agent, and remove the solvent via rotary evaporator to leave the pure acidic component.
2. Isolating the Basic component
   - Use a similar process as the isolation of the acidic component, except basify the solution using 2MNaOH(aq) 2MNaOH(aq) until it gives a pH of 9-10 as determined by pH paper.
3. Isolating the Neutral component
   - The neutral component will be the "leftover" compound in the organic layer. To isolate, wash with brine (×1 ×1) if using diethyl ether or ethyl acetate, dry with a drying agent, and remove the solvent via rotary evaporator to leave the pure neutral component.

Electromembrane Extraction

Electromembrane extraction (EME) is a liquid-phase microextraction concept intended for selective extraction and pre-concentration of target analytes from aqueous samples. Target analytes are typically drug substances, and aqueous samples are typically whole blood and urine. Thus, EME is well suited for pharmaceutical analysis, but environmental and food applications are also frequently seen.

The principle of EME is illustrated in figure. Target analytes are extracted from the sample, through a supported liquid membrane (SLM), and into an acceptor solution. This mass transfer is by electrokinetic migration facilitated by an external electrical field sustained across the SLM. The SLM comprises an organic solvent (5–25 μL) immobilized by capillary forces in the pores of a porous polymeric membrane. This porous polymeric support can be a flat sheet or a hollow fibre membrane as illustrated in figure.

EME Setup and Procedure

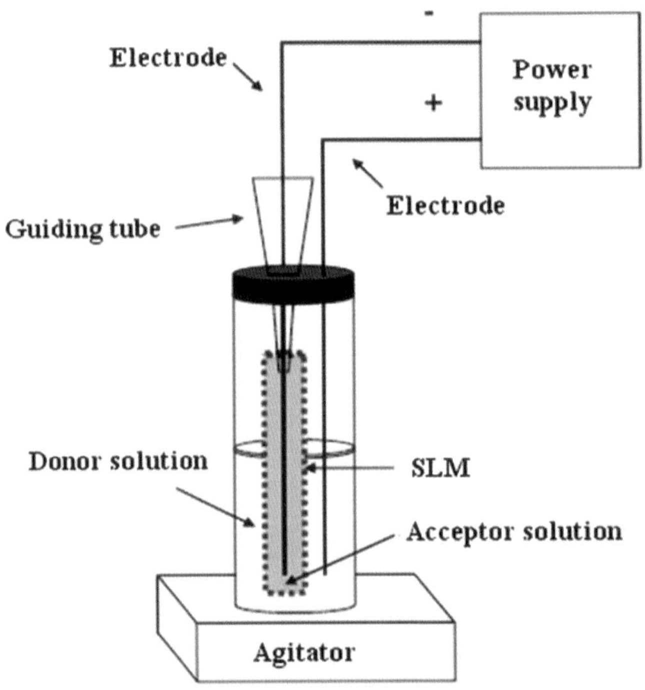

Schematic illustration of the setup for EME.

The setup used for EME is shown in figure. It consists of a glass vial with a screw cap used as a sample compartment, into which sample solution is filled. The HF (usually made up of polypropylene or other porous hydrophobic material) is cut into small pieces of required length and the lower end of the HF of required length is sealed by mechanical pressure, whereas the upper end is connected to a pipette tip as a guiding tube. The HF is dipped for few seconds into the organic solvent serving as the SLM. Excess of solvent in the SLM is gently removed with a medical wipe or by air-blowing with a medical syringe. Through the guiding tube, acceptor solution is filled into the lumen of the HF with a microsyringe. The filled HF and the guiding tube are inserted through the vial cap into the sample solution. One of the electrodes is set into the lumen of the fiber via the guiding tube, whereas the other electrode is led through the cap and directly into the sample solution. The electrodes are connected to the power supply with programmable voltage and the extraction unit is agitated on an agitator. Under the applied voltage, the target analytes migrate from the sample solution into the SLM and transported into the acceptor phase. After the extraction is completed, the acceptor solution is collected by a microsyringe and transferred to a vial for analysis in high-performance liquid chromatography (HPLC), GC, CE or any other technique.

Parameters Influencing Flux of Analytes Across SLM

Composition (Organic Solvent) of SLM, Viscosity and Thickness

The chemical nature of the SLM is highly critical to success in EME. The flux of the analyte is affected by difference in concentration of the analyte across the SLM, which is partially determined by the sample to SLM distribution ratio; this, in turn, is controlled by the type of solvent used as the SLM. In addition, the type of solvent also influences the diffusion coefficient of the analyte, which can also be tuned to increase the selectivity and to obtain good clean up during extraction. There are specific requirements for a solvent to be used as an SLM in EME. The organic phase should have a certain dipole moment or electrical conductivity to support a relatively low current flow in the system, and it should have certain chemical properties to enable phase transfer and electrokinetic migration of the model analytes. Furthermore, the solvent should be immiscible in water to avoid losses from pores of the HF membrane wall and dissolution in the sample solution during stirring.

Extraction of Basic Drugs

Based upon literature reports, nitro-aromatic solvents such as 2-nitrophenyl octylether (NPOE) and nitrophenyl pentyl ether (NPPE) are efficient organic solvents for the elctrokinetic migration of nonpolar (logP > 2) basic drugs through the SLM. It has also been reported that addition of hydrophobic alkylated phosphate reagents to SLM can improve the phase transfer and electrokinetic migration of basic analytes.

Polar (logP < 1) basic drugs were unable to penetrate through the interface between the donor phase and the SLM with pure NPOE. The high polarity of these analytes seemed to counteract the influence of the electric field. To facilitate the transport of polar drugs through SLM, an ion pair reagent such as di-(2-ethylhexyl) phosphate (DEHP) or tris-(2-ethylhexyl) phosphate (TEHP) was added to the organic solvent to form ion pairs with the analytes. An SLM consisting of 10% DEHP and 10% TEHP in NPOE was useful for the extraction of basic drugs with a large logP window.

Extraction of Acidic Drugs

Nitro-aromatic solvents are inefficient for the extraction of acidic drugs. Aliphatic alcohols like 1-octanol and 1-heptanol were found to be efficient for the extraction of acidic drugs. They are insoluble in water and easily impregnate the membrane. In addition, the alcohols showed an appropriate electrical resistance to the applied voltage (i.e., the current was low enough to avoid bubble formation and excessive electrolysis, but still sufficiently high for the system to maintain the properties of the closed circuit). The chemical nature of the organic solvent in combination with a proper potential difference can be used to design very selective extractions of different analyte classes.

The viscosity of the organic solvent is another parameter that significantly influences the diffusion of analytes through SLM: if viscosity is low, more diffusion is observed. The thickness of the SLM also influences the flux of analytes: a thicker membrane results in an increased diffusion path and theoretically reduces the extraction recovery.

Extraction Voltage and Time

In EME, the electrokinetic migration of the analytes across the SLM into the acceptor solution is greatly dependent upon the applied voltage. The flux of analytes (Ji) is affected by the magnitude of the applied voltage. Equation describes how different parameters affect the flux of analytes through the SLM:

$$J_i = \frac{-D_i}{b}\left(1+\frac{v}{\ln\chi}\right)\left[\frac{\chi-1}{\chi-\exp(-v)}\right]\left\{c_1 - c_{10}\exp(-v)\right\}$$

where, Di is the diffusion coefficient for the analyte, h is the thickness of the membrane, ci is the analyte concentration at the SLM/sample interface and c_{io} is the concentration of the analyte at the acceptor/SLM interface. Additionally, v is a function of electrical potential and χ is the ratio of the total ionic concentration in the sample solution to that in the acceptor solution, which is defined as the ion balance.

Generally, the voltage is applied in the range of 5–600 V. At higher voltages, recovery values decrease because of bubble formation at the electrodes and instability of the system. Time is another parameter that can affect the flux of analytes in EME. Both time and voltage directly increase the flux of ions, and thus, increase extraction recovery, but there is an antagonist effect in which they are simultaneously considered; thus, an increase in extraction time limits the voltage and vice versa. Generally, at extraction times above 15 min, the recovery is decreased due to the instability of the electrical current in the system, back-diffusion of the analytes toward the donor phase due to changes in pH from electrolysis and a small loss of artificial liquid membrane.

pH of Donor and Acceptor Phases (Ion Balance Effect)

The pH values of donor and acceptor phases can determine the ion balance in the system. It has been shown that the total ionic concentration on the donor phase to that on the acceptor phase impresses the flux over the membrane . The flux may decrease by increasing this ratio. In EME, for the rapid extraction of basic analytes, pH in the sample and acceptor solutions is low to ionize the analytes, and the analytes are extracted across the SLM with a positive electrode placed in the sample and a negative electrode placed in the acceptor solution. For the extraction of acidic drugs, the extraction system is operated in reversed polarity and the pH of the donor and acceptor solutions was high (alkaline).

Stirring Speed

Stirring speed plays an essential role in increasing the kinetics and efficiency of extraction by increasing the mass transfer and reducing the thickness of the double layer around the SLM.

Generally, a stirring speed is used in the range of 0–1,250 rpm. At higher stirring rates, extraction recovery is decreased due to bubble formation in both the donor and acceptor phases and the leakage of the organic solvent from the SLM. This leads to incompatible analytical results.

Salt Effect

The presence of ionic substances in the donor solution causes an increase in the value of the ion balance in the system, which decreases the flux of analytes across the SLM. Thus, the migration of analytes is more efficient in the absence of salt. However, some studies have shown that the addition of salt to the donor solution increased the extraction efficiency. Basheer et al. Investigated the effect of the addition of salt (NaCl) on the EME of acidic and basic drugs. The optimum extraction was achieved with 30% (w/v) NaCl samples. Alhooshani et al. Investigated the influence of salt on the EME of haloacetic acids. Various amounts of NaCl were added to the donor solution, ranging from 3 to 15%. The addition of NaCl increased the extraction efficiency from ultrapure water to 5% NaCl; above 5%, a decrease in the extraction was observed due to an increase in viscosity and a change in the conductivity of the donor solution.

Temperature

The effect of temperature on extraction has also been reported. It was reported that high temperatures (up to 40° C) can speed up the extraction process and temperatures above 40° C cause partial degradation of the SLM.

Leaching

Even if you don't know the definition for the chemical process of leaching, you've probably witnessed it take place. If you've ever brewed a cup of tea or made instant coffee, for instance, you've seen leaching in action. Leaching is just the process of extracting a substance from a solid material that has come into contact with a liquid.

In leaching, the liquid is very important, as it facilitates the ability to remove, or extract, a given substance from a solid matrix (i.e. material). Let's break this definition down by using the example of brewing tea.

You go into your kitchen and decide you would like to make a hot cup of green tea. Of course you know that a green tea bag, hot water, and a cup is needed to make the tea. Shown in this figure, we can relate each of these components to the definition of leaching:

(1) The tea bag would be our solid matrix,

(2) The green tea would be the substance extracted, and

(3) The hot water would be our liquid source.

Going back to our story on making tea, you proceed with boiling water and add it to a cup. As you steep your green tea bag in the hot water, what do you notice? The water not only changes colors, but more importantly, green tea is extracting from the tea bag into the water. After steeping for a few minutes, you sip from your mug and taste not just hot water, but delicious green tea, too.

Figure: Example of Making Green Tea To Illustrate Leaching

Although making tea is a very well known process, it's also an ideal example of how leaching works. You took a solid matrix, like the tea bag, introduced it to a liquid, in this instance the hot water, and extracted green tea, or a substance, to make a cup of tea.

Process of Leaching

Although some species may be more of an environmental concern than others, the leaching process is indiscriminant such that all constituents (e.g., major or minor matrix components as well as inorganic, organic and radionuclide contaminants) are released under a common set of chemical phenomena which may include mineral dissolution, desorption and complexation, and mass transport processes. In turn, these phenomena are affected by certain factors that can alter the rate or extent of leaching. Among these factors are:

- Internal chemical and physical reactions.
- External stresses from the surrounding environment.
- Physical degradation of the solid matrix due to erosion or cracking.
- Loss of matrix constituents due to the leaching process itself.

The process of leaching includes the partitioning of contaminants between a solid and liquid phase (e.g., assuming local equilibrium) coupled with the mass transport of aqueous or dissolved constituents. Mass transport is the summation of diffusion, hindered diffusion, tortuosity effects, and effective surface area effects through the pore structure of the material to the environment. Important chemical factors, those that influence the liquid-solid partitioning (LSP) of a constituent, include solution pH, redox, the presence of dissolved organic matter, and biological activity. Physical factors, such as relative hydraulic conductivity, porosity and fill geometry, play an important role in determining the rate at which constituents transport through a solid into a passing liquid phase.

The process itself is universal, as any material exposed to contact with water will leach components from its surface or its interior depending on the porosity of the material considered.

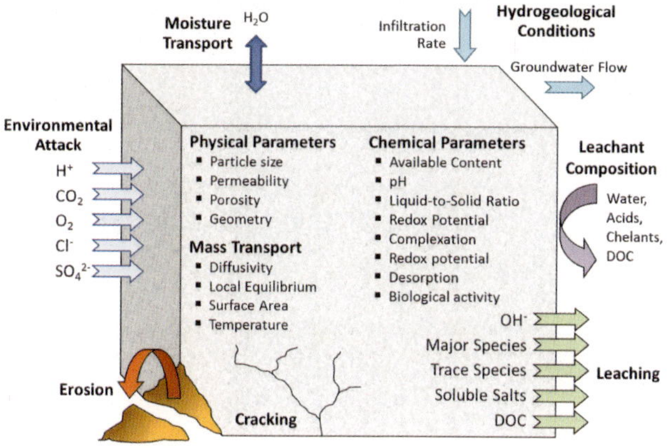

Accelerated Solvent Extraction

Accelerated Solvent Extraction (ASE) eliminates many of the manual steps involved in preparing food samples for analysis, which helps ensure increased reproducibility and accelerates the process significantly. ASE is an automated extraction technique that uses elevated temperatures and pressures to achieve extractions in very short periods of time (a 10-g sample can be extracted in less than 15 minutes using less than 15 mL solvent).

Many of the organic solvents used in extractions boil at relatively low temperatures. This is a limitation to techniques such as Soxhlet or automated Soxhlet the highest temperatures at which extractions take place in these techniques will be the boiling point of the solvent.

If sufficient pressure is exerted on the solvent during the extractions, temperatures above the boiling point can be used. This means that all of the advantages of working at elevated temperature can be realized even with solvents with relatively low boiling points. Operating at elevated pressures also helps the extraction process to happen more quickly. Pumping solvent through a packed bed is easier at elevated pressures; pressurized solvent is forced into the pores of the sample matrix. Hence, the combination of elevated temperatures and pressures allows extractions to occur rapidly and completely.

When extractions are achieved at elevated temperatures, several factors contribute to improved speed, efficiency and reduced solvent use:

1) Solvent strength is higher.

2) Diffusion rates are faster.

3) Solvent viscosity is decreased.

4) Solute-matrix interactions (dipole attractions, van der Waals forces, hydrogen bonding, etc.) are more easily disrupted allowing the analytes to be removed from the matrix.

The net result means performing extractions at elevated temperatures uses less time and with less solvent.

In the food industry, time can be critical. Quality control checks need to be completed quickly and accurately. Whether a laboratory is analyzing the fat content of food for labeling or analyzing food for environmental contaminants, ASE is a powerful tool for preparing these samples in the fastest time possible while ensuring accurate results.

Common Applications of Accelerated Solvent Extraction

Environmental samples

- Pesticides/herbicides
- PAHs/semivolatiles
- PCBs
- Dioxins/furans
- TPH (DRO)
- Explosives
- Air sampling cartridges
- Brominated flame retardants

Pharmaceutical samples

- Active ingredients
- Drug delivery devices/packaging

Polymer samples

- Oils/organic acids
- Xylene solubles
- Polymer additives

Food samples

- Vitamins/antibiotics
- Contaminants
- Natural products
- Pesticides/herbicides
- Pesticide residues

- Fats/lipids
- Herbal/dietary supplements

Consumer products

- Detergents
- Paper and pulp
- Biofuels
- Textiles and fibers.

Supercritical Fluid Extraction

Supercritical fluids have been investigated since last century, with the strongest commercial interest initially focusing on the use of supercritical toluene in petroleum and shale oil refining during the 1970s. Supercritical water is also being investigated as a means of destroying toxic wastes, and as an unusual synthesis medium. The biggest interest for the last decade has been the applications of supercritical carbon dioxide, because it has a nearambient critical temperature (31° C), thus biological materials can be processed at temperatures around 35° C. The density of the supercritical CO_2 at around 200 bar pressure is close to that of hexane, and the solvation characteristics are also similar to hexane; thus, it acts as a non-polar solvent. Around the supercritical region, CO_2 can dissolve triglycerides at concentrations up to 1% mass. The major advantage is that a small reduction in temperature, or a slightly larger reduction in pressure, will result in almost the entire solute precipitating out as the supercritical conditions are changed or made sub critical. Supercritical fluids can produce a product with no solvent residues. Examples of pilot and production scale products include decaffeinated coffee, cholesterol-free butter, low-fat meat, evening primrose oil, squalene from shark liver oil, etc. The solvation characteristics of supercritical CO_2 can be modified by the addition of an entrainer, such as ethanol, however some entrainer remains as a solvent residue in the product, negating some of the advantages of the "residue-free" extraction.

Supercritical fluid extraction (SFE) is the process of separating one component (the extractant) from another (the matrix) using supercritical fluids as the extracting solvent. Extraction is usually from a solid matrix, but it can also be from liquids. SFE can be used as a sample preparation step for analytical purposes, or on a larger scale to either strip unwanted material from a product (e.g. decaffeination) or collect a desired product (e.g. essential oils). Carbon dioxide (CO_2) is the most used supercritical fluid, sometimes modified by co-solvents such as ethanol or methanol. Extraction conditions for supercritical CO_2 are above the critical temperature of 31° C and critical pressure of 74 bar. Addition of modifiers may slightly alter this. Supercritical extraction mostly uses carbon dioxide at high pressure to extract the high value products from natural materials. Unlike other processes, the extraction process leaves no solvent residue behind. Moreover the CO_2 is non-toxic, nonflammable, odorless, tasteless, inert, and inexpensive. Due to its low critical temperature 31°C, carbon dioxide is known to be perfectly adapted in food, aromas, essential oils and nutraceutical industries.

Supercritical Fluid Extraction Process

The system must contain a pump for the CO_2, a pressure cell to contain the sample, a means of maintaining pressure in the system and a collecting vessel. The liquid is pumped to a heating zone, where it is heated to supercritical conditions. It then passes into the extraction vessel, where it rapidly diffuses into the solid matrix and dissolves the material to be extracted. The dissolved material is swept from the extraction cell into a separator at lower pressure, and the extracted material settles out. The CO_2 can then be cooled, recompressed and recycled, or discharged to atmosphere.

Pumps

Carbon dioxide is usually pumped as a liquid, usually below 5° C and a pressure of about 50 bars. The solvent is pumped as a liquid as it is then almost incompressible. As a supercritical fluid, much of the pump stroke will be "used up" in compressing the fluid, rather than pumping it. For small-scale extractions (up to a few grams/minute), reciprocating CO_2 pumps or syringe pumps are often used. For larger scale extractions, diaphragm pumps are most common. The pump heads will usually require cooling, and the CO_2 will also be cooled before entering the pump.

Pressure Vessels

Pressure vessels can range from simple tubing to more sophisticated purpose built vessels with quick release fittings. The pressure requirement is at least 74 bars, and most extractions are conducted at less than 350 bar. However, sometimes, higher pressures will be needed, such as extraction of vegetable oils, where pressures of 800 bars are sometimes required for complete miscibility of the two phases.

The vessel must be equipped with a means of heating. It can be placed inside an oven for small vessels, or oil or electrically heated jacket for larger vessels. Care must be taken if rubber seals are used on the vessel, as the CO_2 may dissolve in the rubber, causing swelling, and the rubber will rupture on depressurization.

Pressure Maintenance

The pressure in the system must be maintained from the pump right through the pressure vessel. In smaller systems (up to about 10 mL/min), a simple restrictor can be used. This can be either a capillary tube cut to length, or a needle valve, which can be adjusted to maintain pressure at different flow rates. In larger systems a backpressure regulator will be used, which maintains pressure upstream of the regulator by means of a spring, compressed air, or electronically driven valve. Whichever is used, heating must be supplied, as the adiabatic expansion of the CO_2 results in significant cooling. This is problematic, if water or other extracted material is present in the sample, as this may freeze in the restrictor or valve and cause blockages.

Collection

The supercritical solvent is passed into a vessel at lower pressure than the extraction vessel. The density, and thus, dissolving power, of supercritical fluids varies sharply with pressure, and hence, the solubility in the lower density CO_2 is much lower, and the material precipitates for collection.

It is possible to fractionate the dissolved material using a series of vessels at reducing pressure. The CO_2 can be recycled or depressurized to atmospheric pressure and vented. For analytical SFE, the pressure is usually dropped to atmospheric, and now the gaseous carbon dioxide is bubbled through the solvent to trap the precipitated components.

Heating and Cooling

This is an important aspect. The fluid is cooled before pumping to maintain liquid conditions, and then heated after pressurization. As the fluid is expanded into the separator, heat must be provided to prevent excessive cooling. For small-scale extractions, such as for analytical purposes, it is usually sufficient to pre-heat the fluid in a length of tubing inside the oven containing the extraction cell. The restrictor can be electrically heated, or even heated with a hair dryer. For larger systems, the energy required during each stage of the process can be calculated using the thermodynamic properties of the supercritical fluid.

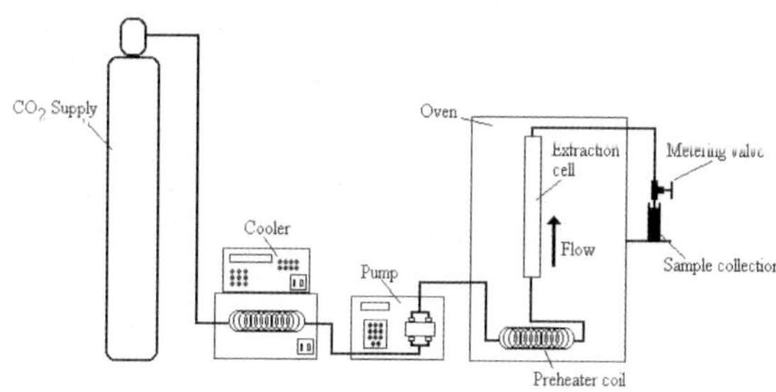

Figure: Schematic diagram of SFE apparatus

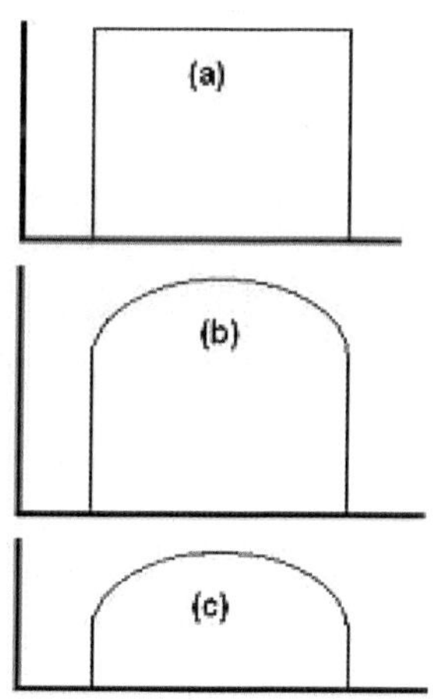

Figure: Concentration profiles during a typical SFE extraction

There are two essential steps to SFE, transport (by diffusion or otherwise) from with the solid particles to the surface, and dissolution in the supercritical fluid. Other factors, such as diffusion into the particle by the SF and reversible release such as desorption from an active site are sometimes significant, but not dealt in detail here. Figure shows the stages during extraction from a spherical particle where at the start of the extraction, the level of extractant is equal across the whole sphere figure a. As extraction commences, material is initially extracted from the edge of the sphere, and the concentration in the center is unchanged. As the extraction progresses, the concentration in the center of the sphere drops as the extractant diffuses towards the edge of the sphere figure c.

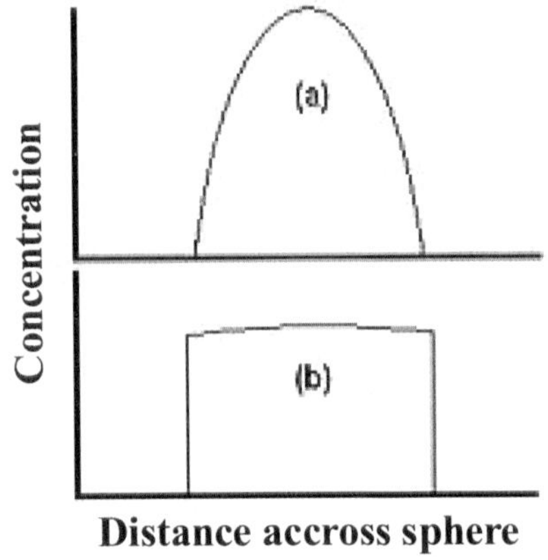

Figure: Concentration profiles for (a) diffusion limited and (b) solubility limited extraction

The relative rates of diffusion and dissolution are illustrated by two extreme cases shows a case where dissolution is fast relative to diffusion. The material is carried away from the edge faster than it can diffuse from the center, so the concentration at the edge drops to zero. The material is carried away as fast as it arrives at the surface, and the extraction is completely diffusion limited. Here, the rate of extraction can be increased by increasing diffusion rate, for example raising the temperature, but not by increasing the flow rate of the solvent. Figure shows a case where solubility is low relative to diffusion. The extractant is able to diffuse to the edge faster than the solvent can carry it away, and the concentration profile is flat. In this case, the extraction rate can be increased by increasing the rate of dissolution, for example by increasing flow rate of the solvent.

The extraction curve of % recovery against time can be used to elucidate the type of extraction occurring. Figure shows a typical diffusion controlled curve. The extraction is initially rapid, until the concentration at the surface drops to zero, and the rate then becomes much slower. The % extracted eventually approaches 100%. Fig. shows a curve for solubility limited extraction. The extraction rate is almost constant, and only flattens off towards the end of the extraction. Figure shows a curve where there are significant matrix effects, where there is some sort of reversible interaction with the matrix, such as desorption from an active site. The recovery flattens off, and if the 100% value is not known, then it is hard to tell that extraction is less than complete.

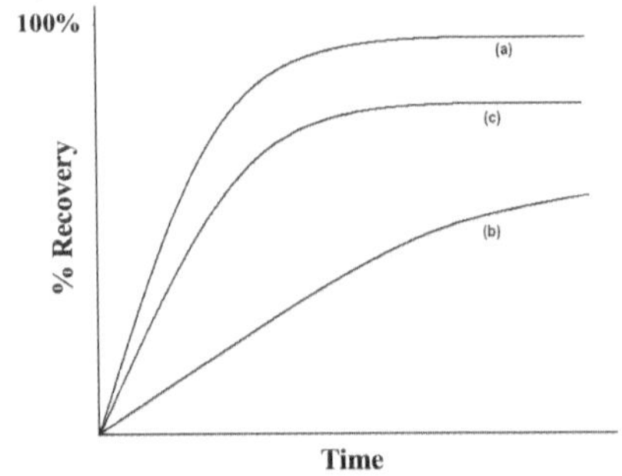

Figure: Extraction profile for different types of extraction

Optimization

The optimum will depend on the purpose of the extraction. For an analytical extraction is to determine, say, antioxidant content of a polymer, then the essential factors are complete extraction in the shortest time. However, for production of an essential oil extract from a plant, then quantity of CO_2 used will be a significant cost, and "complete" extraction is not required; a yield of 70 - 80% perhaps being sufficient to provide economic returns. In another case, the selectivity may be more important, and a reduced rate of extraction will be preferable if it provides greater discrimination. Therefore, few comments can be made, which are universally applicable.

Maximizing Diffusion

This can be achieved by increasing the temperature, swelling the matrix, or reducing the particle size. Matrix swelling can sometimes be increased by increasing the pressure of the solvent, and by adding modifiers to the solvent. Some polymers and elastomers in particular are swelled dramatically by CO_2, with diffusion being increased by several orders of magnitude in some cases.

Maximizing Solubility

Generally, higher pressure will increase solubility. The effect of temperature is less certain, as close to the critical point, increase in the temperature causes a decrease in density, and hence, dissolving power. At pressures well above the critical pressure, solubility is likely to increase with temperature. Addition of low levels of modifiers (sometimes called entrainers), such as methanol and ethanol, can also significantly increase solubility, particularly of more polar compounds.

Optimizing Flow Rate

The flow rate of CO_2 should be measured in terms of mass flow rather than by volume because the density of the CO_2 changes according to the temperature both; before entering the pump heads and during compression. Coriolis flow meters are best used to achieve such flow confirmation. To maximize the rate of extraction, the flow rate should be high enough for the extraction to be completely diffusion limited (but this will be very wasteful of solvent). However, to minimize the amount of solvent used, the extraction should be completely solubility limited (which will take a very long time). Flow rate must therefore be determined depending on the competing factors of time and solvent costs, and also capital costs of pumps, heaters and heat exchangers. The optimum flow rate will probably be somewhere in the region where both solubility and diffusion are significant factors.

Advantages

Environmental Improvement and Reduced Product Contamination

SFE is an alternative to liquid extraction using solvents such as hexane or dichloromethane. There will always be some residual solvent left in the extract and matrix, and there is always some level of environmental contamination from their use. In contrast, carbon dioxide is easy to remove simply by reducing the pressure, leaving almost no trace, and it is also environmentally benign. The use of SFE with CO_2 is approved by the Soil Association for organic products. The CO_2 used is largely a by-product of industrial processes or brewing, and its use in SFE does not cause any extra emissions.

Selectivity

The solvent strength of a supercritical fluid can be varied by change in the pressure and a lesser extent in the temperature.

For example, volatile oils can be extracted from a plant with low pressures (100 bar), whereas liquid extraction would also remove lipids. Lipids can be removed using pure CO_2 at higher pressures, and then phospholipids can be removed by adding ethanol to the solvent.

Speed

It is a fast process and completed in 10 to 60 minutes. It is a diffusion-based process, with the solvent required to diffuse into the matrix, and the extracted material to diffuse out of the matrix into the solvent. Diffusivities are much faster in supercritical fluids than in liquids, and therefore extraction can occur faster. Also, there is no surface tension and viscosities are much lower than in liquids, so the solvent can penetrate into small pores within the matrix inaccessible to liquids.

Purity

A supercritical fluid can be separated from analyte by simply releasing pressure. Supercritical fluids are cheap, inert and nontoxic. Thus, they are readily disposed off after an extraction is completed by allowing them to evaporate into the atmosphere.

Recovery

Recovery of analytes becomes simple.

Raffinate

A raffinate describes any material remaining from a chemical or mechanical process after a desired chemical or material is removed. It may describe a liquid or solid, and may be further processed in additional steps or treated as a waste material. Even if a raffinate is not used to recover additional products, it may require further treatment to avoid creating any environmental issues when sent to a landfill or disposal site.

When crude oil from the ground is processed in a refinery, a mixture of carbon-based petroleum products is produced. Similar products can be created from reactions of natural gas, or methane, to form complex organic, or carbon-containing, molecules. Different processes can be used to extract desired compounds, including distillation by boiling the liquids, or solvent or water extraction from liquids or solids. After any of these processes are complete, a raffinate remains that is often further processed to remove additional materials.

A common raffinate stream in petroleum processing is known as the C4 stream. The term C4 refers to the four carbons in the various molecules being processed. A C4 molecule is described as being in the butane family, and can include butene, isobutane and isobutylene. All have four carbons in their molecules, and are valuable as raw materials for plastics and fuels.

When a C4 stream is processed, each step produces one desired chemical, and the remaining material must be processed further or discarded. Each raffinate becomes the feedstock, or raw material, for the next extraction or distillation step. The remaining or residual product may be tars or heavy petroleum products not suitable for further processing or commercial sale.

Mining uses a combination of different steps to remove the desired metals from the unprocessed ore, or metal-containing earth or rock. Grinding and separating the ore by mechanical shakers or separators can be used to separate some metals from the rock. The rock, now considered a raffinate, is often treated chemically to extract or remove additional metals. In some cases, liquids from the chemical reactions are also raffinates, because they are processed in multiple steps to remove as much valuable metal as possible.

Chemical processing of many materials can create residues or waste products requiring special handling to prevent environmental contamination. Gold and other metal ores contain small amounts of radioactive compounds that can be concentrated in the raffinate stream. These materials may need further processing to minimize the risk to both people and the environment. Petroleum processing creates heavy tar-like compounds that may cause cancer in humans, and may need to be incinerated or processed further before disposal. Some of the chemicals used to process raffinates are hazardous, and need to be removed before disposal.

Solvent Impregnated Resin

In recent years increasing interest in environment protection, economy of energy, as well as process optimization and the continuous progress in fundamental chemistry have produced an important development of new chemical separation techniques. The need of more specific systems for dilute metal recovery from both ecological and economic aspects has led to the development of the synthesis of new extractants, exchangers and adsorbents. These products have improved significantly the selectivity and efficiency of a large number of separation process techniques such as extraction with solvent supported liquid membranes, precipitation, etc. Among these new products, Solvent Impregnated Resins (SIR) has been postulated as a new technological alternative for problems associated with metal separation and recovery.

The solvent impregnated resins have been shown to be effective sorbent for the selective recovery of metal ions from aqueous solutions. They combine not only the advantages of resin ion exchange for processing dilute liquors with specific properties of the extractants, but also a high distribution ratio and selectivity characteristic of the extractants dissolved in a liquid organic phase with the simplicity of equipment and operation characteristic of solid ion-exchange technology. It was well recognized that the impregnated extractants can exhibit strong affinity for the polymeric matrix but still behave as in the liquid state. SIR was introduced in hydro- metallurgical applications by Warshawsky.

Development of chelating materials for solid-phase extraction has gained special attention due to the advantages in the use of these substances in metal ion enrichment. These advantages include high degree of selectivity by controlling pH, versatility, durability, good metal loading capacity and enhanced hydrophilicity. Chelating ligands have been functionalized in several support materials, including commercially available XAD resin series. Amberlite XAD is resin widely used to develop several chelating materials for pre-concentration procedures due to its good physical and chemical properties such as porosity, high surface area, durability, and purity. Amberlite XAD-4 has been often used as a solid sorbent to prepare ligand-loated resin. Recently, the XAD-4 resin, impregnated with many compound as complexing agents has been used for pre-concentration of heavy metals.

Some authors have studied the recovery of different metals using Amberlite XAD-2 resins impregnated with different extractants.

Conventional separation techniques, such as fractional crystallization and fractional precipitation, are inefficient. In addition, solvent extraction techniques have been replaced by solid phase extraction procedures, such as chromatography by extraction resin. Solid phase extraction procedures combine the advantages of high selectivity of solvent extraction and high efficiencies of chromatographic separation, and they separates elements with high purity. Jia, Wang et al. have used an extractant resin, with styrene-divinyl benzene copolymer as support and (HEH [EHP]) as the extractant, and its performance in separating rare earth, such as gadolinium and terbium.

Because of these advantages, extraction resin and their loaded chromatographic separation of metal ions have been under development since 1970. Many papers on the adsorption of metal ions have focused on the use of resins containing organo-phosphorus acids as extractants.

Amberlite XAD-4 has been often used as a solid sorbent to prepare a ligand-loaded resin. Recently, the XAD-4 resin impregnated with many compounds as complexing agents has been used for pre-concentration of heavy metals. The use of biological materials for effective removal and pre-concentration of heavy metals from contaminated waters has emerged as a potential alternative method to conventional treatment techniques of all the pre-concentration methods, bio-sorption by microorganisms immobilized on solid support seems to be the most effective pre-concentration methods, due to their higher recoveries, economical advantages, simplicity and environmental safety. M. Dogru et al. proposed the use of bacillussubtilis immobilized on Amberlite XAD-4 as new bio-sorbent in trace metal determination. This procedure was applied to the determination of Cu^{2+} and Cd^{2+} in aqueous solutions.

T. Saitoh et al. has used the styrene-divinyl benzene XAD-4 for the collection of precious metals from water by impregnated them with trioctylamine (TOA), it was found that this resin was useful for concentrating precious metals such as: platinium, gold and palladium.

Impregnation Procedure

SIR can be modelled as "a liquid complexing agent dispersed homogenously in a solid polymeric medium". The impregnated agent should behave as in the liquid state but exhibit strong affinity to the matrix. The Amberlite XAD resins are widely-used. As stated above, the attractiveness of SIR lies in the possibility of selecting an existing extractant and adapting it to solve a pressing problem. Naturally, most of the SIR studies so far have been concerned with immediate problems, and the extractants chosen have been mostly commercial. They include acidic organo-phosphorous extractant:

a) di(2-ethylhexyl)phosphoric acid (D2EHPA).

b) 2-ethylhexylphosphonic acid mono-2-ethylhexyl ester (IONQUEST 801).

c) di(2,4,4-trimethylpentyl)phosphonic acid (CYANEX 272).

The SIR were prepared as dry method most widely used, the extractant, diluted by a solvent, is placed in contact with the polymer, and the solvent is removed by slow evaporation under vacuum. This method is most successful in the impregnation of hydrophilic extractant, such as amines, ethers, esters, etc. An aliquot of extractant (D2EHPA, TOPO and Diphenylthiourea) was diluted in a pre-calculated amount of solvent (n-hexane for D2EHPA and chloroform for the others). The resulting solution was contacted with fresh resin until all organic solution was absorbed by the resins. Finally, these resins were evaporated to remove the solvent in a vacuum.

Donnan Dialysis Cell and Procedure

Dialysis was carried out in a stirred cell represented in figure. The cell consisted of two compartments made of Teflon. The membrane (or membrane and resin) was sandwiched between the two half cells. The effective volume of each cell was 140 ml. The feed compartment contains the metal solution at a concentration of 10^{-3} M of metal salt; the other compartment noted strip contains variables concentrations of nitric acid. Each compartment was provided with a vertical mechanical stirrer at stirring speed 600 rpm which was previously determined as high enough to minimise the thickness of the boundary layer. The experiments began when starting the stirring motors in the two compartments of the cell. The exposed membrane area was 7 cm^2. All the experiments were performed in a thermostat at 25° C. The experiment duration has been fixed to six hours and 1 ml of solution has been taken up in the regular time (1 h) from the feed and strip compartments in the order of the determination of metal ions flux through the membrane.

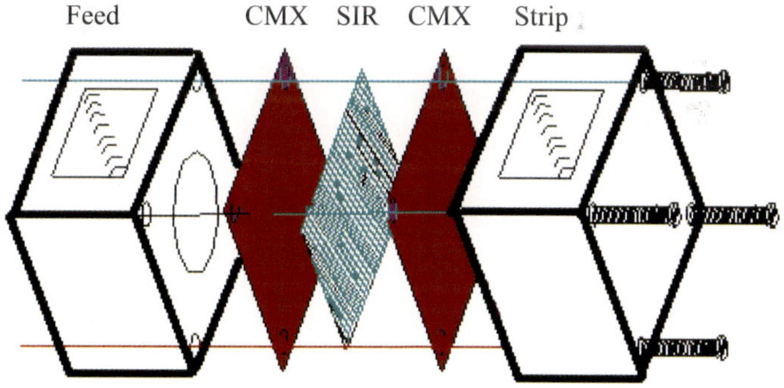

Figure: Donnan dialysis cell (hybrid process coupling the SIR to ion-exchange membranes).

Magnetic Separation

Magnetic separation is a process used to separate materials from those that are less or nonmagnetic. All materials have a response when placed in a magnetic field lthough with most, the effect is too slight to be detected. The few materials that are strongly affected (magnetised) by magnetic fids are known as "Ferromagnetics", those lesser (though noticeably) affected are known as "Paramagnetics".

Ferromagnetics require relatively weak magnetic elds to be attracted and devices to separate these materials usually have magnets that are permanently magnetised ("Permanent" magnets do not require electricity to maintain their magnetic elds). Paramagnetics require stronger magnetic elds and these can only be achieved and maintained by "electro" magnets (large wire coils around an iron frame – current is continuously passed through the coils creating the magnetic eld within the iron. The eld is concentrated across an air gap in the circuit).

Both ferromagnetic (low intensity) and paramagnetic (high intensity) separation devices may be operated with dry solids or with solids in pulp form.

Magnetic separation has two major applications in mineral processing plants:

- The removal of tramp iron (which would deleteriously affect subsequent processes) from an ore stream. This is usually achieved by a low intensity magnet suspended above, or at the head of, a conveyor.
- Separation of magnetic minerals from less magnetic minerals.

This table shows the more common magnetic minerals and the range of eld intensities that may be required to separate them.

	Mineral	**Formula**	**Field Strength***
Ferromagnetic	Magnetic	Fe_3O_4	→ 1
	Pyrrhotite	Fe_7S_8	0.5 - 4
Paramagnetic	Ilmenite	$FeTiO_3$	8 – 16
	Siderite	$FeCO_3$	9 – 18
	Chromite	$FeCr_2O_4$	10 – 16
	Hematite	Fe_2O_3	12 – 18
	Wolframite	$(Fe,Mn)WO_4$	12 – 18
	Tourmaline		16 – 20

*The units given are kilogauss (kG). These are the units most commonly used. The equivalent S.I. unit is the Tesla (T) * 1 Tesla = 10 kilogauss). The extremes of eld strength used are based on experience from a magnetic separation testing laboratory over many years.

Laboratory Wet Drum Magnetic WLIMS

Industrial Mag Separator

References

- Design-principles-liquid-liquid-extraction: aiche.org, Retrieved 15 May 2018
- Solid-phase-extraction, technology: affinisep.com: Retrieved 26 March 2018
- Leaching-definition-process: study.com, Retrieved 25 April 2018
- Leaching-process: vanderbilt.edu, Retrieved 30 June 2018
- What-is-a-raffinate: wisegeek.com, Retrieved 21 June 2018

5

Fractionation

Fractionation is a separation process concerning the segregation of a mixture of solid, liquid, gas or suspension, into a number of smaller factions during a phase transition, in such a way that the composition varies relative to a gradient. This chapter closely examines the key concepts of fractionation and discusses topics like field flow fractionation, equilibrium isotope fractionation and polymer fractionation for an extensive understanding of the subject.

Physical fractionation separates subsets of soil according to physical criteria, such as size and/or density or magnetic susceptibility. This is done after various degrees of dispersion are applied to soil to break bonds between elements of the soil structure. Physical fractionation procedures seek to minimize chemical alterations and aim to separate primary soil particles, primary organomineral or secondary organomineral associations.

Primary organomineral associations relate to the “primary structure of soils as defined by soil texture” (Christensen, 1992). They result from the association of SOM with primary mineral particles and are isolated after complete dispersion of soils. Secondary organomineral associations are aggregates, separated after limited dispersion of soil. In many soils the complete dispersion of particles, especially clay-sized particles, is not possible if organic matter is not oxidized; hence, true primary organomineral complexes cannot be isolated.

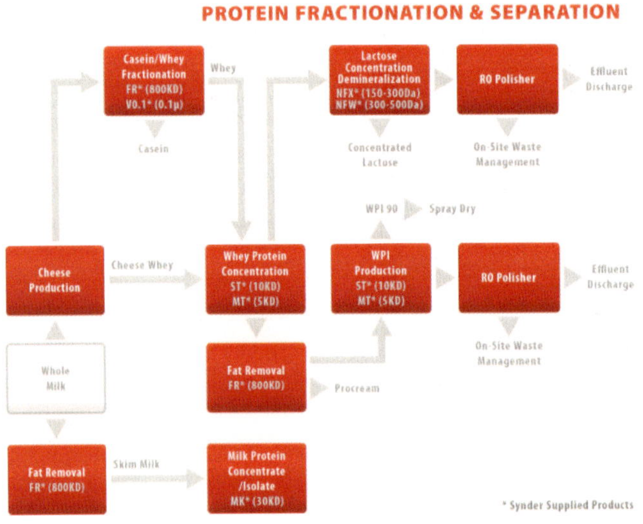

Physical fractionation can pursue different objectives. One is to separate soil organic constituents in their original state (i.e., in the state they have in soils) without any chemical alteration. This is one of the reasons behind the separation of particulate organic matter (POM), which is essentially composed of weakly decomposed plant debris, root fragments, fungal hyphae, spores, seeds, fecal pellets, and faunal bodies. Another objective is to separate SOM in its physical surroundings with the assumption that the association of organic matter with minerals and their location within the

3-D soil architecture controls their dynamics (physicochemical and physical protection). Physical fractionation methods have brought a considerable amount of information on SOM and have led to major advances in the understanding of soil C dynamics and stabilization processes.

Fractionation of Polymers

Methods suitable for segmentation of macromolecules according to their size are based on different solubility of polymers. Experimental processes of fractionation can be divided approximately into two groups:

1. Preparative methods that allow segmentation of polymer into fractions with different molar mass and consequential separation of particular fractions. Curve corresponding to distribution of molar masses in a given polymer is obtained by processing of results. Here belong the techniques of gradual fraction precipitation and fraction dissolution that use to be variously modified.

2. Analytic methods are those, at which polymer is divided into fractions according to molar mass, but fractions are not isolated. Distribution curve is determined on the basis of change observations of a physical property of polymer that is dependant on molar mass of fractions and thus accompanies the segmentation. For instance, gel permeation chromatography (GPC fractionation).

Gradual fraction precipitation is frequently used segmentation method. The principle of this method lies in gradual addition of precipitant into solution of polymer. First separated fraction is the one containing polymer with the biggest molar mass. Polymers with lower and lower molar mass gradually drop out from the solution after addition of more precipitant.

Gradual fraction dissolving (elution fractionation) is method based on gradual polymer extraction by solvent with growing solvent power. For acceleration of extraction, we have to put a layer of polymer in the form of film onto inert support with large area. Polymer deposited on the support is placed into column that could be tempered for dissolving acceleration. Polymer is gradually washed out by eluent (mixture of solvent and precipitant) that is being enriched by solvent. First isolated fraction is the one with the lowest molar mass. Fractions with bigger and bigger molar mass are washed out along with growing solvent power.

Gel permeation chromatography (size exclusion chromatography) is column segmentation technique. Fraction column is filled with granules of gel with porous structure and correctly defined pore dimensions. Solution of polymer is passing through column, where small macromolecules are trapped in gel pores and bigger ones are situated in the space between granules. Big molecules are washed up first during column washing with solvent. Analyzer, that analyzes fractions e.g. by refractive index, is located in the lower part of column. Experimental result of fractionation is then dependence of refractive index on retention volume. For plotting of typical distribution curve, we have to determine by calibration what molar masses correspond to particular retention volumes.

Results of fractionation are most often processed in distribution diagrams. Distribution curve of molar masses of polymer illustrates the share of fractions of macromolecules with certain size (molar mass) in the sample. According to mathematic process of fractionation results processing, we distinguish integral (Fig., line 1) and differential (Fig., line 2) distribution curve for description of polymer's molar mass distribution.

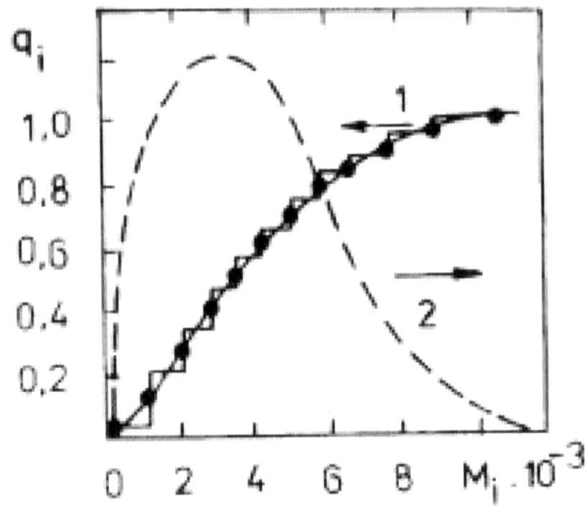

Distribution curves of molar masses: 1: integral; 2: differential

Fractional Distillation

Fractional distillation is a technique for the separation of miscible liquids. It is a distillation technique used for the separation of liquids which dissolve in each other. Fractional distillation is a process of separating the components of a mixture by repeated distillations and condensations.

Fractional Distillation Column

Fractional distillation column is a fractionating column used for the separation of miscible liquids. It is a long glass tube with wide bore. It either filled with glass beads, small stones, porcelain rings or coke. It can also be blown in to a shape with several spheres or pear bulbs. The purpose of the glass beads is that they improve the distillation process.

There can be several trays in the fractionating column in place of glass beads and bulbs. These trays are called theoretical plates. Theses trays or theoretical plates serve the purpose of the glass beads.

For commercial purpose fractionating column with hundredth of theoretical plates are used. The main aim of using the fractionating column is to increase the surface area for cooling of liquid and to provide hurdles to the rising vapours and descending liquid.

Principle of Fractional Distillation

The principle of fractional distillation is that different liquids boil at different temperature. The miscible liquids boil at different temperature and evaporate at different temperature.

When the mixture is heated, the liquid with lower boiling point boils and turns into vapours. So the mixture is heated to a temperature at which one or two components of the mixture will vaporise. Fractional distillation involves repeated distillations and condensations.

Working of Fractional Distillation Process

The fractional distillation apparatus involves *fractionating column, condenser, distilling flask, receiver, and heat source.*

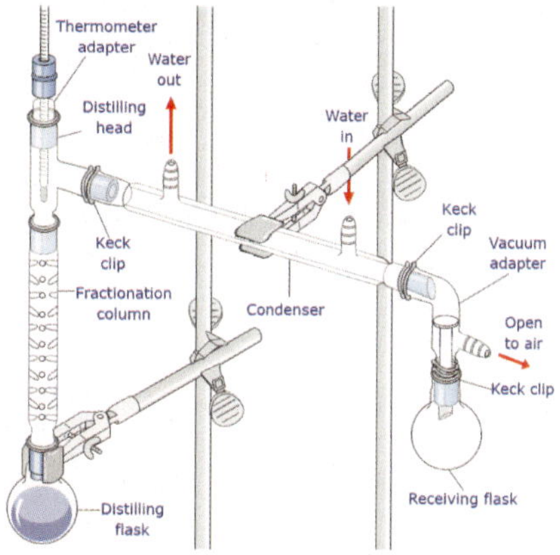

The process starts with taking a mixture of two liquids **X and Y** of which X is more volatile than B. The mixture is taken in the distilling flask and is heated. The temperature rises slowly and the mixture begins boiling. Vapours will start forming. The vapours formed consists of more volatile component X with little amount of less volatile component Y.

Vapours start travelling in the fractionating column. Condensation of vapours of the less volatile liquid Y will start before than those of more volatile liquid X. So by time, the vapours rising in the fractionating column will be more of X and the liquid flowing down will consists of liquid Y.

The process of distillations and condensation is repeated. As a result of successive distillations, the vapours of more volatile component X will reach at the top, from there they are taken into the condenser and finally the substance is transferred to container.

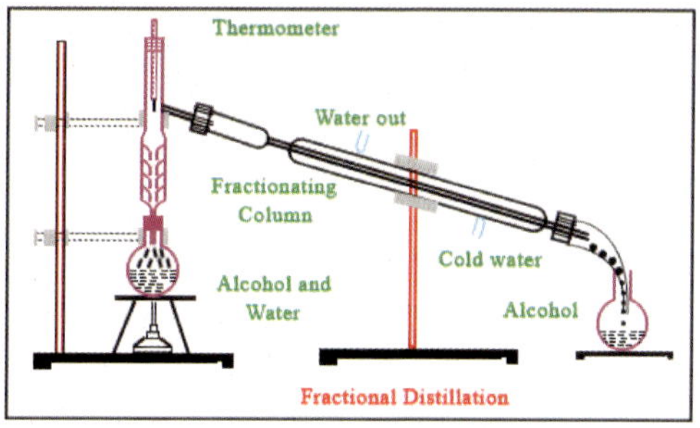

Similarly after a series of distillations, the remaining liquid in the distillation flask gets enriched in components with higher boiling points.

Fractional Distillation Uses

There are number of uses of fractional distillation.

- One of the important use of using fractional distillation is to separate the crude oil into its various components such as gasoline, kerosene oil, diesel oil, paraffin wax, liberating oil.

- Fractional distillation is also used for the purification of water. Water contains many dissolved impurities; these can be removed by this process.
- It is also used for separating acetone and water.
- Industrial use of fractional distillation is in petroleum refineries, chemical plants, natural gas processing and separation of pure gases from mixture of gases.
- It has other industrial uses as it is used for purification and separation of many organic compounds.

Fractional Distillation of Crude Oil

Fractional distillation is used to separate the various components of crude oil. Various components of crude oil such as gasoline, paraffin wax, lubricating oil, diesel oil, fuel oil, naphtha, kerosene are separated. Crude oil is heated and the mixture stars boiling. Hydrocarbon gases are introduced into the column.

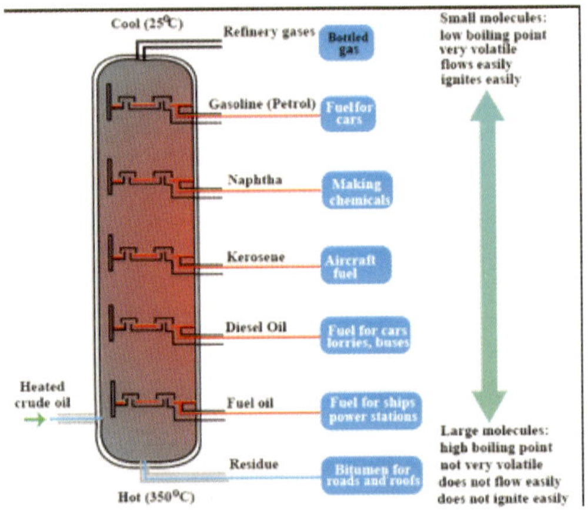

The temperature decreases in going up the fractionating column. The hot vapours rise up in the fractionating column. Therefore vapours with higher boiling point condense in the lower part and the vapours of components with low boiling point condense at the top. The condensed vapours are removed from the sides of the column.

Fractional Crystallization (Chemistry)

Sometimes the solubility of an organic compound is quite similar to that of its impurities. In such a case it is not possible to purify an organic compound by simple crystallization. The crystals thus obtained are not pure. These contain impurities as well. However, these can be purified by fractional crystallization.

Fractional crystallization is done by first preparing a hot, concentrated solution.

When the solution is cooled, the first drop of crystals is obtained. The crystals thus obtained contain more of one of the components which has a lower solubility and less of the component with a higher solubility. On the other hand, the filtrate contains less of components with a lesser solubility and more of components with a higher solubility.

The crystals thus obtained are crystallized again. The crystals now obtained will be less impure as more impurities will be passed on to the mother liquor. The crystallization is repeated a number of times till the crystals are free from impurities.

Similarly the filtrate, i.e. the mother liquor, is crystallized again and again. Each crystallization helps remove some more of the compound which is less soluble. Finally the mother liquor contains a pure compound, which has a slightly higher solubility.

Field Flow Fractionation

Field-flow fractionation (FFF) represents a whole family of methods, separating particles due to differences in size (or the diffusion constant) and *no* other property. In contrast to Analytical Ultracentrifugation, also a method fractionating by particle size, particle density plays no part in FFF: Diffusion constants remain the only target property in FFF.

Field-flow fractionation is a chromatographic method. Separation is achieved by the influence of diffusion on elution time. Elution occurs by means of a laminar flow through the channel, exhibiting a parabolic flow profile. As illustrated in figure, particles elute at a rate dependent on their proximity to the wall. A transverse separation field attempts to displace particles towards an accumulation wall, opposed by the particles back diffusion. In this way, separation occurs dependent on the particles' diffusion properties.

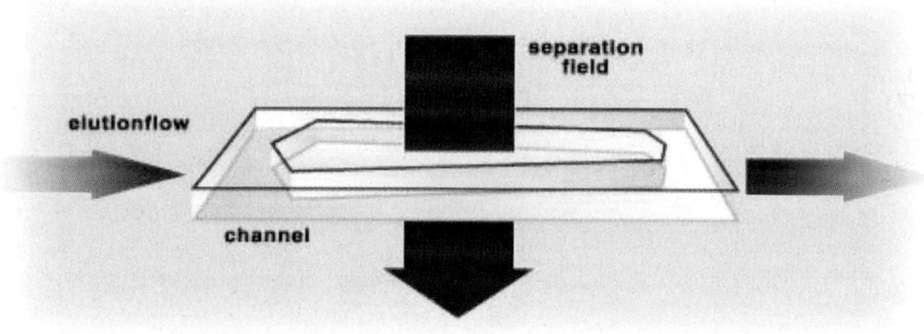

Figure: Principle of Field-flow fractionation

For particle dimensions greater than a nominal 0.5 μm threshold, the centre of mass of a particle will extend farther into the parabolic profile than the diffusion of smaller particles. This is called inversion to the sterical mode of field-flow fractionation. This effect must be taken into account when choosing channel dimensions for the system in question.

Other problems arise when the particles are not spherical. Rod-shaped particles experience some restrictions in their degrees of freedom in the vicinity of the accumulation wall. This effect leads to considerable errors (orders of magnitude) in calculated results.

As the FFF theory is still subject to controverse scientific discussions, a characterisation by means of detection is often preferred to a characterisation by elution time. Consequently, static light scattering detectors are readily applied and also dynamic light scattering is currently introduced into the market as a means of characterisation.

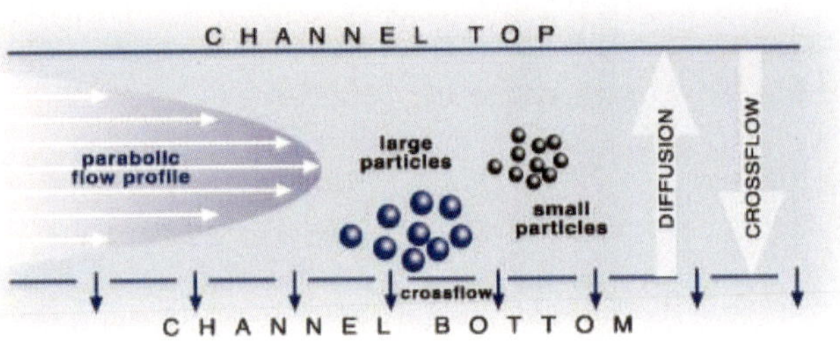

Figure: Influence of the separation field in F-FFF

The separation field is arbitrary, making FFF a large family of methods. Most common are thermal, sedimentational and magnetic fields, or a hydrodynamic field (transverse flow).

Asymmetric Flow Field Flow Fractionation

Asymmetric Flow Field-Flow Fractionation is a one-phase chromatography technique. High-resolution separation is achieved within a very thin flow against which a perpendicular force field is applied. The flow and sample are confined within a channel consisting of two plates that are separated by a spacer foil; the plates are then bolted together. The spacer foil has a typical thickness of 100 – 500 µm.

The upper channel plate is impermeable. The bottom channel plate, on the other hand, is permeable, and made of a porous frit material. An ultra filtration membrane with a typical size barrier of 10 kDa covers the bottom plate to prevent the sample from penetrating the channel.

Within the flow channel a parabolic flow profile is created because of the laminar flow of the liquid: the stream moves slower closer to the boundary edges than it does at the center of the channel flow. When the perpendicular force field is applied to the flowing, laminar stream, the analytes are driven towards the boundary layer the so-called "accumulation wall" of the channel.

Diffusion associated with Brownian motion, in turn, creates a counteracting motion. Smaller particles, which have higher diffusion rates, tend to reach an equilibrium position higher up in the channel, where the longitudinal flow is faster. Thus, the velocity gradient flowing inside the channel separates different sizes of particles.

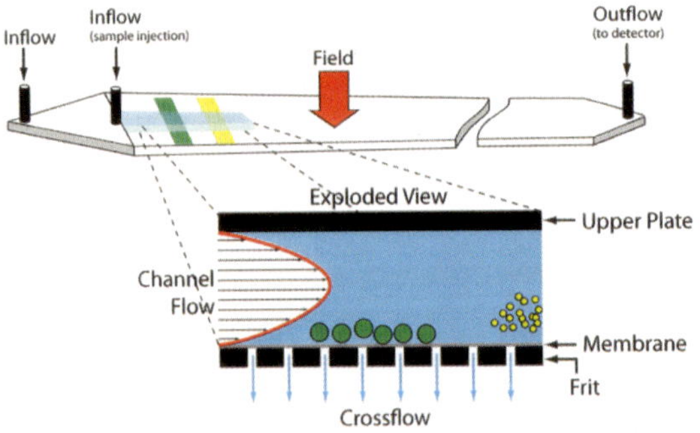

The smaller particles are transported more rapidly along the channel than the larger particles. This results in the smaller particles eluting before the larger ones – the opposite of a Size Exclusion/Gel Permeation Chromatography (SEC/GPC) separation in which the large molecules elute first.

With AF4 separation there is no column media to interact with the samples, so for very high molar mass polymers, there is no need to worry about shearing forces being applied. The entire separation is gentle, rapid, and non-destructive without a stationary phase that may interact, degrade, or alter the sample.

The separation process requires three steps: During the first two steps, injection and focusing, the main flow is split, enters the channel from both ends and is balanced to meet under the injection port. At this point the flow will move only down and permeate through the membrane. When the sample is injected it is focused in a thin band and concentrated towards the membrane. After complete transfer of the sample volume the injection flow is stopped and one typically allows for another minute of focusing before the flow pattern is switched to the elution mode. Now the flow enters only from the inlet port and exits at the outlet which is connected to the detectors. Sample constituents elute separated according to size and are monitored by the array of detectors.

Equilibrium Isotope Fractionation

Properties of Isotopic Molecules

The various isotopes of an element have slightly different chemical and physical properties because of their mass differences. Under the proper circumstances, such differences can manifest themselves as a *mass-dependent* isotope fractionation effect. Nuclear interactions, on the other hand, lead to a *non-mass-dependent* effect in the sense that they depend on the nuclear structure, rather than on the weight difference per se. In the first case, for example, the properties of molecules with ^{17}O will be intermediate between those of molecules with ^{16}O and ^{18}O; this is not necessarily the case for the non-mass-dependent effects. For elements of low atomic numbers, these mass differences are large enough for many physical, chemical, and biological processes or reactions to *fractionate* or change the relative proportions of different isotopes of the same element in various compounds. As a result of fractionation processes, waters and solutes often develop unique isotopic compositions (ratios of heavy to light isotopes) that may be indicative of their source or the processes that formed them.

Two main types of phenomena produce isotopic fractionations: *isotope exchange reactions* and *kinetic processes*. Isotope exchange reactions can be viewed as a subset of kinetic isotope reactions where the reactants and products remain in contact in a closed, well-mixed system such that back reactions can occur and chemical equilibrium can be established. Under such circumstances, isotopic equilibrium can be also established.

Fractionation Accompanying Chemical Reactions and Phase Changes

The strength of chemical bonds involving different isotopic species will usually be different. Molecules containing heavy isotopes are more stable (i.e., have a higher dissociation energy) than molecules with lighter isotopes. Hence, isotopic fractionations between molecules can be explained by differences in their *zero point energies* (ZPE). For example, there is about a 2 kcal/mole difference in ZPE associated with the breaking of the H-H bond compared to the D-D bond. Hence, H-H bonds are broken more easily and D-D bonds are more stable. Chemical reaction rates where such a bond is broken will also show an isotope effect. These are quantum effects that become appreciable at low temperatures and disappear at higher temperatures. The energy differences associated with isotope effects are about 1000 times smaller than the ΔG for chemical reactions, and hence cannot be the driving force for chemical equilibrium.

- Equilibrium fractionations

Equilibrium isotope-exchange reactions involve the redistribution of isotopes of an element among various species or compounds (in a strict sense, this only occurs in a closed, well-mixed system at chemical equilibrium). At isotopic equilibrium, the forward and backward reaction rates of any particular isotope are identical. This does not mean that the isotopic compositions of two compounds at equilibrium are identical, but only that the ratios of the different isotopes in each compound are constant for a particular temperature.

During equilibrium reactions, the heavier isotope generally preferentially accumulates in the species or compound with the higher oxidation state. For example, sulfate becomes enriched in ^{34}S relative to sulfide (i.e., has a more positive various materials ä ^{34}S value); consequently, the residual sulfide becomes depleted in various materials d34S. As a "rule of thumb," among different phases of the same compound or different species of the same element, the more dense the material, the more it tends to be enriched in the heavier isotope. For example, for the various phases of water, at equilibrium, ä $^{18}O_S$ > ä $^{18}O_L$ > ä $^{18}O_V$. Also, the ä ^{13}C and ä ^{18}O values of CO_2 < HCO_3^- < $CaCO_3$.

During phase changes, the ratio of heavy to light isotopes in the molecules in the two phases changes. For example, as water vapor condenses in rain clouds (a process typically viewed as an equilibrium process), the heavier water isotopes (^{18}O and ^{2}H) become enriched in the liquid phase while the lighter isotopes (^{16}O and ^{1}H) remain in the vapor phase. In general, the higher the temperature, the less the difference between the equilibrium isotopic compositions of any two species (because the differences in ZPE between the species become smaller). The fractionation associated with the equilibrium exchange reaction between two substances A and B (i.e., the fractionation of A relative to B) can be expressed by use of the isotope fractionation factor a (alpha):

$$\alpha_{A-B} = R_A / R_B$$

where R = the ratio of the heavy isotope to the lighter isotope (i.e., D/H, $^{18}O/^{16}O$, $^{34}S/^{32}S$, etc.) in compounds A and B.

The value of such an *equilibrium* fractionation factor can be calculated on the basis of spectral data of the isotopic molecular species, at least for simple molecules. The α values generally differ by just a few percent from the equal-energy value of 1.00, except for exchange reactions involving hydrogen isotopes where α values may be as large as 4 at room temperature. The sign and magnitude of a are dependent on many factors, of which temperature is generally the most important. Other factors include chemical composition, crystal structure, and pressure.

The equilibrium fractionation factors (a_{l-v}) for the water liquid-vapor phase transition are 1.0098 and 1.084 at 20° C and 1.0117 and 1.111 at 0° C for ^{18}O and ^{2}H, respectively (Majoube, 1971). In both cases, a_{l-v} > 1, which means that the first phase (the liquid water) is "heavier" than the second phase (e.g., for a_{l-v} = 1.0098, the d^{18}O of water is +9.8‰ higher than the d^{18}O value of vapor at equilibrium). For the ice-water transition (0° C), the values are 1.0035 and 1.0208, respectively (Arnason, 1969).

A useful equation that relates dvalues and fractionation factors is:

$$\alpha_{A\text{-}B} = (1000 + \delta_A) / (1000 + \delta_B) .$$

Other common formulations for fractionation factors include:

$$\alpha^* = 1/\alpha = \alpha_{B\text{-}A} = R_B / R_A$$

and

$$\varepsilon_{A\text{-}B} = (\alpha_{A\text{-}B} - 1) \cdot 1000 .$$

For small values of ε(epsilon), $\varepsilon_{A\text{-}B} \sim \delta_A - \delta_B$. For example, if δ_B = +10‰ and if $\alpha_{A\text{-}B}$ = 1.020, then = 20‰ and $\delta_A \sim$ +30‰. The difference in isotopic composition between two species *A* and *B* is defined as:

$$\varepsilon_{A\text{-}B} \sim \delta_A - \delta_B \sim 1000 \ln \alpha_{A\text{-}B} .$$

Fractionation factors are commonly expressed as "$10^3 \ln \alpha$" because this expression is a very close approximation to the permil fractionation between the materials (ε), especially for the values α of near to unity typical of most elements of interest (O'Neil, 1986), and because the value "$10^3 \ln \alpha$" is nearly proportional to the inverse of temperature (1/T) at low temperatures (°K). Graphical plots of the temperature dependency of α are typically given as $10^3 \ln \alpha$ versus 1/T (Friedman and O'Neil, 1977).

- Kinetic fractionations

Chemical, physical, and biological processes can be viewed as either reversible equilibrium reactions or irreversible unidirectional kinetic reactions. In systems out of chemical and isotopic equilibrium, forward and backward reaction rates are not identical, and isotope reactions may, in fact, be unidirectional if reaction products become physically isolated from the reactants. Such reaction rates are dependent on the ratios of the masses of the isotopes and their vibrational energies, and hence are called kinetic isotope fractionations.

The magnitude of a kinetic isotope fractionation depends on the reaction pathway, the reaction rate, and the relative bond energies of the bonds being severed or formed by the reaction. Kinetic fractionations, especially unidirectional ones, are usually larger than the equilibrium fractionation factor for the same reaction in most low-temperature environments. As a rule, bonds between the lighter isotopes are broken more easily than equivalent bonds of heavier isotopes. Hence, the light isotopes react faster and become concentrated in the products, causing the residual reactants to become enriched in the heavy isotopes. In contrast, reversible equilibrium reactions can produce products heavier or lighter than the original reactants.

Many reactions can take place either under purely equilibrium conditions or be affected by an additional kinetic isotope fractionation. For example, although isotopic exchange between water and vapor can take place under more-or-less equilibrium conditions (i.e., at 100% humidity when the air is still and the system is almost chemically closed), more typically the system is out of chemical equilibrium (i.e., < 100% humidity) or the products become partially isolated from the reactants (e.g., the resultant vapor is blown downwind). Under these conditions, the isotopic compositions of the water and vapor are affected by an additional kinetic isotope fractionation of variable magnitude.

Isotope fractionation factors can be defined as:

$$\alpha = R_p / R_s$$

where R_p and R_s are the ratios of the heavy to light isotope in the *product* and *substrate* (reactant), respectively. An isotope enrichment factor, e, can be defined as:

$$e_{p\text{-}s} = (\alpha - 1) \cdot 1000$$

If the reactant concentration is large and fractionations are small,

$$e_{p\text{-}s} \sim \Delta = d_p - d_s$$

where Δ(del) is another term for the enrichment factor. Note that Equations for kinetic fractionations are the same as equations for equilibrium fractionations, except for the differences in subscripts. One should be especially careful with the superscripts, subscripts, and units of all fractionation factors; different authors may define them differently. The use of $_p$ and $_s$ (or $_r$) for kinetic fractionations like Equation 2.10 reflects the unidirectional nature of these reactions.

The same formulations apply not only when part of the system is removed by a chemical or biological reaction, but also when material escapes by diffusion or outflow (e.g., by effusion through an aperture). In the latter cases the term *transport* fractionation factor may be preferred. The transport fractionation, like the equilibrium factors, is temperature dependent. However, unlike true kinetic fractionation factors, which can be quite appreciable, transport fractionations have only slight (positive) temperature coefficients.

Mass-independent Fractionation

The oxygen and sulfur isotope systems have the most intensively studied and dramatic massindependent chemical isotope fractionations of any elements. Prominent mass-independent fractionation phenomena in these systems are found in primitive meteoritic materials in ozone in the terrestrial stratosphere, and in sulfur extracted from ancient (2.4 Ga or older) terrestrial sediments. Although the mechanisms behind each of these prominent mass-independent fractionation signatures are not yet fully understood, the most widely favored hypotheses are that both the meteoritic oxygen isotope anomaly and the Archean–earliest Proterozoic sulfur isotope anomaly originate from abundance-dependent photochemical fractionation effects, in particular the phenomenon of selfshielding, whereas the oxygen isotope signature of stratospheric ozone is controlled by an isotope effect on the lifetimes of newly formed, vibrationally and rotationally excited ozone molecules.

Mass-Independent Fractionation of Sulfur Isotopes

Although Hulston & Thode (1965) found that $^{33}S/^{32}S$ and $^{34}S/^{32}S$ ratios in most terrestrial and meteoritic samples closely follow a single mass fractionation line, Farquhar et al. (2000) discovered that terrestrial samples of sulfide and sulfate minerals formed near Earth's surface in the Archean and earliest Proterozoic eons before 2.4 Ga exhibit pronounced mass-independent fractionation signatures. They were able to reproduce similar fractionations by exposing sulfur-bearing molecules in the vapor phase to UV photons energetic enough to drive photochemical breakdown. The most likely candidate reactant for the Archean fractionation reaction is sulfur dioxide (SO_2), and the fractionating mechanism is thought to be an atmospheric self-shielding process similar to the nebular process proposed for CO. Sulfur gases, including SO_2, are very rare and generally short lived in the modern oxidizing atmosphere of Earth, and the surficial sulfur reservoir is dominated by sulfate and sulfide in solution and chemical sediments. The modern stratosphere's ozone layer also strongly absorbs UV light, greatly attenuating photochemical breakdown of sulfur species in most of the atmosphere. More detailed chemical arguments, experiments, and reaction modeling indicate that the strong signature of SO_2 self-shielding observed in Archean sulfate and sulfide can only occur in an atmosphere in which O_2 is a trace gas ($pO_2 \approx 10^{-5}$ bar or less). The Archean sulfur isotope record, and the disappearance of the mass-independent signature after this epoch, thus provides key evidence that the first appearance of O_2 as a significant component of air occurred at ~2.4 Ga. Mass-independent fractionation of sulfur in modern environments has also been noted, related to sulfate aerosols lofted to the stratosphere by extremely large volcanic eruptions.

Mass-Independent Fractionation of Ozone in Earth's Stratosphere

Mass-independent fractionation of ozone (O_3) was first observed in laboratory synthesis experiments using an electric discharge (Heidenreich & Thiemens 1986). In these experiments O_3 is enriched in ^{17}O and ^{18}O, with $\delta^{17}O \approx \delta^{18}O$ up to +30% depending on the gas pressure and the presence of an inert gas. A similar signature of ozone in the stratosphere was first tentatively observed in balloon-collected samples by Mauersberger (1987), and it is now known that stratospheric ozone can show enrichments in 17O- and 18O-bearing molecules of 100 and more, with mass-49 (17O-bearing) and mass-50 (mostly 18O-bearing) enrichments following a roughly 1:1 relationship. This unusual isotopic signature influences the isotopic composition of the O_2 residue and is transferred to other molecules in the stratosphere through exchange reactions, leading to mass-independent signatures in several tropospheric molecular species, such as O_2 and CO_2.

Unlike in meteoritic oxygen and Archean sulfur, the mass-independent fractionation in ozone is thought to stem mainly from the formation reaction. Photolysis reactions are likely to be secondary in importance. When O_2 and O atom react to form ozone, the O_3 product molecule is initially highly excited vibrationally and rotationally, with enough energy to dissociate back into the reactant state. Stabilization of the product depends on inelastic collisions with other molecules (typically N_2, in the case of air) that remove enough energy to prevent dissociation. The net formation rate of ozone thus depends critically on the intrinsic dissociation lifetime of newly formed ozone molecules, relative to the rate of collision with other molecules. Substitution of ^{17}O or ^{18}O into an ozone molecule strongly affects the dissociation lifetime, and the location of rare isotopes on the central or distal position of the ozone molecule is also important. These considerations lead ozone to have high $^{17}O/^{16}O$ and $^{18}O/^{16}O$ ratios relative to atomic and molecular oxygen. The effect diminishes at high gas pressure (where collision rates with surrounding molecules become large and lifetime differences are less important). In detail, however, quantitatively reproducing laboratory experiments and environmental isotopic signatures with theoretical models has proven difficult, and studies are ongoing.

Apparent mass-independent ^{17}O signatures in oxygen, derived from the fractionation of stratospheric ozone and from tropospheric chemical processes with varying mass fractionation exponents due to photosynthesis and respiration are of interest as tracers of oxygen in the ambient environment. For instance, Luz proposed the use of ^{17}O measurements to track photosynthetic productivity. More recently, ^{17}O in chemical sediment and biominerals has been used as a proxy for atmospheric CO_2 abundance, tracking the O_3-CO_2 exchange process.

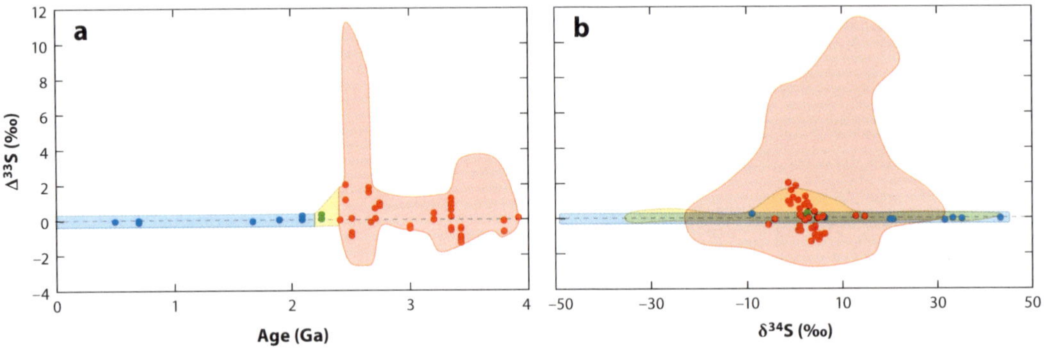

Transient Kinetic Isotope Fractionation

First-order kinetic isotope fractionation and the Rayleigh equation First-order (FO) kinetics have found a wide application, especially in the last two decades, in environmental disciplines related to nutrient biogeochemical cycling in soils. The principle of FO kinetics in isotope modelling is based on the uncoupled kinetic reactions

$$S \xrightarrow{k} P, \quad S' \xrightarrow{k'} P' \qquad (1)$$

which describe simultaneous transformation of the isotopologue substrates S and S′ into the corresponding products P and P′ at rates k and k′. Under the hypothesis that the system is closed, i.e., $S(t) + P(t) = S_0$ and $S'(t) + P'(t) = S'_0$, where S_0 and S'_0 are the initial substrate concentrations, the first-order kinetics describing the rates of change of S, S′, P and P′ are

$$\frac{dS}{dt} = -\frac{dP}{dt} = -kS, \frac{dS'}{dt} = -\frac{dP'}{dt} = -k'S' \qquad (2)$$

with analytical solutions

$$S(t) = S_0 e^{-kt}, S'(t) = S'_0 e^{-k't}, \quad (3)$$

$$P(t) = S_0\left(1 - e^{-kt}\right), \quad P'(t) = S'_0\left(1 - e^{-k't}\right). \quad (4)$$

Equation (2) can be used to define the instantaneous isotope product composition $R_{iP} = dP'/dP$, and the accumulated isotope composition $R_P = P'/P$ and $R_S = S'/S$ for the product and substrate, respectively. The isotope compositions R_{iP} and R_S are used to define the isotope fractionation factor α.

$$á = \frac{R^{iP}}{R_S} = \frac{dP'}{dP} / \frac{S'}{S}, \qquad (5)$$

which becomes

$$\alpha = \frac{dP'}{dP} / \frac{S'}{S} = k' / k = \text{const}, \qquad (6)$$

upon substitution of Eqs. (2). Integration of Eq. (6) for $dP' = -dS'$ and $dP = -dS$ yields the Rayleigh equation describing $R_S(t)$ over time

$$R_S(t) \simeq R_{S,o} f(t)^{\alpha-1}, \qquad (7)$$

Where $R_{S,o}$ is the initial substrate isotope composition, and $f(t) = S(t)/S_o$ is the fraction84of remaining substrate. Because the system is closed (i.e., $S_o = S(t) + P(t)$), the isotope composition of accumulated, R_P, and instantaneous, R_{iP}, products can be written as

$$R_P(t) \simeq R_{S,o} \frac{1 - f(t)^{\alpha}}{1 - f(t)}, \qquad (8)$$

$$R_{iP}(t + dt) = \alpha R_S(t) \qquad (9)$$

It is common to describe the isotope composition relative to a standard, R_{std}, as δ=87($R/R_{std}-1$)1000. Therefore, equations (7), (8), and (9) become

$$\delta_S(t) \simeq \delta_{S,o} + \epsilon \ln[f(t)], \qquad (10a)$$

$$\delta_P(t) \simeq \delta_{S,o} - \epsilon \frac{f(t)\ln[f]}{1 - f(t)}, \qquad (10b)$$

$$\delta_{iP}(t + dt) \simeq \delta_S(t) + \epsilon, \qquad (10c)$$

With $\epsilon = (\alpha - 1)1000$ the enrichment factor. Equations (2) and the Rayleigh approximations in equations (10) have been used to interpret the isotopic signature in several biochemical reactions. The advantages in using FO kinetics lie in the model91simplicity and the ability to derive exact solutions for $S(t)$, $S'(t)$, $P(t)$, $P'(t)$ and α, and approximate analytic solutions for the δ values as in Eqs. (10). However, FO kinetics and the Rayleigh equations are limited in isotope applications to biochemistry because substrate consumption is normally associated with microbial biomass dynamics and specific enzymes that compete for the substrates (i.e., the reactions are coupled) to form the products.

Quasi Steady State Michaelis-Menten-Monod Kinetic Isotope Fractionation

A richer description with respect to first-order kinetics was achieved by the 1913's Michaelis-Menten model where the substrate, *S*, was assumed to attach to an enzyme, *E*, to form a complex, C. In that work, the complex was assumed to be in100fast equilibrium with *S* and *E* and to dissociate yielding the product, *P*, and releasing free, unchanged enzyme. Within our purpose of modeling isotopic effects, this approach leads to the two coupled reactions for S and S'.

$$\mathrm{S+E} \underset{k_2}{\overset{k_1}{\rightleftharpoons}} \mathrm{C} \xrightarrow{k_3} \mathrm{P+E}, \qquad \mathrm{S'+E} \underset{k'_2}{\overset{k'_1}{\rightleftharpoons}} \mathrm{C'} \xrightarrow{k'_3} \mathrm{P'+E}, \tag{11}$$

Where k_i and k'_i are the rate constants corresponding to the isotopologue reactions.

In these reactions all stoichiometric coefficients have been taken equal to 1 while the concentrations of S, S', C, C', P, and P' have units of mass per gram of soil. For example, using the N nitrogen atom as the tracer of the denitrification reaction that produces N_2O from NO_3^-, S, C, and P are expressed in $\left[\mathrm{mg\ N\ kg_{soil}^{-1}}\right]$. Identically, for the isotopologue reaction that produces $^{15}N^{14}NO$ from $^{15}NO_3^-$ and $^{14}NO_3^-$, S', C', and P' are also expressed in $\left[\mathrm{mg\ kg_{soil}^{-1}}\right]$. The isotopologue reaction that consumes $^{15}NO_3^-$ and ^{15}NO and produces $^{15}N^{15}NO$ is not included in reactions of eq. (11) due to its scarsity. The same approach is used for the reaction that produces N_2 from N_2O.

The two reactions can be conveniently described by choosing the concentrations S, S', C, C' as independent variables. Assuming that the system is closed to mass transfer, the following mass conservation laws can be written

$$S_o = S(t) + C(t) + P(t), \quad (12a)$$
$$S'_o = S'(t) + C'(t) + P'(t), \quad (12a)$$
$$E_o = E(t) + C(t) + C'(t). \quad (12c)$$

Where S_o, S'_o, and E_o are the initial substrate and enzyme concentrations. The mass balance for the substrate S has been simplified under the assumption that the transfer of light isotopes from the substrate S towards the heavy product P' does not appreciably affect the concentration of S. For example, the rate of consumption of $^{14}NO_3^-$ substrate is 1/2 the rate of production of $^{15}N^{14}NO$; because the isotopic ratio $^{15}N^{14}NO/^{14}N_2O$ is in the order of magnitude of 10^{-2}, the consumption of ^{14}N from $^{14}NO{-}3$to form $^{15}N\ ^{14}NO$ is in the order of $5{\cdot}10^{-3}$ of the concentration of $^{14}NO_3^-$. Equation (12c) establishes that the two substrates S and S' compete to bound to the enzyme to form the complexes C and C', and is responsible for the coupling of the two reactions in Eq. (11).

In the approach presented here the enzyme concentration is assumed to be linearly proportional by a factor z to the biomass concentration B as

$$E(t) = z\,B(t) \qquad (13)$$

Because B is assumed to increase in response to the release of P and P' over time according to the Monod model, equation (13) implies that the enzyme is synthesized at the same rate as biomass growth and deteriorates at the same rate as microbial cells die. This approach, though simplifying the enzyme dynamics, improves the original Michelis-Menten formulation by which E was considered constant and not linked to any microbial biomass dynamics.

Using the mass conservation laws in equations (12), the kinetic equations for each component in the system of equation (11), including the biomass, can be written as follows

$$\frac{dS}{dt} = -k_1 SE + k_2 C, \qquad (14a)$$

$$\frac{dS'}{dt} = -k_1' S'E + k_2' C', \qquad (14b)$$

$$\frac{dC}{dt} = k_1 SE - (k_2 + k_3)C, \qquad (14c)$$

$$\frac{dC'}{dt} = k_1' S'E - (k_2' + k_3')C', \qquad (14d)$$

$$\frac{dP}{dt} = -\frac{dS}{dt} - \frac{dC}{dt} = k_3 C, \qquad (14e)$$

$$\frac{dP'}{dt} = -\frac{dS'}{dt} - \frac{dC'}{dt} = k_3' C', \qquad (14f)$$

$$\frac{dE}{dt} = z\frac{dB}{dt} - \frac{dC}{dt} - \frac{dC'}{dt}, \qquad (14g)$$

$$\frac{dB}{dt} = Y\left(\frac{dP}{dt} + \frac{dP'}{dt}\right) - \beta B \qquad (14h)$$

Where equation (14h) is the Monod equation describing microbial biomass dynamics with Y the yield coefficient expressing the biomass gain per unit of released product and β expressing the biomass death rate.

A simplified expression for this system of eight ordinary differential equations in eight unknowns can be obtained assuming that complexation of C and C' is very fast at the early stage of the reactions, and that, afterwards, their concentrations do not change appreciably in time. This assumption, known as Haldane-Briggs, is commonly referred to as quasi steady state, and implies that C and C′ are small and that

$$\frac{dC}{dt} \simeq 0, \quad \frac{dC'}{dt} \simeq 0. \qquad (15)$$

Substituting equations (15) into equations (14e), (14f), and (14g), the quasi-steady-state assumption also implies

$$\frac{dS}{dt} \simeq -\frac{dP}{dt}, \quad \frac{dS'}{dt} \simeq -\frac{dP'}{dt}, \quad \frac{dE}{dt} \simeq z\frac{dB}{dt}. \qquad (16)$$

Introducing equations (12) into equations (14c) and (14d) under the quasi-steady-state assumption, the kinetic equations for C and C' become

$$\frac{dC}{dt} = k_1 S(zB - C - C') - (k_2 + k_3)C \simeq 0, \quad (17a)$$

$$\frac{dC'}{dt} = k_1' S'(zB - C - C') - (k_2' + k_3')C' \simeq 0, \quad (17b)$$

and can be solved for C and C′ to yield

$$C \simeq \frac{zBS}{S+K(1+S'/K')}, \quad C' \simeq \frac{zBS'}{S'+K'(1+S/K)}, \tag{18}$$

With $K= (k_2+k_3)/k_1$ and $K'= (k'_2+k'_3)/k'_1$ the Michaelis-Menten parameters (or half saturation concentrations). In the classic approach to solve the Michaelis - Menten kinetics with the Haldane-Briggs' quasi-steady-state assumption, the total enzyme concentration (free plus bound in the complexes) E_t= constant appears in place of the transient product zB in the numerators of equation (18) of our approach. Introducing the approximate solutions for C and C′ of equation (18) into equation (14e), (14f), and (14h), and using equations (16), we obtain

$$\frac{\mathrm{d}P}{\mathrm{d}t} \simeq k_3 \frac{zBS}{S+K(1+S'/K')}, \tag{19a}$$

$$\frac{\mathrm{d}P'}{\mathrm{d}t} \simeq k'_3 \frac{zBS'}{S'+K'(1+S/K)}, \tag{19b}$$

$$\frac{\mathrm{d}S}{\mathrm{d}t} \simeq -\frac{\mathrm{d}P}{\mathrm{d}t}, \tag{19c}$$

$$\frac{\mathrm{d}S'}{\mathrm{d}t} \simeq -\frac{\mathrm{d}P'}{\mathrm{d}t}, \tag{19d}$$

$$\frac{\mathrm{d}B}{\mathrm{d}t} \simeq \frac{k_3 YzBS}{S+K(1+S'+K')} + \frac{k'_3 S'YzB}{S'+K'(1+S/K)} = -\beta B, \tag{19e}$$

which are normally known as the Michaelis-Menten kinetics with biomass dynamics following Monod kinetics. From here on we will call this approach with the acronym QSS-MMM kinetics to highlight that the Michaelis-Menten-Monod kinetics are solved156under the assumption of quasi steady state.

Introducing dP and dP' of equations (19a) and (19b) into equation (5) for the fractionation factor α we obtain

$$\alpha = \frac{\mathrm{d}P'}{\mathrm{d}P} \Big/ \frac{S'}{S} \simeq \frac{k'_3}{k_3} \frac{S+K(1+S'/K')}{S'+K'(1+S/K)} = \frac{k'_3 K}{k_3 K'} = \text{constant} \tag{20}$$

In this case, however, equation (19a) and (19b) cannot be integrated analytically so no approximation for the δ^{15}N can be derived as was done for the Rayleigh approximation equations.

The approximate solutions of equations (19) to the system of differential equations in equations (14), achieved using the Haldane-Briggs' quasi-steady-state assumption, imply that CK ≃ SE and $C'K' \simeq S'E$ over the entire reaction course. The circumstances under which these conditions are valid are that C and C′ have to be small and time invariant (Eqs. (15)). For C and C' to be small the conditions C ≪ S and C′ ≪ S′ have to be satisfied. Consequently, the K and K' values have to be large numbers on the same order of magnitude as S and S′(i.e., S~K and S′~K′), while the free enzyme concentration E has to be a small number on the same order of magnitude as C and C′(i.e., E~C~C′). While these conditions may be encountered in common enzymatic reactions, the boundaries of validity in isotopic applications become more restrictive. If the refers to rare isotopologues with natural abundance (i.e., $S \gg S'$, $C \gg C'$, and $K \simeq K'$) and if $S \sim K \simeq K'$, then it must be $S' \ll K'$. This relation poses a first limitation to the quasi-steady-state assumption which requires $S' \sim K'$. In addition, if $E \sim C$, then it must be $E \gg C'$ where as $E \sim C'$ is expected for assuming quasi steady state. This second condition that applies to E, C, and C′ strongly limits the quasi-steady-state assumption because in our modeling development we have assumed that the enzyme concentration is not constant but varies with biomass concentration. In instances where the biomass concentration increases, and in so doing the free enzyme concentration increases too, the system may depart from the condition $E \sim C \sim C'$(i.e., $E > C \gg C'$), worsening the degree of applicability of quasi steady state. Conversely, in instances where the free enzyme concentration decreases, the system may meet only one of the conditions $E \sim C$ or $E \sim C'$.

Transient Kinetic Isotope Fractionation

Regardless of the fact that isotopic effects measured in natural-abundance biochemical reactions are normally very small (i.e., $\alpha \approx 1$), the isotopic effects modeled with either first-order kinetics or quasi-steady-state Michaelis-Menten-Monod kinetics have as outcome a constant isotopic effect. This constancy implies that, for stable isotopologues, biochemical reactions always produce depleted products while enriching the remaining substrates owing to the preference of enzyme systems to use lighter isotopologue sub-strates. In the case of first-order kinetics and quasi-steady-state Michaelis-Menten-Monod kinetics this is a straightforward conclusion from Eqs. (6) and (20), as the substrate is directly transformed into a product (i.e., $dP = -dS$ and $dP' = -dS'$) or, equivalently, the complexation rate is null (i.e., $dC = dC' = 0$). If however, the quasi-steady-state Haldane-Briggs hypothesis is not used to solve the Michaelis-Menten-Monod kinetics (i.e., $dC \neq 0$ and $dC' \neq 0$), a different conclusion is reached. The isotoplogue substrates S and S′, while competing to bond to the enzyme without conditions imposed on the concentrations discussed above, will form reversible complexes that will either accumulate, decompose back to substrates, or release products. Along each of these pathways, the isotopic fractionation changes the isotope composition of the total substrate $S(t) + S'(t)$, complex $C(t) + C'(t)$, and product $P(t) + P'(t)$ over time. This fractionation sequence is further fed by the feedback introduced by the concentration of free enzyme, which is synthesized by microbes that grow at the rate at which the products are released (i.e., proportional to k_3 and k'_3) and affects the rate at which the substrates bind in the complexes (i.e., proportional to k_1 and k'_1). The quasi-steady-state assumption applied to C and C′ is invoked to achieve a simplified expression of these kinetics but does not allow quantitative treatment of these effects, which, although small in absolute terms, have the same order of magnitude (one part per thousand) as the fractionation factor. It is therefore meaningful to challenge the quasi-steady-state assumption and question whether doing so could bring to light nonlinear effects during transient complexation in microbially-induced isotope fractionation modelled with Michaelis-Menten-Monod kinetics.

By relaxing the quasi-steady-state assumption in equation (15), the system kinetics in equations (14) cannot be re-written in a simplified form as equations (19). We name the full set of differential equations in equations (14) with the acronym TR-MMM to underline that it is solved for the transient kinetics.

An expression for α can be derived by combining Eqs. (14e) and (14f) as follows α

$$\alpha(t) = \frac{dP'}{dP} / \frac{S'}{S} = \frac{dS' + dC'}{dS + dC} / \frac{S'}{S} = \frac{k_3' C'}{k_3 C} \frac{S}{S'} = \frac{k_3'}{k_3} \frac{Rc(t)}{Rs(t)}. \tag{21}$$

This expression for α shows that in the most general case of biological kinetic isotope fractionation, the isotopic effect may change with the substrate concentrations S and S′ and complex concentrations C and C′, and can be re-written in terms of isotope compositions R_C and R_S. Equation (21), derived without imposing conditions on complex dynamics, generalizes the fractionation factor derived for quasi-steady-state Michaelis- Menten-Monod kinetics in euation (19). Because equations (14e) and (14f) do not have an analytic solution, the Rayleigh approximation equations cannot be derived as for the case of quasi-steady-state Michaelis-Menten-Monod kinetics.

Polymer Fractionation

Polymers are increasingly used as high-performance materials and crucial in the areas of medicine, cosmetics and technique. For sophisticated applications it is often harmful that the polymers have a broad molecular mass distribution or contain impurities. By means of fractionation we are in the position to remove unwanted material on a technical scale enabling the optimization of the polymer properties.

Modifiable Properties:	
• Viscosity	• Refractive index
• Density	• Degradation
• Crystallinity, glass temperature	• Permeability through skin /kidney
• Purity	• Inflammable behavior
• Volatility	• Oligomer content
• Shelf life	• etc.

The fractionation method is based on liquid/liquid phase separation and is therefore suitable for almost all soluble polymers. The phase separation is induced by lowering the solvent quality of an initially homogenous polymer solution. Two phases are formed: one containing the long chain material (gel-phase) and the other one contains mainly the short chain material and the low molecular mass impurities (sol-phase).

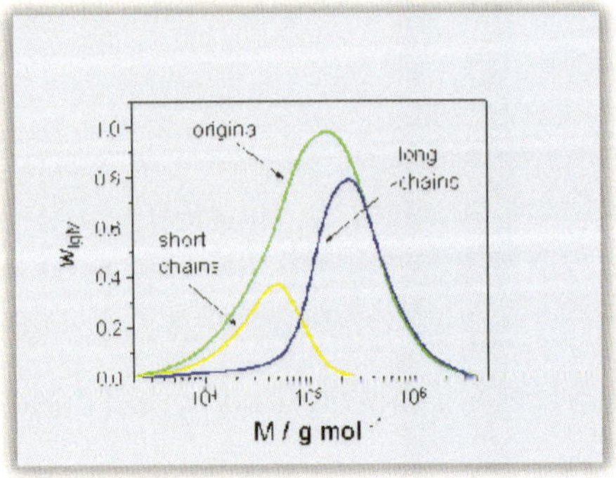

By the choice of the fractionation parameters (temperature, compositions) the “cut” through the molecular mass distribution can be varied. Narrow molecular mass distributions can be achieved by repeated fractionation.

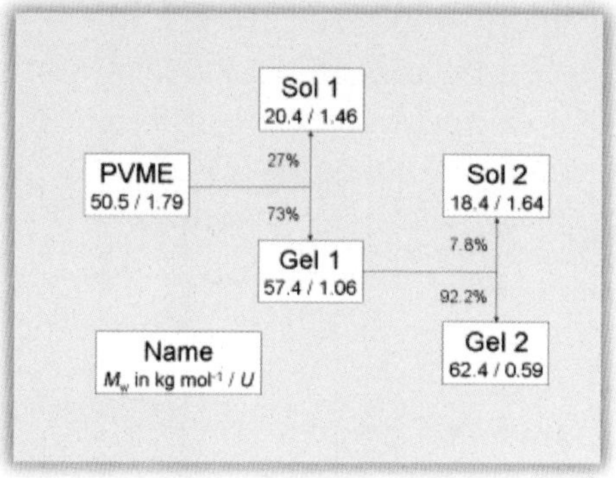

The reasons for fractionation are as divers as the different applications of polymers are. The following examples are therefore taken from very different areas.

- Recovery/purification of polymers produced by means of biotechnology

  Fractionation enables the removal of low molecular mass material of polymers stemming from biologiocal sources, yielding products with high purities.

- Application in medicine

  The short chains as well as the long chains of hydroxyethyl starch are troublesome for its use as blood plasma expander. The short chains leave the body through the kidney very fast and are therefore inefficient. The long chains on the other hand remain in the body, are stored in the skin and induce itching.

  Another sample is hyaluronic acid wich is used in eye surgeries as a viscoelastic agent or for the treatment of arthrosis as a natural lubricant. By means of fractionation the properties of hyaluronic acid can be tuned for the application.

- Harmful contents in photoresists

  Photoresist, like Novolak, are used for the production of electronic circuits in the micro- or submicroscale. For this application the oligomers are disturbing, because they are still sufficient volatile and can be deposited at wrong places. Therefore they impede a further miniaturization. By removing the oligomeric contents this problem can be solved.

- Filter dust at the membrane production

  During the production of filtration membranes (e.g. from cellulose acetate or- nitrate) very often filter dust is formed, which is blocking the pores of the membrane and decreasing its performance. Furthermore the filter dust can leave the membrane and enter the filtrate. The filter dust consists of the short chains of the membrane material and is transported during the production according to the phase inversion process towards the surface of the forming membrane. At this place it precipitates and blocks the pores. A removal of the short chains reduces the formation of filter dust markedly.

- Production of samples with different molecular masses

  By means of the fractionation of broadly distributed samples one can obtain fractions differing in the molecular mass. This is for example necessary, if relations between polymer properties and its molar mass shall be examined, but no suitable samples are availabe. A familiar example for a molecular mass dependent property is the hydrodynamic radius of polymers in solution: The Kuhn-MarkHouwink relation shows the dependence between the intrinsic viscosity and the chain length.

- Shelf life of polymers

  Certain components can reduce the shelf life of poylmers. This can be e.g. impurities or the short chain material of the polymer itself (as they act as plasticizer or because the end groups have a negative influence). By means of fractionation these components can be removed or reduced to an accaptable amount.

- Fractionation according to chemistry

  Copolymers (like Polystyrolacrylnitril, SAN) often possess not only a distribution of the molecular mass but also in chemical composition. Under certain circumstances it is possible to fractionate these polymers according to chemistry and therefore opimize their properties.

- Fractionation according to architecture

  During the sythesis of polymers with complex molecular architecture, like branched polymers or stars, it often happens that linear chains or stars with lower number of arms are formed as unwanted side products. By means of fractionation these components can be removed.

- Standards for analytics

  By means of repeated fractionation of a broadly distributed sample it is possible to obtain samples with narrow distributions. These fractions can be used as standards, e.g. for the gel permeation chromatography (GPC/SEC).

References

- Fractionation, earth-and-planetary-sciences: sciencedirect.com, Retrieved 28 April 2018
- Protein-fractionation-separation, dairy-industry: synderfiltration.com, Retrieved 25 March 2018
- Fractional-distillation, inorganic-chemistry: tutorvista.com, Retrieved 20 June 2018
- Field-flow-fractionation, other-methods-colloidal-analytics: nanolytics.de, Retrieved 17 July 2018
- Flow-field-flow-fractionation-theory: wyatt.com, Retrieved 19 April 2018

6

Membrane Technology

Membrane technology is the technology, concerning the transport of substances between different fractions through a permeable membrane. It is commonly used for the mechanical separation of liquid or gaseous streams. The topics elaborated in this chapter on mass transfer, concentration polarization, pervaporation, membrane gas separation, ultrafiltration, membrane distillation, reverse osmosis, etc. will help in providing a better perspective about membrane technology.

Membrane technology has become a dignified separation technology over the past decennia. The main force of membrane technology is the fact that it works without the addition of chemicals, with a relatively low energy use and easy and well-arranged process conductions.

Membrane technology is a generic term for a number of different, very characteristic separation processes. These processes are of the same kind, because in each of them a membrane is used. Membranes are used more and more often for the creation of process water from groundwater, surface water or wastewater. Membranes are now competitive for conventional techniques. The membrane separation process is based on the presence of semi permeable membranes.

The principle is quite simple: the membrane acts as a very specific filter that will let water flow through, while it catches suspended solids and other substances.

There are various methods to enable substances to penetrate a membrane. Examples of these methods are the applications of high pressure, the maintenance of a concentration gradient on both sides of the membrane and the introduction of an electric potential.

Membranes occupy through a selective separation wall. Certain substances can pass through the membrane, while other substances are caught. Membrane filtration can be used as an alternative for flocculation, sediment purification techniques, adsorption (sand filters and active carbon filters, ion exchangers), extraction and distillation.

There are two factors that determine the affectivity of a membrane filtration process; selectivity and productivity. Selectivity is expressed as a parameter called retention or separation factor (expressed by the unit l/m2·h). Productivity is expressed as a parameter called flux (expressed by the unit l/m2·h). Selectivity and productivity are membrane-dependent.

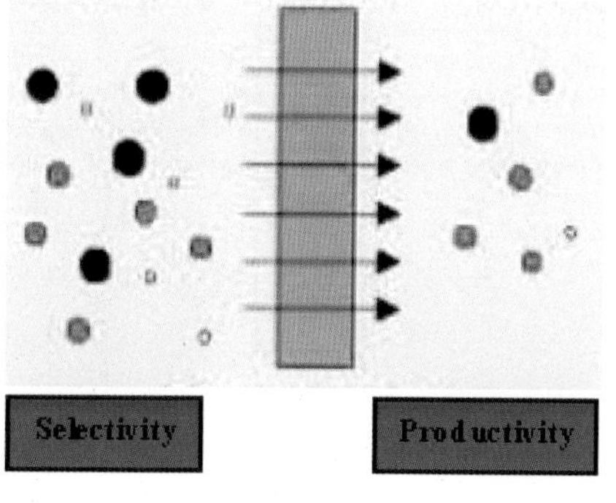

Membrane filtration can be divided up between micro and ultra filtration on the one hand and nano filtration and Reverse Osmosis (RO or hyper filtration) on the other hand.

When membrane filtration is used for the removal of larger particles, micro filtration and ultra filtration are applied. Because of the open character of the membranes the productivity is high while the pressure differences are low.

When salts need to be removed from water, nano filtration and Reverse Osmosis are applied. Nano filtration and RO membranes do not work according to the principle of pores; separation takes place by diffusion through the membrane. The pressure that is required to perform nano filtration and Reverse Osmosis is much higher than the pressure required for micro and ultra filtration, while productivity is much lower.

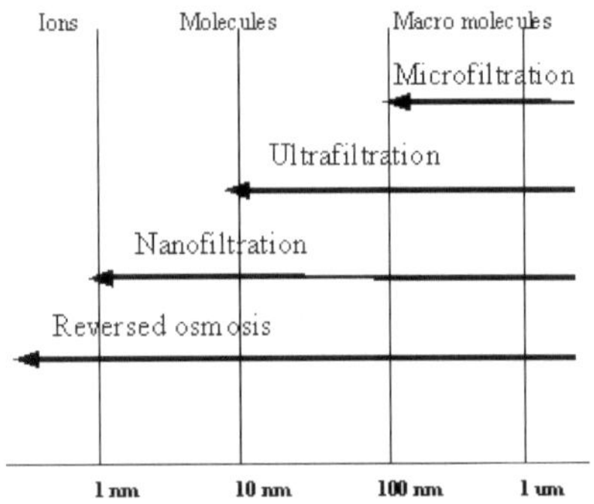

Membrane filtration has a number of benefits over the existing water purification techniques:

- It is a process that can take place while temperatures are low. This is mainly important because it enables the treatment of heat-sensitive matter. That is why these applications are widely used for food production.
- It is a process with low energy cost. Most of the energy that is required is used to pump liquids through the membrane. The total amount of energy that is used is minor, compared to alternative techniques, such as evaporation.
- The process can easily be expanded.

Process Management of Membrane Filtration Systems

Membrane filtration systems can be managed in either dead-end flow or cross-flow. The purpose of the optimisation of the membrane techniques is the achievement of the highest possible production for a long period of time, with acceptable pollution levels.

Membrane Systems

The choice for a certain kind of membrane system is determined by a great number of aspects, such as costs, risks of plugging of the membranes, packing density and cleaning opportunities. Membranes are never applied as one flat plate, because this large surface often results in high investing costs. That is why systems are built densely to enable a large membrane surface to be put in the smallest possible volume. Membranes are implemented in several types of modules. There are two main types, called the tubular membrane system and the plate & frame membrane system.

Tubular membrane systems are divided up in tubular, capillary and hollow fiber membranes. Plate & frame membranes are divided up in spiral membranes and pillow-shaped membranes.

Membrane Fouling

During membrane filtration processes membrane fouling is inevitable, even with a sufficient pre-treatment. The types and amounts of fouling are dependent on many different factors, such as feed water quality, membrane type, membrane materials and process design and control.

Particles, biofouling and scaling are the three main types of fouling on a membrane. These contaminants cause that a higher workload is required, to be able to guarantee a continuous capacity of the membranes. At a certain point the pressure will rise so much that it is no longer economically and technically accountable.

Membrane Cleaning

There are a number of cleaning techniques for the removal of membrane fouling. These techniques are forward flushing, backward flushing, air flushing and chemical cleaning, and any combination of the methods.

Membrane Reactor

A membrane reactor is really just a plug-flow reactor that contains an additional cylinder of some porous material within it, kind of like the tube within the shell of a shell-and-tube heat exchanger. This porous inner cylinder is the membrane that gives the membrane reactor its name.

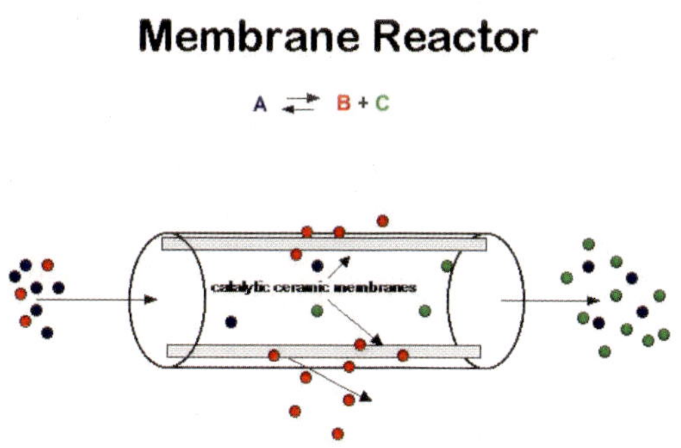

A mixed feed of A and B enters the membrane reactor. C is produced in the reactor, and B diffuses out through the membrane pores. There are multiple ceramic membranes, but only two are shown for simplicity.

The membrane is a barrier that only allows certain components to pass through it. The selectivity of the membrane is controlled by its pore diameter, which can be on the order of Angstroms, for microporous layers, or on the order of microns for macroporous layers.

Use of Membrane Reactor

Membrane reactors combine reaction with separation to increase conversion. One of the products of a given reaction is removed from the reactor through the membrane, forcing the equilibrium of the reaction "to the right" (according to Le Chatelier's Principle), so that more of that product is produced.

Membrane reactors are commonly used in dehydrogenation reactions (e.g., dehydrogenation of ethane), where only one of the products (molecular hydrogen) is small enough to pass through the membrane. This raises the conversion for the reaction, making the process more economical.

Kinds of Membrane Reactors

- Membrane reactors are most commonly used when a reaction involves some form of catalyst, and there are two main types of these membrane reactors: the inert membrance reactor and the catalytic membrane reactor.
- The inert membrane reactor allows catalyst pellets to flow with the reactants on the feed side (usually the inside of the membrane). It is known as an IMRCF, which stands for Inert Membrane Reactor with Catalyst on the Feed side. In this kind of membrane reactor, the membrane does not participate in the reaction directly; it simply acts as a barrier to the reactants and some products.
- A catalytic membrane reactor (CMR) has a membrane that has either been coated with or is made of a material that contains catalyst, which means that the membrane itself participates in the reaction. Some of the reaction products (those that are small enough) pass through the membrane and exit the reactor on the permeate side.

Mass Transfer

Mass transport across membranes involves different driving forces. The wide range of behavior resulting from these various driving forces can be described compactly via the framework provided by non-equilibrium thermodynamics (NET). NET makes use of four postulates above and beyond those of equilibrium thermodynamics:

- The equilibrium thermodynamic relations apply to systems that are not in equilibrium, provided that gradients are not too large (local equilibrium).
- All fluxes in the system may be written as linear relations involving all the driving forces;
- No coupling of fluxes and forces occurs if the difference in tensorial order of the flux and force is an odd number (Curie's Postulate).
- In the absence of magnetic fields, the matrix of the coefficients in the flux-force relations is symmetric (Onsager's reciprocal relations).

The description of mass transfer when multiple driving forces are present is given by the Maxwell-Stefan equations. Eventhough the derivation of these equations is not in the scope of this script, it is important to understand how these equations are derived:

- The starting point for derivation of Maxwell-Stefan equations is the entropy balance equation for irreversible processes (entropy remains a valuable state function in non-equilibrium under the local equilibrium assumption).

Entropy Balance

$$\rho\frac{D\hat{S}}{Dt}=-\left(\nabla.\bar{J}_s\right)+g_s$$

where,

- ρ is the density of the mixture;
- $\bar{J}_s$ is the entropy flux vector;
- g_s is the entropy generation per unit volume.

A lagrangian approach is used where the evolution of a specific control volume element is followed in time.

The derivation of the entropy balance follows

$$\frac{DS}{Dt}=\underbrace{\frac{dS|_e}{Dt}}_{\text{Rate of exchange with exterior}}+\underbrace{\frac{d\,S|_i}{Dt}}_{\text{Rate of internal production}}$$

Appling the divergence theorem over a generic control volume of surface $\sum$

$$\left(\int_V \nabla\cdot\bar{F}\,dV=\oint_{\partial V}\bar{F}\cdot d\bar{S}\right)$$

$$\left(\int_V \rho\frac{D\hat{S}}{Dt}\right)=-\int_\Sigma \bar{J}_s\cdot\bar{n}\,d\sum+\int_V g_s dV=-\int_V \nabla\cdot\bar{J}_s dV+\int_V g_s\,dV$$

Due to the arbitrary chosen control volume the local balance, eq. 1.6, is obtained equating the arguments of the integrals.

By using the conservation laws for mass, momentum and energy, together with the equilibrium relation $dU=TdS-pdV+\sum_i\frac{\mu_i}{M_i}dn_i$, it is possible to express the entropy flux $\bar{J}_s$ and entropy generation gs as a sum of products of fluxes and forces (gs) or sum of fluxes $\left(\bar{J}_s\right)$.

The result of this procedure brings to the Maxwell-Stefan generalized equations:

$$\nabla x_\alpha=\sum_{\beta=1}^{N\neq\alpha}\frac{x_\alpha x_\beta}{Đ_{\alpha\beta}}\left(\bar{v}_\alpha-\bar{v}_\beta\right)=\underbrace{-x_\alpha\nabla\text{In a}_\alpha}_{\text{Concentration diffusion term}}+$$

$$-\frac{1}{cRT}\left[\underbrace{\left(\phi_\alpha-\omega_\alpha\right)\nabla p}_{\text{Pressure diffusion term}}\underbrace{-\rho_\alpha\bar{g}_\alpha+\omega_\alpha\sum_{\beta=1}^{N}\rho\beta\bar{g}\beta}_{\text{Forced diffusion term}}\right]+$$

$$\underbrace{-\sum_{\beta=1}^{N\neq\alpha}\frac{x_\alpha x_\beta}{Đ_{\alpha\beta}}\left(\frac{D_\alpha^T}{\rho_\alpha}-\frac{D_\beta^T}{\rho_\beta}\right)\nabla T}_{\text{Thermal diffusion term}}$$

where,

- $Đ_{\alpha\beta}$ are the Maxwell-Stefan diffusivities;
- D_i^T is the thermal diffusivities of component *i*.

For binary diffusion in gases or liquids

$$\bar{J}_\alpha = cĐ_{\alpha\beta}\nabla x_\alpha = -cĐ_{\alpha\beta}\left[x_\alpha \nabla \text{In } a_\alpha + \right.$$

$$\left. + \frac{1}{cRT}\left((\phi_\alpha - \omega_\alpha)\nabla p - \rho\omega_\alpha\omega_\beta\left(\bar{g}_\alpha - \bar{g}_\beta\right) + k_T \nabla \text{ In T}\right)\right.$$

The contribution of external forces (gravity field, electrostatic field and mechanical forces) can be written as

$$\bar{g}_\alpha = \bar{g} - \frac{z_\alpha F}{M_\alpha}\nabla\phi + \delta_{\alpha,m}\frac{1}{\rho_m}\nabla p$$

The term $\delta_{\alpha,m}\frac{1}{\rho_m}\nabla p$ represents the mechanical constraint that keeps the membrane stationary against external pressure gradients and internal viscous drag (typically a wire mesh or equivalent structure is used).

Membranes consist of an insoluble, selective permeable matrix m and one or more permeating species α, β.

The following constraints hold:

- Negligible curvature: $\delta << R_{curv}$ (mass transport in unidirectional and perpendicular to the membrane surface).
- Immobility of the matrix: $v_m = 0$.
- Pseudo steady behavior: $\frac{\partial_{c_\alpha}}{\partial t} = 0$ (diffusional terms within the membrane are short compared to those in the adjacent solution.
- No thermal diffusion.

It follows that the Maxwell-Stefan equations can be slightly simplified:

- The term $w_\alpha \nabla p$ can be set to zero as the membrane is stationary.
- The thermal diffusion term can be neglected.
- Since the membrane is, to an excellent approximation, incompressible and of uniform chemical composition, the equivalent body force will be uniform and of magnitude $\bar{g} = \frac{1}{p_m}\nabla p$.

The resulting Maxwell-Stefan equation can be therefore written as:

$$\sum_{\beta=1}^{N}\frac{RT}{Đ_{\alpha\beta c}}\left(x_\beta \bar{N}_\alpha - x_\alpha \bar{N}_\beta\right) =$$

$$= -x_\alpha RT \nabla \ln a_\alpha - z_\alpha F \nabla V + \left(\frac{\delta_{\alpha,m}}{\rho_m} - \bar{V}_\alpha \right) \nabla p$$

where $\bar{N}_\alpha$ and $\bar{N}_\beta$ represent the molar fluxes, $\delta_{\alpha,m}$ the Kronecker delta and $\bar{V}_\alpha$ the partial molar volume of α.

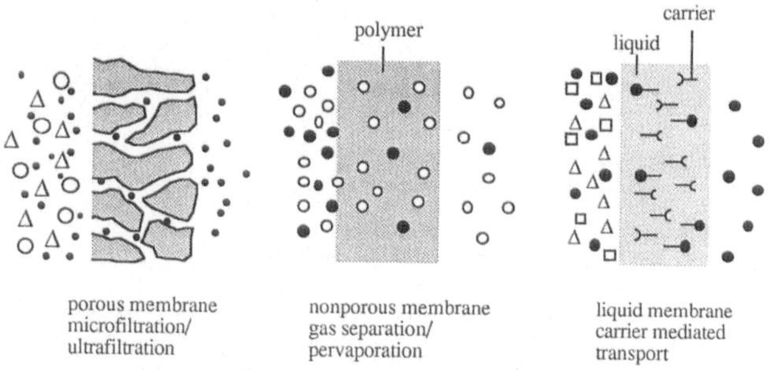

Figure: Schematic representation of the basic types of membrane

Along with the Maxwell-Stefan equation for membrane transport, a generally applicable set of boundary conditions has to be defined: these conditions are obtained by requiring the "total potential" of each species to be continuous across the boundary of the membrane and the solution bathing the membrane (local equilibrium). Since conditions within the membrane are very difficult, or even impossible, to determine, this equation is primarily useful to obtain a qualitative understanding of membrane behavior. What is the way to proceed then? The approach is to use simplified models (based on the Maxwell-Stefan equation introducing simplifications) that allow to describe a membrane quantitatively with a good agreement with respect to experimental data. The model will be therefore simplified according to the type of membrane considered. There are many transport mechanisms involved depending on the type of selective barrier:

- *Porous membranes* (mean pore size diameter 5000 - 1 nm)

  In porous membranes molecules are transported by a pressure-driven convective flow through tiny membrane pores (pores size bigger that 10–9 m). A difference in steric hindrances between the components' molecules and the membrane material leads therefore to different compounds permeability's. The permeant concentration within a porous membrane is uniform and the only driving force across the membrane is the pressure gradient.

- *Homogeneous membranes* (mean pore size diameter less than 1 nm)

  In dense membranes molecules of the different compounds first dissolve into the membrane matrix and then diffuse through the membrane under a concentration gradient. The permeability of each species is therefore affected by the solubility of each compound into the membrane material (thermodynamic aspect) and by the rate at which each component diffuses through the membrane (kinetics aspect).The average pore diameter in dense polymer membranes is within the thermal motion of the polymer chains from which the membrane is made of.

- Facilitated diffusion/charged membranes

  In a carrier-mediated diffusion membrane the movement of molecules across the membrane occurs via very specific carrier-molecules that are embedded within the membrane. The perm selectivity towards a component depends mainly on the specificity of the carrier molecule. Through the use of specially tailored carriers, extremely high selectivity's can be obtained. The component to be removed can be gaseous or liquid, ionic or non-ionic. To some extent the functionality of this kind of membranes approaches that of a cell.

- Active transport

  In membranes where active transport occurs the movement of molecules (against some concentration gradients or other form of resistance) across the membrane requires energy. Unlike passive transport, which uses natural entropy of molecules moving down a gradient, active transport uses external sources of energy (electrochemical gradients, etc.) to perform a separation.

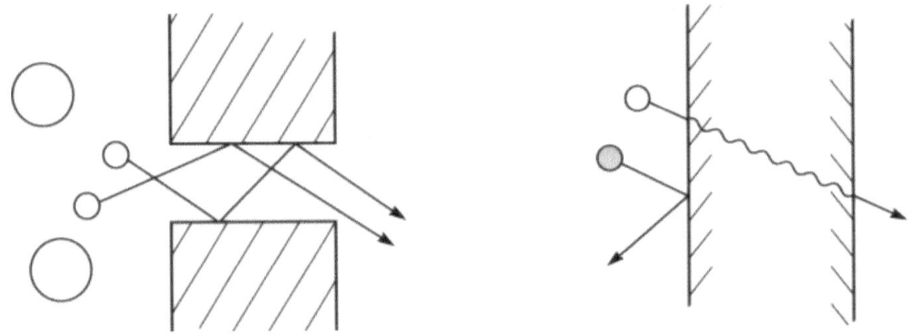

(a) Microporous membranes separate by molecular filtration.

(b) Dense solution-diffusion membranes separate because of differences in the solubility and mobility of permeants dissolved in the membrane material.

Figure: Membranes mass transfer models

Models for Mass Transfer Through Membranes

Two basic models for mass transfer through the membrane will be considered:

- The hydrodynamic model for porous membranes.
- The solution-diffusion model for dense membranes.

Through a manipulation of the simplified Maxwell-Stefan equation (eq. 1.10) it is possible to express the molar flux of component α as:

$$\bar{N}_\alpha = \bar{J}_\alpha = -cĐ_{\alpha\beta}\left[x_\alpha \nabla \ln a_\alpha + \frac{1}{cRT} c_\alpha \bar{v}_\alpha \nabla p - \rho_\alpha \bar{g}_\alpha + \omega_\alpha \sum_{\beta=1}^{N} \rho_\beta \bar{g}_\beta\right]$$

In the case of binary mixtures the left term in eq. 1.10 can be written as

$$\sum_{\beta=1}^{N} \frac{1}{D_{\alpha\beta}c}\left(x_\beta \bar{N}_\alpha - x_\alpha \bar{N}_\beta\right) = \frac{1}{D_{\alpha\beta}c}\bar{N}_\alpha = -\frac{1}{D_{\alpha\beta}c}\bar{N}_\beta$$

due to the fact that $x_\alpha + x_\beta = 1$ and $\bar{N}_\alpha + \bar{N}_\beta = \bar{0}$.

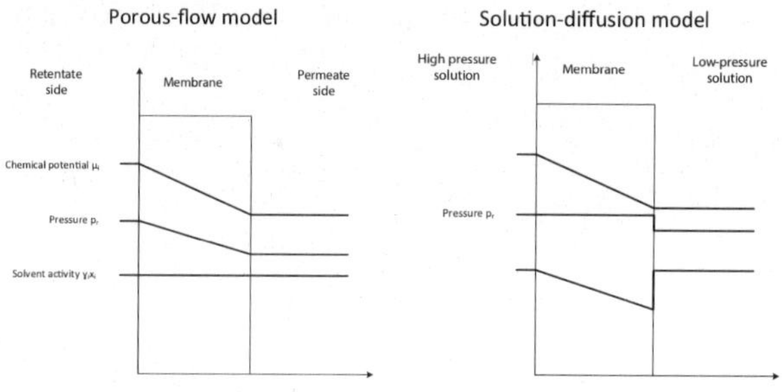

(a) *Porous-flow model.* (b) *Solution-diffusion permeation model.*

Figure: Comparison between the hydrodynamic permeation and solutiondiffusion permeation models for mass transfer through the membrane.

Hydrodynamic Model

In the framework of porous membranes the mass transport mechanism is model as a pure convective motion of molecules across the membrane due to a superimposed pressure gradient. Eq. can be reconsidered in light of the simplification introduced by this model.

$$\bar{J}_\alpha = -c\mathit{Đ}_{\alpha\beta}\left[\frac{1}{cRT}c_\alpha\bar{\upsilon}_\alpha\nabla p - \rho_\alpha\bar{g}_\alpha\right]$$

It follows that

$$\bar{J}_\alpha = -\frac{\mathit{Đ}_{\alpha\beta}}{RT}\bar{\upsilon}_\alpha c_\alpha\nabla p = -c_\alpha K'\frac{dp}{dz}$$

Equation is named Darcy equation where the following terms represents

- $\frac{dp}{dz}$: pressure gradient;
- c_α: concentration of component A in the medium;
- K': permeability of the medium.

Solution-Diffusion Model

In the solution-diffusion model, transport occurs only by diffusion. The component that needs to be transported must first dissolve in the membrane. The molar flux of component α can be therefore expressed using eq.

$$\bar{J}_\alpha = -c\mathit{Đ}_{\alpha\beta x_\alpha}\nabla \text{ In } a_\alpha$$

since

- No pressure gradient exists within the membrane;
- No external forces act on the membrane.

In the solution-diffusion model, the pressure within the membrane is constant at the high-pressure value (p_r), and the gradient in chemical potential across the membrane is expressed as a smooth gradient in solvent activity $(\gamma_s x_s)$ as shown in figure. By introducing the definition of activity, $a_\alpha = \gamma_\alpha x_\alpha$, in equation.

$$\bar{J}_\alpha = -c\text{Đ}_{\alpha\beta}\left(\frac{\partial \ln a_\alpha}{\partial \ln x_\alpha}\right)_{T,p} \nabla x_\alpha = -c\text{Đ}_{\alpha\beta}\left[1+\left(\frac{\partial \ln \gamma_\alpha}{\partial \ln x_\alpha}\right)_{T,p}\right]\nabla x_\alpha$$

Under the hypothesis of perpendicular flux it is possible to write

$$\bar{J}_\alpha = -cD_{\alpha,\beta}\nabla x_\alpha = -cD_{\alpha,\beta}\frac{dx_\alpha}{dz}\hat{n}$$

where,

- $D_{\alpha,\beta} = \text{Đ}_{\alpha\beta}$ for an ideal mixture;
- $D_{\alpha,\beta} = \text{Đ}_{\alpha\beta}\left[1+\left(\frac{\partial \ln \gamma_\alpha}{\partial \ln x_\alpha}\right)_{T,p}\right]$ for a non-ideal mixture.

Equation is the simplified Fick equation with diffusivities accounting for mixtures non-ideality.

The flux equations respectively for porous membranes and homogeneous membranes have been obtained. Making use of the expression of the flux in the continuity equation

$$\frac{\partial \rho}{\partial t} + \nabla\cdot(\rho\bar{v}) = 0 \text{ Overall continuity equation}$$

$$c\frac{Dx_\alpha}{Dt} = -\nabla\cdot\bar{J}_\alpha + \left(R_\alpha - x_\alpha\sum_\beta^N R_\beta\right)$$

$$\rho\frac{D\bar{v}}{Dt} = -\nabla p + \mu\nabla^2\bar{v} + \rho\bar{g} \text{ Momentum}$$

$$\rho\hat{c}_p\frac{DT}{Dt} = -\nabla\cdot\bar{q} - \left(\frac{\partial \ln \rho}{\partial \ln T}\right)_p\frac{Dp}{Dt} - r : \nabla\bar{v} \text{ Energy}$$

All the latter equations needs to be solved simultaneously in order to obtain the *T*, *p*, ω distributions.

Considering the continuity equation for a species α in molar term

$$c\frac{D_{x_\alpha}}{Dt} = -\nabla\cdot\bar{J}_\alpha + \left(R_\alpha - x_\alpha\sum_\beta^N R_\beta\right)$$

For a steady state process, with fixed membrane and no reactions inside the membrane

$$\nabla\cdot\bar{J}_\alpha = 0$$

Also, considering the mass transfer to happen only perpendicularly to the membrane

$$\frac{dJ_\alpha}{dz} = 0$$

Considering the flux equations for porous and homogeneous membranes and substituting them into the continuity equation

$$\frac{d}{dz}\left(cD_{\alpha,\beta}\frac{dx_\alpha}{dz}\right) = 0 \text{ Solution-Diffusion model}$$

$$\frac{d}{dz}\left(cD_{\alpha,\beta}\frac{dp}{dz}\right) = 0 \text{ Pore - flow model}$$

In the framework of the solution-diffusion model the aim is to link the flux equation and the continuity equation to the conditions in the mixture bathing the membrane. It is therefore necessary to consider the conditions at the membrane interfaces: local equilibrium assumption between the mixture and the membrane surface on both sides of the membrane (permeate or retentate) has been made.

$$\frac{d}{dz}\left(cD_{\alpha,\beta}\frac{dx_\alpha}{dz}\right) = 0$$

$$z = 0;\ x_\alpha = x^I_{\alpha,R}$$

$$Z = \delta;\ x_\alpha = x^I_{\alpha,P}$$

General solution $x(z) = A_2 z + A_1$, where A_1, A_2 can be retrieved from the boundary conditions obtaining

$$J_\alpha = cD_{\alpha,\beta}\frac{x^I_{\alpha,R} - x^I_{\alpha,P}}{\delta}$$

where,

- $x^I_{\alpha,R} c^I_{\alpha,R,} x^I_{\alpha,P,} c^I_{\alpha,P}$: molar fractions and compositions for component α within the membrane but either at the feed or permeate interface;
- δ: membrane thickness.

Determination of the Species Compositions at the Membrane Interfaces

$x^I_{\alpha,R}$ and $x^I_{\alpha,P,}$ have to be expressed as a function of the conditions outside the membrane. The adopted procedure follows:

- Local equilibrium at the membrane interface;
- Obtaining $x^I_{\alpha,R}$ and $x^I_{\alpha,P,}$ from isofugacity condition;
- Substituting $x^I_{\alpha,R}$ and $c^I_{\alpha,P}$ into the flux equation.

The general form for liquid and gas fugacities in a liquid or gas mixture are:

1. Liquid Fugacity

$$f_\alpha^L(T,p,\bar{x}) = f_\alpha^L(T,p)x_\alpha\gamma_\alpha(T,p,\bar{x})$$

$$= \left[f_\alpha^V(T,p^*(T))\exp\int_{p^*}^{p}\frac{\bar{\upsilon}_{\alpha,l}}{RT}dp\right]x_\alpha\gamma_\alpha$$

$$= p^*(T)\left[\exp\int_0^{p^*}\frac{(z-1)}{p}dp\right]\left[\exp\int_{p^*}^{p}\frac{\bar{\upsilon}_{\alpha,l}}{RT}dp\right]x_\alpha\gamma_\alpha$$

$$= p_\alpha^*(T)\phi_\alpha^V(T,p_\alpha^*)\mathfrak{F}_\alpha x_\alpha\gamma_\alpha$$

where,

- f_α^V is the pure vapor fugacity of component α;
- $\phi_\alpha^V = \exp\int_0^{p^*}\frac{(z-1)}{p}dp$ is the pure vapor fugacity coefficient of component α;
- $\mathfrak{F}_\alpha = \exp\int_{p^*}^{p}\frac{\bar{\upsilon}_{\alpha,l}}{RT}dp$ is the Poyting factor and $\bar{\upsilon}_{\alpha,l}$ the pure liquid molar volume

2. Gas Fugacity

$$f_\alpha^V(T,p,\bar{y}) = py_\alpha\phi_\alpha(T,p,\bar{x})$$

where,

- y_α is the vapor molar composition of species α;
- ϕ_α is the mixture vapor fugacity coefficient of component α.

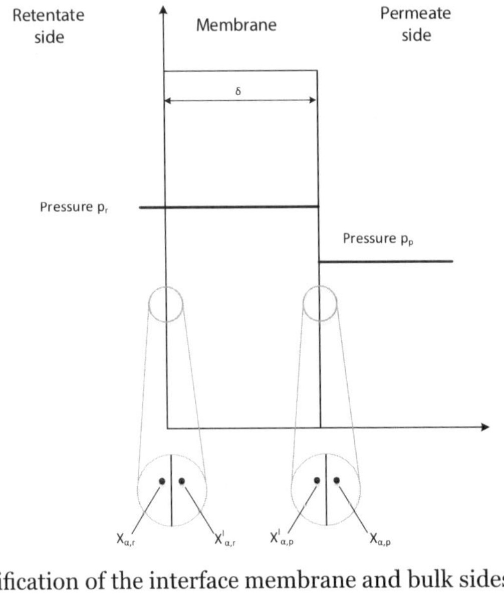

Figure: Identification of the interface membrane and bulk sides compositions

Concentration Polarization

Concentration polarization occurs in the portion of the polarization of a cell produced by concentration changes resulting from passage of current through the electrolyte.

In electrochemistry, concentration polarization is the polarization of an electrolytic cell resulting from changes in the electrolyte concentration due to the passage of current through the electrode/solution interface. Concentration polarization commonly occurs during the cathodic processes, depending on the reduction of dissolved oxygen, since it is usually in low concentration.

Concentration polarization also refers to the concentration gradient of salts on the high-pressure side of the reverse-osmosis membrane surface created by the less-than-immediate redilution of salts left behind as water permeates through a membrane.

In concentration polarization the electrode is polarized due to a concentration effect that occurs at the electrode surface. The cause of the changes in concentration (emergence of concentration gradients in the solution adjacent to the electrode surface) is the difference in the rate of electrochemical reaction at the electrode and the rate of ion migration in the solution to or from the surface.

For cathodic processes, the most important concentration effect is the diffusion of dissolved oxygen to the metal surface and the subsequent reduction of oxygen by the half-cell reaction:

$$O_2(g) + 2H_2O(l) + 4e^- = 4OH^-(aq)$$

For anodic processes, concentration polarization occurs due to the slow diffusion away from the metal surface of the metal cations, which are produced by anodic dissolution. For example, the anodic reaction:

$$Fe(s) = Fe^2 + (aq) + 2e^-$$

will produce a concentration effect if Fe^{2+} ions are slow to diffuse away from the metal surface.

Concentration polarization effects can also occur in the corrosion of metals in acid solutions. When a chemical species participating in a corrosion process is in short supply, the mass transport of that species to the corroding surface can become rate controlling in order to maintain the balance of consumption and delivery of that species.

Concentration polarization strongly affects the performance of the separation process. Lower rate of separation under the same external driving force means increased power consumption.

Concentration polarization is inherent to all types of membrane separation processes. It can be used in:

- Gas separations
- Pervaporation
- Membrane distillation

- Reverse osmosis
- Nanofiltration
- Ultrafiltration
- Microfiltration separations

This technique results in better mixing of the solution and in reducing the thickness of the diffusion boundary layer.

Concentration polarization usually predominates when the concentration of the active species is low.

Anion Exchange Membrane

An Anion Exchange Membrane (AEM) is a semipermeable membrane generally made from ionomers and designed to conduct anions while being impermeable to gases such as oxygen or hydrogen. This is their essential function when used in a Membrane Electrode Assembly (MEA) of a Direct Methanol Fuel Cell (DMFC) or Direct-Ethanol Fuel Cell (DEFC): separation of reactants and transport of anions.

Typical Properties of Anion Exchange Membranes Include

- Alkali Resistance
- Resistant to Oxidation
- Chlorine Resistance
- Diffusion Dialysis for Acid Recovery

Alkaline Anion Exchange Membrane Fuel Cell

Alkaline anion exchange membrane fuel cells (AAEMFCs) have received ever-increasing attentions due to the enhanced electrochemical kinetics and the absence of precious metal electrocatalysts, and thus great progress has been made in recent years. The alkaline anion exchange membrane based direct alcohol fuel cells, one type of alkaline anion exchange membrane fuel cells utilizing liquid alcohols as fuel that can be obtained from renewable biomass feedstocks, is another attractive point due to its ability to provide electricity with cogeneration of valuable chemicals. Significant development has been made to improve the selectivity towards high added-value chemicals and power output in the past few years.

There are several advantages of the AEMFC technology over the currently commercialized proton exchange membrane fuel cells (PEMFCs). Technically, AEMFCs are similar to PEMFCs, with the main difference being that the solid membrane is an alkaline AEM instead of an acidic PEM. With an AEM in an AEMFC, the OH^- anion is transported from the cathode to the anode, opposite to the H^+ conduction direction in a PEMFC. Although in principle both technologies are similar, the use of an AEM creates an alkaline pH cell environment, and therefore, the AEMFC offers several potential advantages over the mature PEMFC technology. Among them, the following are particularly noteworthy:

a. Enhanced oxygen reduction catalysis, allowing for the use of less expensive, Pt-free or precious group metal (PGM)-free catalysts;

b. Extended range of cell and stack materials stable in the fuel cell environment;

c. A wider choice of fuels in addition to pure hydrogen; and

d. A wider range of less expensive polymer chemistry (fluorinated raw materials are not necessary).

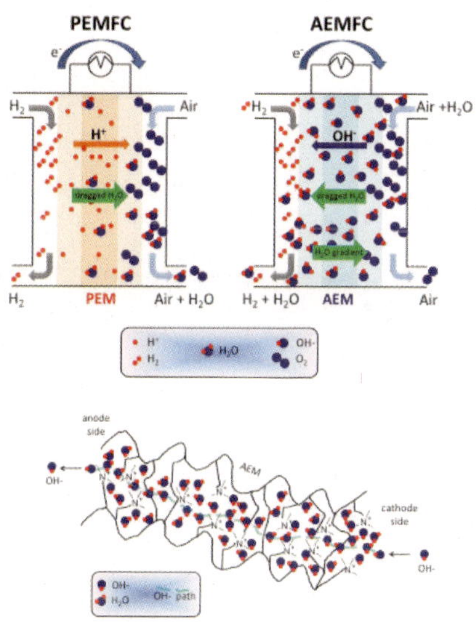

Figure: Schematic of an AEMFC as compared to a PEMFC (top scheme), and of an AEM based on a quaternary ammonium pendant functional group (bottom scheme).

As mentioned before, the use of an AEM in the fuel cell creates an alkaline environment inside the cell. Therefore, the AEM is the key differentiating element in the AEMFC device. AEMs are solid polymer electrolytes that conduct anions, such as OH^- and carbonate anions, as they contain positively charged cationic groups bounded to (or part of) a polymer backbone. Recent years have seen extensive research on the preparation and properties of AEMs. Numerous studies on new chemistries have been reported. The different cationic functional groups and polymer backbones used in typical AEMs for fuel cells are well summarized in the literature. Initially, most studies on AEMFCs focused on the development of suitable AEMs, and only a small subset of them reported data on cell performance and performance stability.

Before reviewing the state of the art in cell performance and performance stability, it should be remarked that in an AEMFC, ions and water are not transported in the same directions as in PEMFCs. During the operation of an AEMFC, when ambient air is used, several anions are present in the cell: HCO_3-, $CO_3=$, and OH^-. However, in an AEMFC operated at high current densities, the most common anion species present across the membrane is the OH^-, generated according to the electrochemical oxygen reduction reaction (ORR) at the cathode:

$$\frac{1}{2} O_2 + H_2O + 2\,e^- \rightarrow 2\,OH^-$$

The OH− is transported to the anode where, if hydrogen is used as fuel, the following oxidation reaction takes place:

$$2\,OH^- + H_2 \rightarrow 2H_2O + 2\,e^-$$

In the overall reaction, both PEMFC and AEMFC produce water as a byproduct. However, in contrast to PEMFC technology, in an AEMFC water is generated at the anode (twice as much as in a PEMFC, per electron), while at same time water is a reactant at the cathode. This distinctive water transport scenario, together with the high alkaline medium in AEMFCs, represent a unique feature of AEMFCs. Figure shows a schematic drawing of these transport characteristics in an AEMFC.

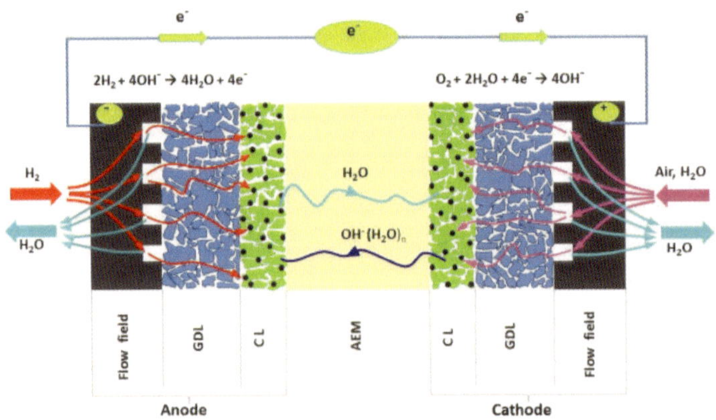

Fig. Schematic diagram of an H2-AEMFC and the main species transported through the cell.

Membrane Gas Separation

Separation processes constitute an integral part of chemical industries for isolation of products and recovery of reactants. Membrane methods of separation are modern techniques which could reduce capital investment, operating cost and process safety. Separation of binary gas mixtures is an important unit operation in a large number of industries to recover valuable gases and enable pollution control. Such separations are currently undertaken by conventional technology viz., cryogenic distillation, adsorption on solid surface and solvent absorption. Now membranes are increasingly finding application for gaseous separations and have exhibited tremendous potential. Membrane technology could compete as well as compliment conventional separation techniques due to its multidimensional character, which involves rapid mass transfer rate and high selectivity towards a specific gas. Inherent advantages of membrane gas separation process are as follows:

- Simplicity of operation and installation.
- Feasibility under mild conditions, for eg. air separation can be carried out at lower pressure and ambient temperature rather than extreme conditions in cryogenic distillation.
- Smaller footprint and flexibility of operation due to compactness of modules with huge reduction in consumption of electricity and fuel.
- No extra agents are required as compared to corrosive and polluting amine solvents in absorption process.

- Continuous mode of operation with partial or complete recycle of retentate/permeate.
- Possibility of integration with other separation units to constitute effective hybrid processes for achieving improved economy and desired purity levels.
- Membranes can be "tailored" to adapt to a specific separation task.

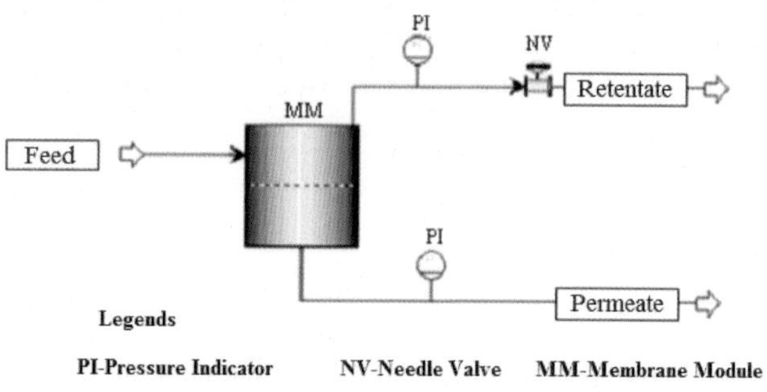

Figure. Schematic of Membrane Gas Separation Process

A simple sketch of gas separation by membrane is provided in Fig. The feed gets split into permeate stream and an unpermeate stream also known as retentate. Depending upon the application, either the permeate or retentate could be the final product. The membrane preferably allows one component of the feed stream to pass through it due to preferential interaction and sorption combined with higher diffusivity of the same component owing to differences in molecular size. The kinetic diameters of a few of the common gases along with their other properties are provided in Table. The lower the kinetic diameter, the higher the diffusivity.

The development of a potential polymer is the key area of research in membrane technology. The aim in the development of new membranes is to increase the permeability and selectivity or increasing the permeability without compromising the selectivity or enhancing the selectivity at constant permeability. Several researchers have been working to synthesize and use different varieties of membranes for separation of $O_2/N_2$from air, CO_2/CH_4 from landfill, biogas or natural gas, CO_2/N_2 in power plants, H_2 recovery in ammonia manufacture or water gas shift reaction and even olefin–paraffin separation in refinery. Membranes can be prepared from different polymers such as polycarbonates, polyimides, polyamides, polyetherimides, or polysulfones by various techniques.

Table. Properties of Common Gases

Gas	State	Molecular weight	Kinetic diameter Å	Specific gravity at 70 °F (1 atm)	Critical temp (°F)	Critical pressure (psig)	Specific volume (cf/lb)
H_2	Compressed Gas	2.02	2.89	0.069	-399.96	190.8	192
N_2	Compressed Gas	28.01	3.64	0.967	-232.4	492.9	13.8
H_e	Compressed Gas	4.003	2.6	0.138	-450.3	33.2	96.7
O_2	Compressed Gas	32.0	3.46	1.105	-181.1	736.9	12.1
CO_2	Liquefied Gas	44.01	3.3	1.52	87.8	1071	8.74
C_3H_6	Liquefied Gas	42.08	4.4	1.501	197.5	666.3	9.05
C_3H_8	Liquefied Gas	44.1	3.96	1.55	206.2	617.4	8.5
CH_4	Compressed Gas	16.04	3.8	0.555	-115.8	673	23.7

CO	Compressed Gas	28.01	3.76	0.97	-220.4	507.4	13.8
Ar	Compressed Gas	39.95	3.4	1.38	-188.1	710	9.74

Theory of Gas Permeation

The driving force for gas separation is partial pressure gradient which is the product of total pressure and mole fraction. Most membranes used for gas separation are non-porous.

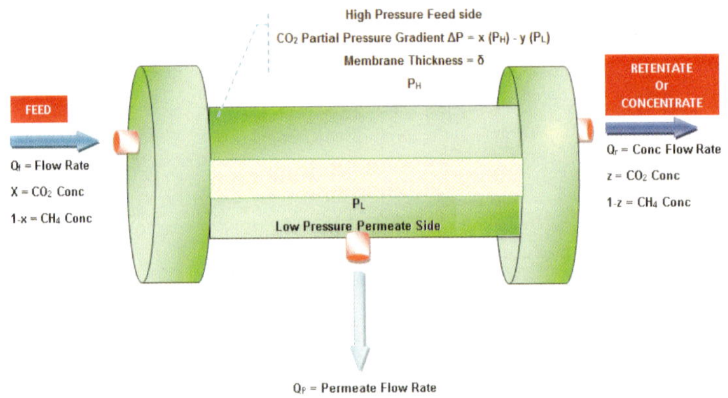

Figure: Principle of Gas Permeation and Separation

Separation takes place according to the morphology of the membrane which is based on different transport mechanisms. Porous polymer membranes have a pore size ranging from 0.5 to 10 μm and non-porous membranes with a polymer chain gap of 10^{-3} to 10^{-4} μm induced by thermal vibration. Depending on the porosity of the membrane matrix, various transport mechanisms including Poiseuille (viscous) flow, Knudsen's diffusion, molecular sieving, capillary condensation and solution–diffusion mechanism for gas transport through membranes are proposed as illustrated in Fig. Porous membranes are used to separate the uranium hexafluoro (UF6) isotope gas mixtures to isolate the radioactive U235 isotope.

Mass Transfer Through Porous Membranes

Knudsen's diffusion occurs in a porous membrane, whose pore sizes are smaller than the mean free path of the gas molecules. Figure reveals the transport of gases through porous membranes by Knudsen diffusion mechanism. Gas molecules interact with the pore walls much more frequently than colliding with one another which allows lighter molecules to preferentially diffuse through pores. Knudsen's diffusion principally takes place in membranes with a pore diameter in the range of 50-100Å [ref hand book].Molecular sieving is primarily based on the precise size exclusion and shape discrimination amongst various gas molecules by the micropores of < 7Å in diameter as per figure.

Mechanisms of Permeation in Nonporous Membranes

Transport of gases through nonporous polymeric membranes is usually described by solution–diffusion mechanism (Fig.) which consists of three steps, (a) sorption of preferential permeate (solubility) at the upstream surface of the membrane, (b) activated diffusion through the membrane and (c) desorption at the downstream side. Solution-diffusion mechanism is driven by a difference in thermodynamic activities existing across the membrane and interacting forces working between membrane material and permeating molecules. The activity difference causes a concentration gradient that leads to diffusion in the direction of decreasing activity as depicted in Fig.). Carbon molecular sieve (CMS) membranes and zeolites are the typical membranes dominated by molecular sieving mechanism and give high separation performance. Ratio of the gas molecular size to micropore diameter controls the permeation rate. Permeability (P) is the capability of a membrane to permeate gas molecules:

$$P = D \times S$$

The ability of a membrane to separate two molecules, A to B, is the ratio of their permeability, called as the membrane selectivity α.

$$\alpha = \frac{P_A}{P_B}$$

For a binary gas mixture the selectivity can also be determined from a molar concentration of the two gases in feed and permeate:

$$\alpha = \frac{y(1-x)}{x(1-y)}$$

where y is the permeate concentration of the fast permeating gas and x is its feed concentration.

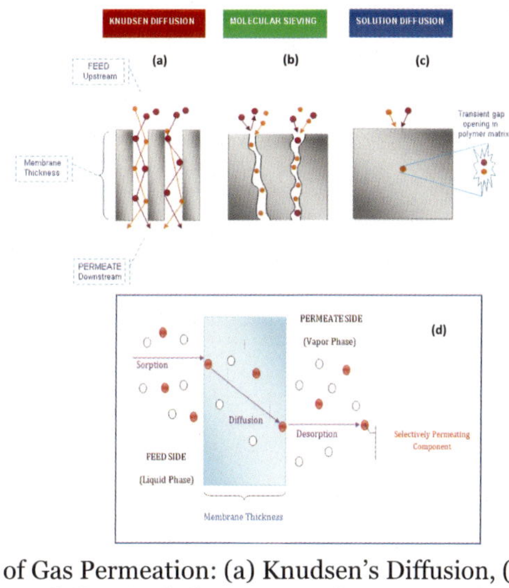

Figure. Mechanisms of Gas Permeation: (a) Knudsen's Diffusion, (b) Molecular Sieving, (c) Solution-Diffusion and (d) Detailed Overview of Mass Transfer by Solution-Diffusion

Gas permselectivity of a membrane can be enhanced by facilitated transport which involves reversible complexation and transport of permeant molecules across a membrane through utilization of a carrier species. The carrier reacts with a permeant molecule to form a labile complex. Within the membrane, the carrier shuttles the permeant across the membrane boundaries to downstream side where the carrier-gas molecule complex breaks allowing the gas to flow into the permeate stream. When a feed mixture contains only one species with which the carrier will react, the transport of that particular species will be "facilitated" or carrier-mediated across the membrane as illustrated in Fig. The driving force in facilitated transport is a concentration gradient of permeant-carrier complexes across the membrane. It has been found that olefin transport can be facilitated by transition metal ions such as Cuprous and Silver ions which possess vacant d-orbitals due to formation of a $\pi-$ bond complex.

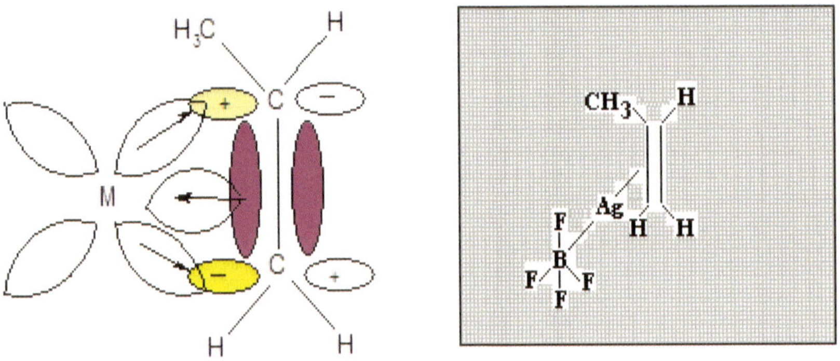

Fig. Mechanism of Facilitated Transport

Gas separation using facilitated transport is most commonly done with "immobilized liquid" membranes, prepared by dissolving the carrier in an appropriate solvent to impregnate an electrically neutral but rigid microporous polymer matrix. Once formed, surface tension forces serve to hold the carrier molecules inside the membrane.

Pervaporation

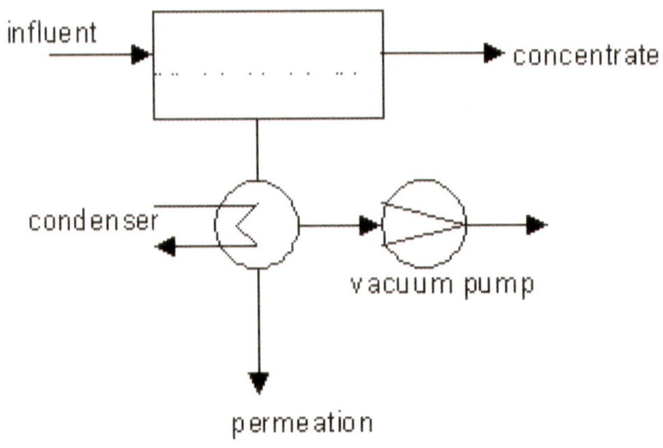

Method and Installation Description

Pervaporation is a membrane process comparable to distillation, and combines permeation and vaporisation. Pervaporation is used to separate liquid mixes. The used membrane is a dense non-porous membrane or a very finely-porous ceramic membrane that displays an affinity towards the component one wants to remove. Specifically for this process, the permeating component is converted into an evaporation phase, due to the low (partial) vapour pressure on the permeate-side. This low vapour pressure is normally achieved by placing a (slight) vacuum on the permeate-side of the membrane. In most cases, the permeate is re-condensed. The pervaporation process contains 3 steps:

- Selective sorption in the membrane on the influent-side;
- Selective diffusion through the membrane;
- Desorption in the gas phase on the permeate-side.

There is also an alternative process, referred to as vapour permeation. Vapour permeation involves the supply being introduced in vapour form before it comes into contact with the membrane. This alternative is particularly interesting if the supply comes from an earlier process step in vapour form.

Pervaporation is not based on a liquid-vapour balance like distillation, but is based on differences in sorption and diffusion in the various components in the supply. This it is an interesting alternative for separating azeotropic mixes or mixes with similar boiling points. Further, it offers an energy advantage because only the permeating component must be converted in the gas phase.

Good hydrophilic pervaporation membranes are available on the market. The use of pervaporation to drain solvents is clearly on the increase. Recently, the technique was also used to drain bio-ethanol. Good hydrophobic commercial membranes are also available on the market for removing volatile organic components from wastewater. They are also suitable for removing volatile components like hydrocarbons from air. Membranes for separating 2 organic components are still in the development phase.

The main advantage of pervaporation is the major potential to save energy. Savings of above 50% are easy to realise. However, the membranes and membrane installations, (incl. need for vacuum) are currently relatively expensive. Thanks to considerable energy saving, the investment is still economically viable for complex distillations.

Further, pervaporation avoids contamination of the required flow, in contrast to, for example, azeotropic distillation where entrainers or extra components are used to eradicate the azeotrope.

If the supply contains suspended matter or dissolved salts, then membrane pollution may be encountered. In this case, an effective pre-treatment or vapour permeation can be implemented.

Application

Pervaporation is a practical and financially viable alternative to distillation when draining solvents, particularly where azeotropes must be degraded or liquids with similar boiling points must be separated. In such cases, a hybrid process (distillation combined with pervaporation) is normally the most economically viable.

Hydrophilic pervaporation or vapour permeation can also be used to remove water from a reaction mix and thus favourably alter the reaction-balance. This is, for example, used for esterification reactions.

To remove volatile organic substances (incl. chlorinated hydrocarbons and BTEX) from wastewater, pervaporation is competitive with active carbon adsorption and stripping.

Throughout the world, there are various large-scale implementations, with the majority addressing drainage of alcohol (ethanol and isopropanol). The capacity of most installations lies between 50 and 300 l/day. Thus, these are rather small installations.

Here are a few examples of sectors where pervaporation can be implemented: the chemicals industry, pharmaceutical industry, petrochemical sector, refineries, tank and drum cleaning and paint, gloss, varnish or printing-ink production.

Boundary Conditions

In order to realise sufficient flux during pervaporation, work is normally carried out at temperatures between 50 and 100°C. Typical fluxes averaged for a pervaporation process are around 1 kg/lhm2.

Pervaporation should be implemented to remove the minority component from a mix. In this case, the required membrane surface and required energy will be at their lowest.

Most pervaporation membranes work optimally when the permeating component in the supply is below a particular maximum level.

Components with high boiling points make pervaporation more difficult. They restrict selectivity and can block the membrane.

Effectiveness

Pervaporation is primarily used for the removal of volatile organic compounds (incl. VOX, BTEX, MAK, methyl tert-butyl ether, phenols, styrene, PAHs and apolar KWS) from wastewater. For this purpose, the technique is competitive with active carbon adsorption ad stripping.

Pervaporation easily removes components from water to below 1%. Higher concentrate purification levels are possible but require a relatively large membrane surface, because the encountered fluxes are relatively low (low driving force).

Support Products

If the correct pre-treatment is implemented (e.g. micro-filtration) in the presence of suspended matter, and if a vapour permeation process is used for the presence of salts, then no problems are expected during membrane cleaning - thus no chemicals are needed.

Environmental Issues

Pervaporation is recognised as an energy-efficient alternative to difficult distillations. The concentrate is normally the product (incl. for drainage). Due to the high selectivity of the membranes, the permeate can often be discharged or re-used.

Membrane Fouling

Contamination of membranes causes a higher energy use, a higher cleaning frequency and a shorter life span of the membrane. Membrane contamination is usually called fouling.

The International Union of Pure and Applied Chemistry defines fouling as follows:

The process that results in a decrease in performance of a membrane, caused by the deposition of suspended or dissolved solids on the external membrane surface, on the membrane pores, or within the membrane pores. (Koros 1996).

When clean water is filtered, the membrane material is the only resistance caused (R_m). The flux is than called the clean water-flux. As a result of the accumulation of particles on the membrane through the filtration of water with a certain level of suspended solids, a cake will form on the membrane (R_c; particles). When particles block the membrane pores, this is called pore plugging (R_{pb}; scaling). Resistance as a consequence of adsorption in or on the membrane is called biofouling (R_a).

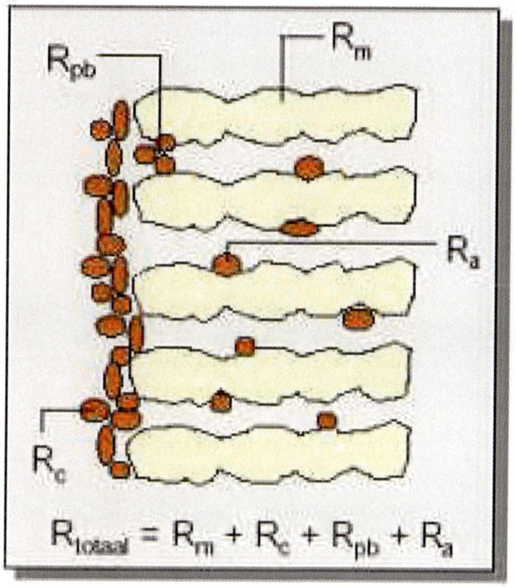

Resistances:

R_m = membrane resistance

R_a = adsorption, biofouling

R_{pb} = pore plugging

R_c = cake layer

Particles, biofouling and scaling are three main groups of pollutants that can be distinguished from membrane fouling. These will cause the need of a higher workload, to keep the filtration capacity at a certain level. At a certain point the pressure has increased so much that it is no longer economical.

For the control of membrane fouling there are many different techniques.

One way to predict fouling is by using the Silt Density Index (SDI) of the feed water. The SDI, which is based on experience, can be defined as the time that is needed to filtrate an amount of water with a noted concentration of salts through a standard 0.45 mm micro filtration membrane. When the SDI is high, one can conclude that the feed water contains a high amount of membrane plugging matter.

Types of Membrane Fouling

There are four categories of membrane fouling. They are (a) inorganic fouling/scaling, (b) particle/colloids fouling, (c) icrobial fouling, and (d) organic fouling. A brief description on the nature of fouling, relevant water quality as ndicators, and control measures are summarized below for each type of membrane fouling.

Inorganic Fouling

Inorganic fouling or scaling is caused by the accumulation of inorganic precipitates such as metal hydroxides, and "cales" on membrane surface or within pore structure. Precipitates are formed when the concentration of chemical species exceeding their saturation concentrations. Scaling is a major concern for reverse osmosis (RO) and nanofiltration (NF). RO and NF membranes reject inorganic species. Those species form a concentrated layer in the vicinity of membrane-liquid interface - a phenomenon referred to "concentration polarization". For microfiltration (MF) and ultrafiltration (UF), inorganic fouling due to concentration polarization is much less profound, but can exist most likely due to interactions between ions and other fouling materials (i.e., organic polymers) via chemical bonding.

Particulate/Colloid Fouling

Algae, bacteria, and certain natural organic matters fall into the size range of particle and colloids. However, they are ifferent from inert particles and colloids such as silts and clays. To distinguish the different fouling phenomena, particles and colloids here are referred to biologically inert particles and colloids that are inorganic in nature and are riginated from weathering of rocks.

Microbial/Biological Fouling

Microbial fouling is a result of formation of biofilms on membrane surfaces. Once bacteria attach to the membrane, they start to multiple and produce extracellular polymetric substances (EPS) to form a viscous, slimy, hydrated gel. EPS typically consists of heteropolysaccharides and have high negative charge density. This gel structure protects bacterial cells from hydraulic shearing and from chemical attacks of biocides such as chlorine.

Organic Fouling

Organic fouling is profound in membrane filtration with source water containing relatively high natural organic matters (NOM). Surface water (lake, river) typically contains higher NOM than ground water, with exceptions. For source water high in NOM, organic fouling is believed to be the most significant factor contributed to flux decline (Mallevialle et al., 1989; Lahoussine-Turcaud et al, 1990). Microfilters usually remove insignificant amount of organic matter, as measured by dissolve organic carbon (DOC). DOC as an indicator for organic fouling is probably neither proper nor adequate. Efforts to identify the effects of subgroups of NOM on membrane fouling have yet been able to draw definitive conclusions.

Factors Contributing to Membrane Fouling

1. Feed stream properties. This includes characteristics such as concentration, pH and ionic strength, dissolved solid content, suspended solid content, etc.

2. Membrane material and its physical-chemical properties. Membrane pore size, porosity and pore distribution, and membrane surface charge and hydrophobicity in certain pH and ionic strength conditions can contribute to fouling.

3. Processing variables. Transmembrane pressure, temperature, and cross-flow velocity have a big impact on membrane fouling.

Approaches to Minimize Membrane Fouling

- Optimize pH and ionic strength of the feed solution to minimize the adsorption or deposition of the feed materials.
- Select an appropriate pre-filtration procedure or other means to remove large molecules, since the presence of larger molecules or particles could cause a steric hindrance to the passage of smaller molecules through the membrane.
- Select a membrane with an optimum pore size to result in good separation performance as well as optimized permeate flux.
- Optimize the operating conditions. This includes increasing transmembrane pressure to maximize flux without introducing more fouling potential.
- Increase the cross-flow velocity, which generally results in an improvement in permeate flux.

Membrane Distillation

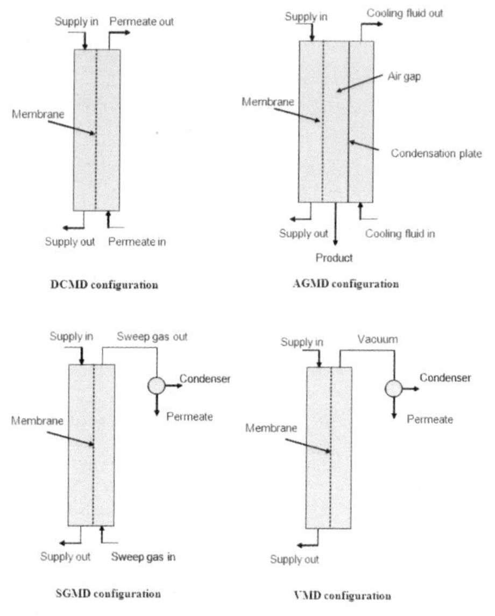

Method and Installation Description

Membrane distillation (MD) is a separation process where a micro-porous hydrophobic membrane separates two aqueous solutions at different temperatures. The hydrophobicity of the membrane prevents mass transfer of the liquid, whereby a gas-liquid interface is created. The temperature gradient on the membrane results in a vapour pressure difference, whereby volatile components in the supply mix evaporate through the pores (10 nm – 1 μm) and, via diffusion and/or convection of the compartment with high vapour pressure, are transported to the compartment with low vapour pressure where they are condensated in the cold liquid/vapour phase. For supply solutions that only contain non-volatile substances, such as salts, water vapour will be transported through the membrane whereby demineralised water is obtained on the distillation-side and a further concentrated salt flow on the supply-side.

The manner in which the vapour pressure difference is generated across the membrane is determined by the specific module configuration. In the most commonly used configuration, direct contact membrane distillation (DCMD), the permeate-side consists of a condensation liquid (often clean water) that is in direct contact with the membrane. Alternatively, the evaporated solvent can be collected on a condensation surface that can be separated from the membrane via an air gap (AGMD) or a vacuum (VMD), or can be discharged via a cold, inert sweep gas (SGMD). In the latter two cases, condensation of vapour molecules takes place outside the membrane module. Theoretically, the type of driving force, together with the water-repelling nature of the membrane, permits full retention of non-volatile components, such as ions, macro-molecules and colloidal particles.

Specific Advantages and Disadvantages

Advantages

The potential advantages of MD, in comparison with conventional separation processes, are found primarily in the lower working temperature and pressure, and thus the lower energy costs and less stringent mechanical properties. In contrast to distillation and RO, supply solutions can be separated at a temperature well below the boiling point and under atmospheric pressure. Typical supply temperatures around 30-60°C permit re-use of residual heat flows, and the use of alternative energy sources such as sun, wind and geothermics.

In addition, in comparison with RO, MD is less susceptible to flux limitations caused by concentration polarisation, whereby a higher concentration of matter is obtained on the supply-side. Theoretically, MD offers 100% retention for non-volatile dissolved substances, whereby there is no limit on the supply concentration.

In comparison with traditional distillation, MD possesses typical basic advantages of membrane separation, namely simple up-scaling, simple operations, possibility for high membrane surface/volume ratios, possibility to treat flows with heat-sensitive components and/or a high suspended particle-content at atmospheric pressure and a temperature below the boiling point of the supply.

Disadvantages

The main factors that determine the feasibility of MD as an industrial separation technique, and thus still form barriers for commercial implementation, are the relatively low permeate flux in comparison with pressure-based membrane processes, flux reductions caused by concentration polarisation, membrane pollution, water loss due to conduction through the polymeric membrane and/or pore wetting, the high cost of MD modules and the high thermal energy consumption, for which the economic costs are determined by the MD configuration and the specific application.

Application

MD can be used for the production of distilled water and for concentrating aqueous solutions. De-salting sea water and the production of clean water from brackish water are the most common MD applications because non-volatile ionic particles are almost completely retained.

Further, MD offers potential for water purification in the pharmaceutical, chemical and textile industries, for the concentration of fruit juices and milk processing, in bio-medical applications like the removal of water from blood, in separating azeotropic aqueous mixes (alcohol-water), in the concentration of cooling liquids (glycols), non-volatile acids and oil-in-water emulsions and applications where high temperature processing causes thermal degradation of the process flow. With the discharge of less harmful waste flows in mind, MD offers greater potential in the textile industry where wastewater is contaminated with dyes.

Effectiveness

MD can be used for the separation/concentration of non-volatile components, like ions, acids, colloids and macro-molecules from aqueous flows, for the removal of traces of volatile organic compounds VOCs), like benzene, chloroform and tri-chloroethylene from water, and for the extraction of other organic components like alcohols from diluted aqueous solutions.

Ultrafiltration

Ultrafiltration (UF) is a membrane filtration process similar to Reverse Osmosis, using hydrostatic pressure to force water through a semi-permeable membrane. The pore size of the ultrafiltration membrane is usually 103 - 106 Daltons. Ultrafiltration (UF) is a pressure-driven barrier to suspended solids, bacteria, viruses, endotoxins and other pathogens to produce water with very high purity and low silt density.

Ultrafiltration (UF) is a variety of membrane filtration in which hydrostatic pressure forces a liquid against a semi permeable membrane. Suspended solids and solutes of high molecular weight are retained, while water and low molecular weight solutes pass through the membrane. Ultrafiltration is not fundamentally different from reverse osmosis, microfiltration or nanofiltration, except in terms of the size of the molecules it retains.

A membrane or, more properly, a semi permeable membrane, is a thin layer of material capable of separating substances when a driving force is applied across the membrane. Once considered a viable technology only for desalination, membrane processes are increasingly employed for removal of bacteria and other microorganisms, particulate material, and natural organic material, which can impart color, tastes, and odors to the water and react with disinfectants to form disinfection byproducts (DBP).

As advancements are made in membrane production and module design, capital and operating costs continue to decline. The pressure-driven membrane processes discussed in this fact sheet are microfiltration (MF), ultrafiltration (UF), nanofiltration (NF), and reverse osmosis (RO).

Ultrafiltration (UF) is used to remove essentially all colloidal particles (0.01 to 1.0 microns) from water and some of the largest dissolved contaminants. The pore size in a UF membrane is mainly responsible for determining the type and size of contaminants removed. In general, membrane pores range in size from 0.005 to 0.1 micron. UF membrane manufacturers classify each UF product as having a specific molecular weight cutoff (MWC), which is a rough measurement of the size of contaminants removed by a given UF membrane. A 100,000 MWC UF membrane means that when water containing a given standard compound with a molecular weight of around 100,000 daltons is fed to the UF unit, nearly all of the compound will not pass through the membrane.

Substances with a molecular weight of 100,000 daltons have a size of about 0.05 microns to about 0.08 microns in diameter. UF membranes are used where essentially all colloidal particles (including most pathogenic organisms) must be removed, but most of the dissolved solids may pass through the membrane without causing problems downstream or in the finished water. UF will remove most turbidity from water.

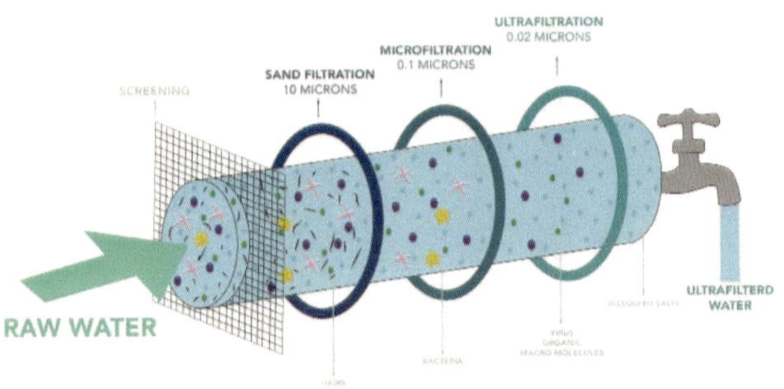

Ultra-filter membranes are offered in various configurations. Possible membrane configurations include:

- Pipe-shaped membranes: capillary, hollow fibre or tubular;
- Plate-shaped membranes: flat plate or spiral.

In addition to the specific membrane configurations, one can also identify a few set-ups. The 2 most commonly used methods are dead-end and cross-flow set-ups. The names refer to the way in which the supply is sent to the membrane. In dead-end UF, the supply is sent directly to the membrane. The pollution layer will thus form on the supply side of the membrane surface. This layer contains all particles that have been separated on the basis of their size (sieve effect). This layer is periodically rinsed away by briefly re-sending the produced liquid (permeate) through the membrane in an opposite direction to the production flow. This helps to loosen the hardened layer, and makes it ready for disposal. This is referred to as a semi dead-end set-up.

This re-rinsing may not be enough to remove the layer from the surface if the hardened layer is too strongly compressed or if the bond with the membrane is very strong. In this case, chemical cleaning must be implemented with, for example, bleach, peroxide, acid and alkali or detergent.

The figure below shows a dead-end set-up, where M = the membrane.

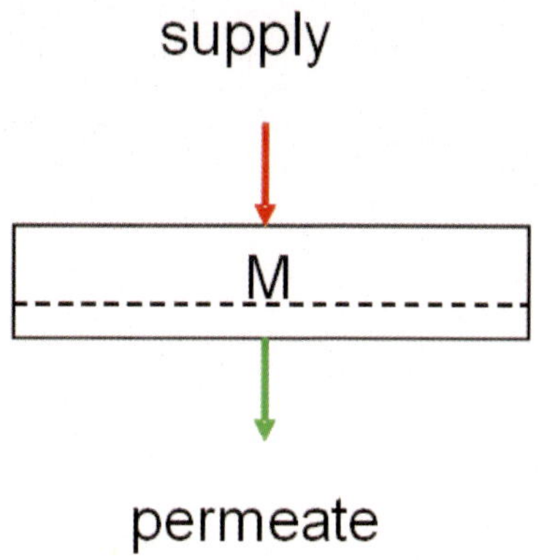

In a cross-flow set-up (see figure below), the liquid is passed along the membrane surface at a particular speed. The permeate is able to pass through the membrane and the larger particles are left behind in a concentrated flow (the retentate). The hardened layer in this set-up is continuously removed by the cross-flowing supply flow.

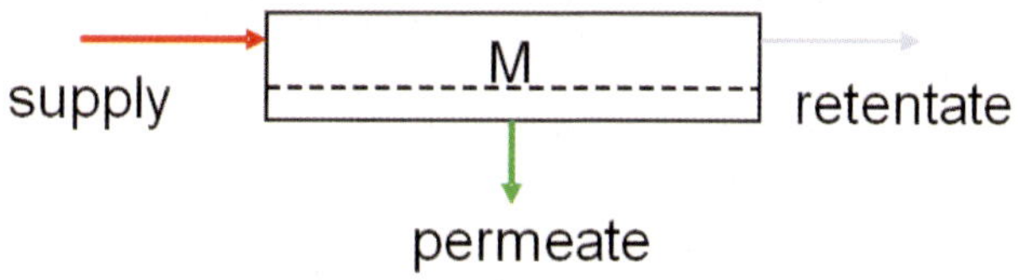

Specific Advantages and Disadvantages

The advantages of UF are:

- Low operating pressure required (higher than MF);
- Lower energy consumption than nano-filtration or reverse osmosis;
- Few manual actions required;
- Relatively cheap;
- Good permeate yield depending on the supply water and membrane choice;
- Disinfection through removal of bacteria. To a certain extent, UF allows viruses, phage, colloids and macro molecules to be removed.

Possible disadvantages of UF are:

- Only removes suspended matter and bacteria;
- Sensitive to oxidative chemicals (e.g. nitric acid, sulphuric acid, peroxide and persulphate in high concentrations); NaOCl exposure determines the life-span of the membrane and is typically 150.000 to 500.000 ppmh and pH dependent;

- Damage may occur when trying to prevent hard and sharp particles > 0.1 mm;
- Membrane damage at pressure > 3 bar.

Application

UF is extremely suitable as pre-treatment in the preparation of drinking water or process water. It has excellent properties for removing suspended matter and bacteria. Further, UF is implemented in the:

- Dairy industry;
- Food industry;
- Metal industry (oil/water emulsion separation);
- Textile industry;
- Pharmaceutical industry;
- Chemicals industry;

UF is also implemented:

- As pre-treatment for nano-filtration of reverse osmosis;
- For applications involving re-use of wastewater flows.

A membrane bioreactor combines biological purification with ultra-filtration or micro-filtration.

Boundary Conditions

Membranes must be protected against hard particles larger than 0.1 mm. They can be removed by regular pre-filtration. Further, supply flows, pH conditions and temperature conditions must be compatible with the membrane material.

Effectiveness

UF can be implemented for removing the following parameters:

- Suspended matter (>99%);
- Harmful micro-organisms (e.g. bacteria, protozoa, algae, fungi) (>99%).
- PAHs.

Further, UF can also be implemented to break down emulsions.

Diafiltration

Diafiltration is a technique that uses ultrafiltration membranes to completely remove, replace, or lower the concentration of salts or solvents from solutions containing proteins, peptides, nucleic acids, and other biomolecules. The process selectively utilizes permeable (porous) membrane filters to separate the components of solutions and suspensions based on their molecular size. An ultrafiltration membrane retains molecules that are larger than the pores of the membrane while smaller molecules such as salts, solvents and water, which are 100% permeable, freely pass through the membrane.

Benefits of Diafiltration

Conventional techniques used for salt removal or buffer exchange such as membrane dialysis and column-based gel filtration can be effective but have limitations. Dialysis procedures can take up to several days, require large volumes of water for equilibration and risk product loss through manual manipulation of the dialysis bags. Gel filtration results in a dilution of the sample and often requires an additional ultrafiltration step to concentrate it back. Adding steps to a process can risk sample loss or possible contamination. With diafiltration, salt or solvent removal as well as buffer exchange can be performed quickly and conveniently. Another big advantage of diafiltration is that the sample is concentrated on the same system, minimizing the risk of sample loss or contamination.

Nanofiltration

The membrane separation process known as nanofiltration is essentially a liquid phase one, because it separates a range of inorganic and organic substances from solution in a liquid – mainly, but by no means entirely, water. This is done by diffusion through a membrane, under pressure differentials that are considerable less than those for reverse osmosis, but still significantly greater than those for ultrafiltration. It was the development of a thin film composite membrane that gave the real impetus to nanofiltration as a recognised process, and its remarkable growth since then is largely because of its unique ability to separate and fractionate ionic and relatively low molecular weight organic species.

The membranes are key to the performance of nanofiltration systems. They are produced in plate and frame form, spiral wound, tubular, capillary and hollow fibre formats, from a range of materials, including cellulose derivatives and synthetic polymers, from inorganic materials, ceramics especially, and from organic/inorganic hybrids.

Recent developments of membranes for NF have greatly extended their capabilities in very high or low pH environments, and in their application to non-aqueous liquids. The plastic media are highly cross-linked, to give long-term stability and a practical lifetime in more aggressive environments. NF membranes tend to have a slightly charged surface, with a negative charge at neutral pH. This surface charge plays an important role in the transportation mechanism and separation properties of the membrane.

As with any other membrane process, nanofiltration is susceptible to fouling, and so nanofiltration systems must be designed to minimise its likelihood – with proper pretreatment, with the right membrane material, with adequate cross-flow velocities to scour the membrane surface clear of accumulated slime, and by use of rotating or vibrating membrane holders.

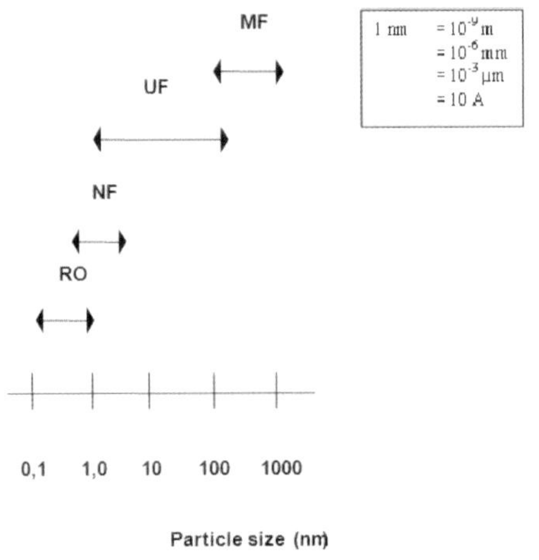

Method and Installation Description

As micro-filtration and ultra-filtration, the sieve effect is one of the separation principles, as are solvent diffusion and electrostatic repulsion. The sieve effect is based on the difference between the particle size and the pore diameter. The pore size of the nano-filtration membrane is characterised by a cut-off value. This cut-off value is consistent with the molecular weight of the smallest molecule that can be 90% restricted by the top layer of the membrane (2 μm thick). The cut-off value is expressed in Dalton (Dalton = weight in grams of mole of the molecule). A typical nano-filtration membrane lies in the range of 150-500 Dalton, depending on the molecular structure.

Nano-filtration membranes possess pores with a size of approximately 1nm. Nano-filtration membranes are characterised based on their retention of loaded and unloaded particles. The retention of a nano-filtration membrane can be determined via experimental filtration tests with pre-selected molecules. A simple salt solution (NaCl or Na2SO4) is selected for loaded particles. Polysaccharides (dextrines) or polyethylene glycols (PEG) with various molecular weights are selected for unloaded particles. The salt retention of a typical nano-filtration membrane is considerably lower than, for example, reverse osmosis, while salt retention is zero for ultra-filtration.

A nano-filtration membrane is also ion-selective. This is the ability to distinguish various ions from one another. Because a nano-filtration membrane collects solid loaded groups in its membrane structure, electrostatic repulsion/attraction forces may occur between the components in the liquid and the (nano-filtration) membrane surface, which results in a certain degree of ion selectivity. Based on the sieve effect (pore size 1 nm) and the molecular size of chlorides (0.12 nm in size) and sulphates (0.23 nm in size), these ions are expected to diffuse through the membrane. Despite this, the retention for chlorides is maximum 90 % and that of sulphates is minimum 90%.

A nano-filtration membrane can be tubular, spiral or flat in shape. A spiral module consists of polyamide membrane layers wound in spiral form. At the edge of the membrane, the wound layers are sealed using a cap. A permeate collection tube is located in the centre of the wound module. All the clean water is passed through the spiral windings and collects in this tube.

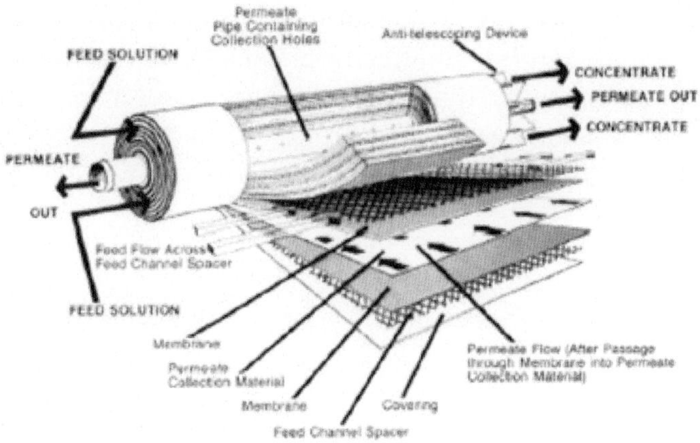

Here are a few specific advantages and disadvantages of NF:

Advantages

- Lower discharge volumes, lower retentate concentrations than RO for low value salts.
- Reduction salt content and dissolved matter content (TDS) in brackish water.
- Reduction in heavy metals.
- Reduction in nitrates and sulphates.
- Reduction in colour, tannins and turbidity.
- Softens hard water when specific softening membranes are used.
- "chemical-free", e.g. needs no salt or chemicals during operation.
- pH of water after nano-filtration is normally non-aggressive.
- Disinfection.

Disadvantages

- Higher energy consumption than UF and MF (0.3 to 1 kWh/m^3).
- Pre-treatment is needed for some heavily polluted waters (pre-filtration 0.1 - 20 microns). This is always the case with spiral-wound membranes.
- Limited retention for salts and univalent ions.
- Nano-filtration membranes are a little more expensive than reverse osmosis membranes.
- Membranes are sensitive to free chlorine (life-span of 1000 ppmh). An active carbon filter or a bi-sulphite treatment is recommended for high chlorine concentrations.

Applications

Industrial applications of nanofiltration are quite common in the food and dairy sector, in chemical processing, in the pulp and paper industry, and in textiles, although the chief application continues to be in the treatment of fresh, process and waste waters.

In the treatment of water, NF finds use in the polishing at the end of conventional processes. It cannot be used for water desalination, but it is an effective means of water softening, as the main hardness chemicals are divalent. At first sight, NF would not seem to have much place in MBR processes, because the higher transmembrane pressure differentials needed for NF are not available in most bioreactor systems, but there are some specialised uses for MBRs in which NF is finding a place.

NF membranes are also used for the removal of natural organic matter from water, especially tastes, odours and colours, and in the removal of trace herbicides from large water flows. They can also be used for the removal of residual quantities of disinfectants in drinking water.

Food industry applications are quite numerous. In the dairy sector, NF is used to concentrate whey, and permeates from other whey treatments, and in the recycle of clean-in-place solutions. In the processing of sugar, dextrose syrup and thin sugar juice are concentrated by NF, while ion exchange brines are demineralised. NF is used for degumming of solutions in the edible oil processing sector, for continuous cheese production, and in the production of alternative sweeteners.

There are probably as many different applications in the whole chemical sector (including petrochemicals and pharmaceuticals) as in the rest of industry put together. Many more are still at the conceptual stage than are in plant use, but NF is a valuable contributor to the totality of the chemicals industry. The production of salt from natural brines uses NF as a purification process, while most chemical processes produce quite vicious wastes, from which valuable chemicals can usually be recovered by processes including NF. The high value of many of the products of the pharmaceutical and biotechnical sectors allows the use of NF in their purification processes.

The paper pulp industry uses a very great quantity of water in its production processes, a quantity that the industry is striving to reduce, mainly by "closing the water cycle" – a system in which the purification properties of NF have a major role.

All of these specifically mentioned applications have been water-based, but nanofiltration is not restricted to the treatment of aqueous suspensions. Indeed one of the largest NF plants was installed at a petroleum refinery for the dewaxing of oils. Boam and Nozari, in their review of organic solvent nanofiltration, point out that many organic system separation processes are quite highly energy intensive, and that, by contrast, OSN can be quite an energy saving alternative.

In aqueous systems, nanofiltration uses hydrophilic polymeric materials, such as polyether-sulphone, polyamides and cellulose derivatives. These materials, in contact with organic solvents, quickly lose their stability. Special membranes have therefore been developed to provide the same kind of performance as in aqueous systems, and they are now used for solvent exchange, solvent recovery and separation, for catalyst recovery and for heavy metal removal.

Reverse Osmosis

Reverse Osmosis, commonly referred to as RO, is a process where you demineralize or deionize water by pushing it under pressure through a semi-permeable Reverse Osmosis Membrane.

Osmosis

To understand the purpose and process of Reverse Osmosis you must first understand the naturally occurring process of Osmosis.

Osmosis is a naturally occurring phenomenon and one of the most important processes in nature. It is a process where a weaker saline solution will tend to migrate to a strong saline solution. Examples of osmosis are when plant roots absorb water from the soil and our kidneys absorb water from our blood.

Below is a diagram which shows how osmosis works. A solution that is less concentrated will have a natural tendency to migrate to a solution with a higher concentration. For example, if you had a container full of water with a low salt concentration and another container full of water with a high salt concentration and they were separated by a semi-permeable membrane, then the water with the lower salt concentration would begin to migrate towards the water container with the higher salt concentration.

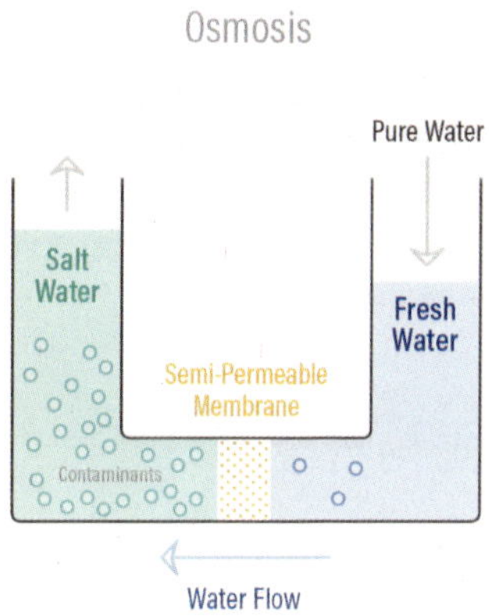

A semi-permeable membrane is a membrane that will allow some atoms or molecules to pass but not others. A simple example is a screen door. It allows air molecules to pass through but not pests or anything larger than the holes in the screen door. Another example is Gore-tex clothing fabric that contains an extremely thin plastic film into which billions of small pores have been cut. The pores are big enough to let water vapor through, but small enough to prevent liquid water from passing.

Reverse Osmosis is the process of Osmosis in reverse. Whereas Osmosis occurs naturally without energy required, to reverse the process of osmosis you need to apply energy to the more saline solution. A reverse osmosis membrane is a semi-permeable membrane that allows the passage of water molecules but not the majority of dissolved salts, organics, bacteria and pyrogens. However, you need to 'push' the water through the reverse osmosis membrane by applying pressure that is greater than the naturally occurring osmotic pressure in order to desalinate (demineralize or deionize) water in the process, allowing pure water through while holding back a majority of contaminants.

Below is a diagram outlining the process of Reverse Osmosis. When pressure is applied to the concentrated solution, the water molecules are forced through the semi-permeable membrane and the contaminants are not allowed through.

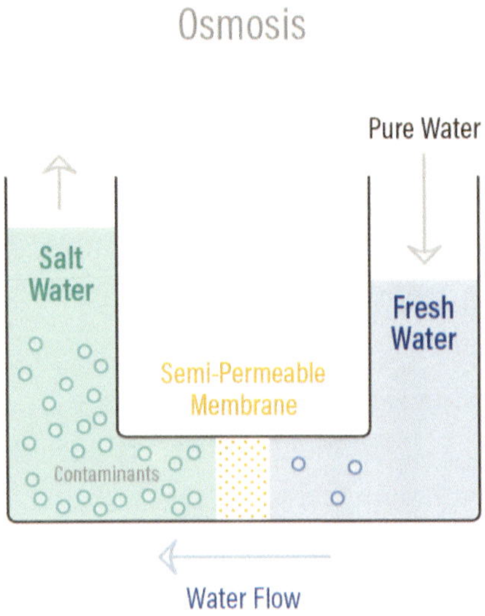

Working of Reverse Osmosis

Reverse Osmosis works by using a high pressure pump to increase the pressure on the salt side of the RO and force the water across the semi-permeable RO membrane, leaving almost all (around 95% to 99%) of dissolved salts behind in the reject stream. The amount of pressure required depends on the salt concentration of the feed water. The more concentrated the feed water, the more pressure is required to overcome the osmotic pressure.

The desalinated water that is demineralized or deionized, is called permeate (or product) water. The water stream that carries the concentrated contaminants that did not pass through the RO membrane is called the reject (or concentrate) stream.

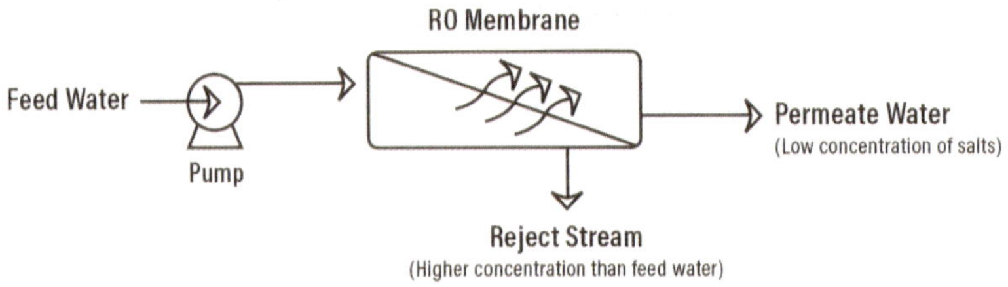

As the feed water enters the RO membrane under pressure (enough pressure to overcome osmotic pressure) the water molecules pass through the semi-permeable membrane and the salts and other contaminants are not allowed to pass and are discharged through the reject stream (also known as the concentrate or brine stream), which goes to drain or can be fed back into the feed water supply in some circumstances to be recycled through the RO system to save water. The water that makes it through the RO membrane is called permeate or product water and usually has around 95% to 99% of the dissolved salts removed from it.

It is important to understand that an RO system employs cross filtration rather than standard filtration where the contaminants are collected within the filter media. With cross filtration, the solution passes through the filter, or crosses the filter, with two outlets: the filtered water goes one way and the contaminated water goes another way. To avoid build up of contaminants, cross flow filtration allows water to sweep away contaminant build up and also allow enough turbulence to keep the membrane surface clean.

Removal of Contaminants from Water by Reverse Osmosis

Reverse Osmosis is capable of removing up to 99%+ of the dissolved salts (ions), particles, colloids, organics, bacteria and pyrogens from the feed water (although an RO system should not be relied upon to remove 100% of bacteria and viruses). An RO membrane rejects contaminants based on their size and charge. Any contaminant that has a molecular weight greater than 200 is likely rejected by a properly running RO system (for comparison a water molecule has a MW of 18). Likewise, the greater the ionic charge of the contaminant, the more likely it will be unable to pass through the RO membrane. For example, a sodium ion has only one charge (monovalent) and is not rejected by the RO membrane as well as calcium for example, which has two charges. Likewise, this is why an RO system does not remove gases such as CO_2 very well because they are not highly ionized (charged) while in solution and have a very low molecular weight. Because an RO system does not remove gases, the permeate water can have a slightly lower than normal pH level depending on CO_2 levels in the feed water as the CO_2 is converted to carbonic acid.

Reverse Osmosis is very effective in treating brackish, surface and ground water for both large and small flows applications. Some examples of industries that use RO water include pharmaceutical, boiler feed water, food and beverage, metal finishing and semiconductor manufacturing to name a few.

Applications and uses of RO

The objectives of an RO plant for industrial use are distributed in the following way: 50% in desalination of seawater and brackish water; 40% in the production of ultrapure water for the electronic, pharmaceutical and energy production industries; 10% as decontamination systems for urban and industrial water.

- Desalination of brackish water

  The salinity of this type of water is 2000 mg/L – 10000 mg/L. In its treatment, pressures of 14 bar – 21 bar are used to achieve rejection coefficients greater than 90% and to obtain water with saline concentrations of lower than 500 mg/L, which are the values recommended by the WHO as a requirement of potability.

  Treatment plants use membrane modules rolled in spirals. It is estimated that the costs for this type of plant are in the region of 0.25 $US/L of treated water/day, with the operating costs being equivalent.

- Desalination of seawater

  Depending on the geographical area, the salinity of this type of water is 30000 mg/L – 40000 mg/L. To meet the potability requirements, polyamide hollow fiber membranes are used which allow rejection coefficients to be achieved of greater than 99.3% with working pressures of 50 bar – 70 bar.

The operating costs of this type of treatment plant are estimated to be 1 – 1.25$US/ L of treated water/day, which means that this treatment system is not competitive in comparison with other systems, such as multistage evaporation processes, if water needs exceed 40000 m3 of treated water/day.

- Production of ultrapure water

  RO allows water of the quality demanded by the electronic industry to be obtained from drinking water (concentration of dissolved solids < 200 mg/L).

  The main problem with this type of installation is the bio-contamination of the membranes, which is why the installation of sterilization systems that use UV radiation is necessary. Table 11 compares the characteristics demanded for drinking water with those for ultrapure water.

- Wastewater treatment

  The use of reverse osmosis in wastewater treatment is limited by the high operating costs due to the problems of contamination of the membranes. In the case of industrial wastewater, RO is used in those industries where it is possible to improve the efficiency of the process by recuperating valuable components that can be recycled in the production process: galvanoplasty industries and paint for metal structures, or where reuse of the treated water represents an important reduction in the consumption of water, e.g. textile industry.

In the case of urban water, RO is a treatment that would be indicated as a tertiary treatment, if it were possible for water to be obtained with a quality that made it appropriate for consumption, at a cost of 0.5 – 0.75 $US/m3.

The main problem for the consolidation of this type of treatment is the social reaction. Nevertheless, in areas of Japan and California, where there are extreme water limitations, RO plants are being used to treat water that comes from the biological treatment of domestic water, with the water treated by RO being used to refill groundwater resources.

Forward Osmosis

Forward Osmosis (FO) is a remarkable new technology that turns muddy, contaminated water, into a clean and refreshing drink. It also removes difficult toxic waste and pollutants from industrial waste streams.

- Osmosis is a natural equilibrium process.
- Two liquids are separated by a special proprietary membrane that allows only water to pass.
- Water passes from one chamber to the next by diffusion until the solids concentration on both sides of the membrane are equal.

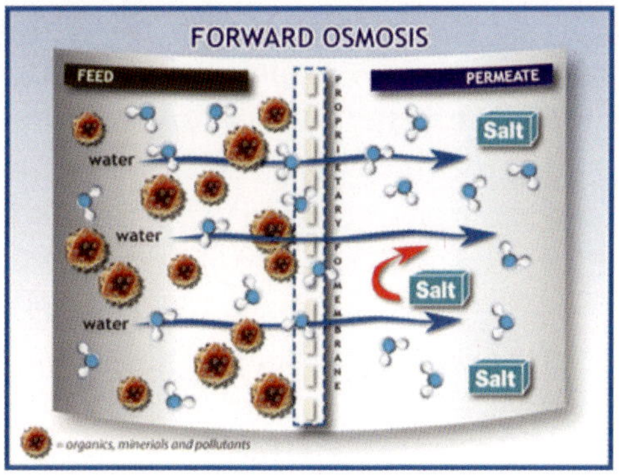

Forward Osmosis Membranes reject organics, minerals and other solids — similar to reverse osmosis (RO) — but resist typical fouling problems.

It is the process of two liquids seeking equilibrium that creates the "power" in Forward Osmosis. In the example above, the permeate side of the membrane contains a salt "draw" solution which has a higher osmotic potential than the dirty feed water on the other side of proprietary membrane. That higher osmotic potential in the salt "draw" solution is the power source nature provides to drive the filtration process. The water moves through the membrane, is filtered in the process and dilutes the higher concentration salt water on the permeate side.

Forward Osmosis Enjoys Several Unique Technological Advantages:

- FO overcomes fouling limitations inherent in pressure driven membrane separations allowing proper filtration and concentration of difficult products and waste streams.
- FO can process feed streams; a source of dirty or polluted water containing high levels of suspended solids.
- In separations driven by hydraulic pressure (such as RO) all components of a feed stream are unselectively forced against the membrane surface.
- In FO, osmotic pressure, which is hydraulic pressure created by Osmosis as water passes through a semi-permeable membrane into a more concentrated stream, selectively draws molecules through the membrane avoiding membrane fouling and compaction.
- FO is a process normally occurring in nature that, when harnessed for water filtration and liquid transfer, requires little or no electricity or external power source.

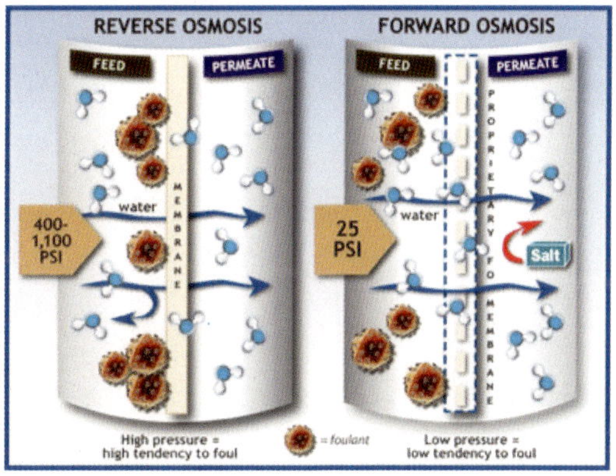

In this example, the process of Reverse Osmosis (RO) is compared with Forward Osmosis (FO). RO typically requires high pumping pressures, which require high energy usage to force the feed solution against a very tight membrane material and force the filtered water through the membrane to the clean water side. During that process all solutes and pollutants on the dirty water side are forced against the membrane material, under high pressure, which compounds RO's sensitivity to fouling and need for relatively clean feed streams. Conversely, FO operates at very low pressure, simply adequate to circulate the fluids, and the natural osmotic pressure in the draw solution pulls water through the membrane, leaving solids and foulants behind in the concentrated feed solution.

Perstraction

Perstraction is a separation process where a separation process is used with one side contacting a solvent, the other side's contacting the desired migrating product. For our purposes, the membrane would have a water and alcohol broth mix on one side hoping to move the alcohol over to the solvent side. That would eliminate the high-energy heat needed to vaporize alcohol out of the broth water.

But one still has to extract the alcohol from the solvent.

The yeasts and bacteria that digest and form alcohols live in broth of water and the prepared biomass feedstock. As they live and grow the microorganisms excrete the desired alcohols. When the alcohol content gets past the organism's tolerance level it slows growth and eventually kills the microorganism. This part of the cycle is a huge problem, which is solved by running batches and distilling off the alcohols before or just at the concentration when microorganism activity stops.

Perstraction in a developed and refined form would allow continuous production. Then adding biomass and water as alcohol is taken off could be practical. The reduction in fuel use to heat the mash broth and distill the product off would be greatly diminished so making alcohols much more energy competitive and more eco friendly.

The range of perstraction in fuels includes methanol, ethanol and butanol. Recently a study was conducted about using perstraction in a whey permeate (a kind of milk sugar) process to make butanol. The experiments offer that recovery of the butanol would be more economical than using batch distillation, better membranes need developed while its now known that silicalite membranes developed for pervaporation can be used, and that the process as done in the lab can be adapted in existing solvent ferment systems without significant changes.

This means that perstraction will be something we're going to need to understand. Biomass and microorganisms at high productivity could displace a huge chunk of fossil oil and gas driving down and moderating prices. As this technology gets better, stabilizes on a few competitive models and integrates into plant engineering plans, biomass to fuels could become much more common than the 10% ethanol in gasoline. Perhaps it will apply in oil extraction from algae someday also.

There is a lot of room for growth in biomass to fuel and perstraction could well be an important part of the system. There is a huge opportunity in the extraction of alcohols from fermentation broths and those that offer the lowest cost to set up and run coupled to microorganisms that eat about anything.

Pressure Exchanger

Pressure exchangers can be used in place of conventional turbine solutions for energy recovery in RO (reverse osmosis) seawater desalination plants.

The pressure exchanger transfers the hydraulic energy in the brine directly to the incoming seawater, referred to as feedwater.

Operating Principle with Two Pressure Vessels

- The pressure exchanger comprises two pressure vessels which are operated simultaneously. The check valve unit is installed at one end of the pressure vessels and the rotary valve at the other.
- The rotary valve is connected to the plant piping through which the brine flows. The check valve unit is connected to the feedwater piping. Each side is equipped.
- With one high-pressure and one low-pressure pipe for the seawater and the brine.
- The pressure exchanger is operated in cycles, with each cycle being defined by a 180° rotation of the rotary valve. Each cycle has a low-pressure phase in which the fluid flows from the check valve unit to the rotary valve and a high-pressure phase in which the fluid flows from the rotary valve to the check valve unit. The pressure vessels are filled with seawater in the low-pressure phase, in the high-pressure phase the seawater is pumped into the plant.
- The rotary valve is driven by a servomotor (gearbox). Actuating the valve changes the pressure in the vessels and the flow's direction. As the pressure vessels operate alternately, one conveys low-pressure fluid and the other high-pressure fluid. The fluids in the vessels always flow in opposing directions. Each pressure vessel may be equipped with a piston separating the fluids (seawater and brine).
- Pressures of between 50 and 80 bar are required to desalinate seawater. A high-pressure pump provides the inlet pressure required for the reverse osmosis process. Concentrated brine resulting from this process leaves the membrane modules under high pressure and flows through a pressure exchanger where this pressure is transferred directly to the incoming seawater (feedwater). A recirculation pump then pumps the feedwater to the membranes.
- The pressure exchanger's control unit records and processes all operating conditions in the higher-level reverse osmosis (RO) system in order to optimise operation. It ensures that the water columns in the pressure exchanger vessels are accelerated and decelerated so gently that actuating the rotary valves of the pressure exchanger does not cause pulsations.

- All components including high-pressure and recirculation pumps are made from seawater-resistant materials.

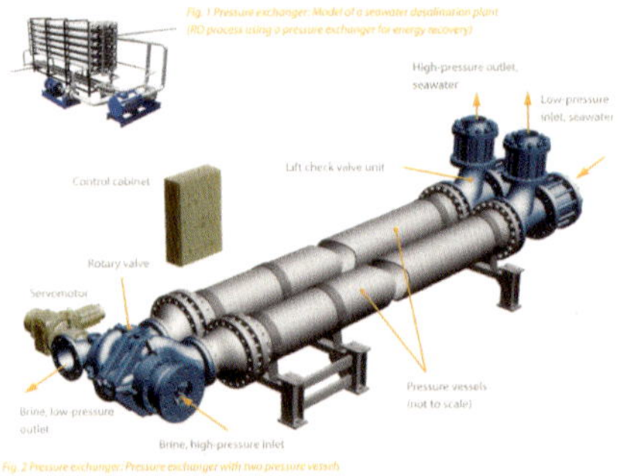

Fig. Pressure exchanger: Model of a seawater desalination plant Fig. Pressure exchanger: Pressure exchanger with two pressure vessels

Vibratory Shear-enhanced Process

While membrane-based separations of liquids from solids have enjoyed increasing popularity over the last 20 years, the technology has an inherent Achilles heel that affects all membrane devices: fouling. This long-term loss in throughput capacity is due primarily to the formation of a boundary layer that builds up naturally on the membranes surface during the filtration process. In addition to cutting down on the flux performance of the membrane, this boundary or gel layer acts as a secondary membrane reducing the native design selectivity of the membrane in use. This inability to handle the buildup of solids has also limited the use of membranes to low-solids feed streams.

Crossflow

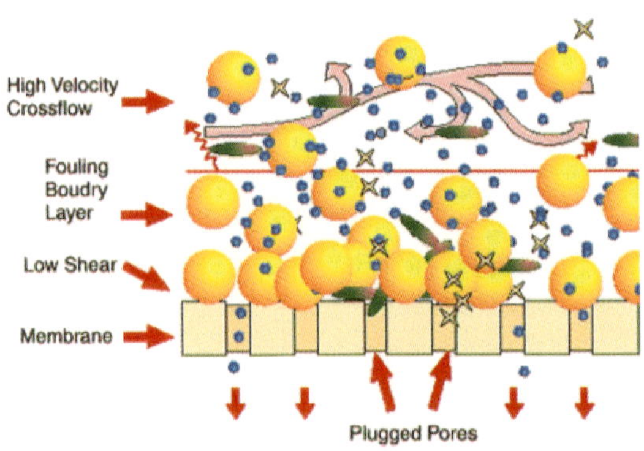

To help minimize this boundary layer buildup, membrane designers have used a method known as tangential-flow or cross-flow filtration that relies on high velocity fluid flow pumped across the membranes surface as a means of reducing the boundary layer effect. In this method, membrane elements are placed in a plate-and-frame, tubular, or spiral-wound cartridge assembly, through which the substance to be filtered (the feed stream), is pumped rapidly.

In cross-flow designs, it is not economic to create shear forces measuring more than 10-15 thousand inverse seconds, thus limiting the use of cross-flow to low-viscosity (watery) fluids. In addition, increased cross-flow velocities result in a significant pressure drop from the inlet (high pressure) to the outlet (lower pressure) end of the device, which leads to premature fouling of the membrane that creeps up the device until permeate rates drop to unacceptably low levels.

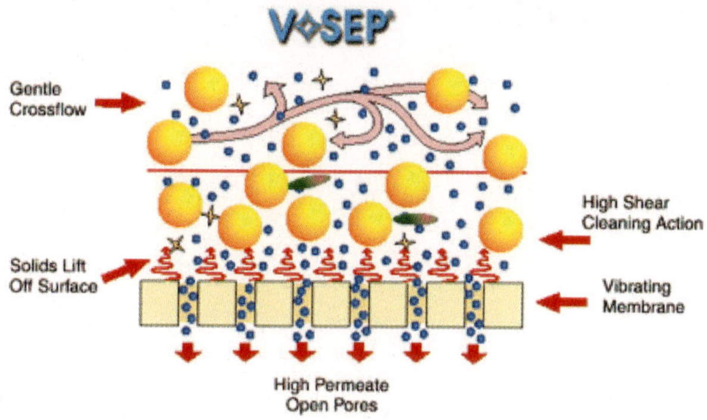

New Logic, however, has developed an alternative method for producing intense shear waves on the face of a membrane. The technique is called Vibratory Shear Enhanced Processing (VSEP). In a VSEP System, the feed slurry remains nearly stationary, moving in a leisurely, meandering flow between parallel membrane leaf elements. Shear cleaning action is created by vigorously vibrating the leaf elements in a direction tangent to the faces of the membranes.

The shear waves produced by the membrane's vibration cause solids and foulants to be lifted off the membrane surface and remixed with the bulk material flowing through the membrane stack. This high shear processing exposes the membrane pores for maximum throughput that is typically between 3 and 10 times the throughput of conventional cross-flow systems.

The VSEP membrane filter pack consists of leaf elements arrayed as parallel discs and separated by gaskets. The disc stack resembles records on a record changer with membrane on each side.

The disk stack is oscillated above a torsion spring that moves the stack back and forth approximately 7/8 inches (2.22 centimeters). This motion is analogous to the agitator of a washing machine but occurs at a speed faster than that which can be perceived by the human eye.

The oscillation produces a shear at the membrane surface of about 150,000 inverse seconds (equivalent to over 200 G's of force), which is approximately ten times the shear rate of the best conventional cross-flow systems. More importantly, the shear in a VSEP System is focused at the membrane surface where it is cost effective and most useful in preventing fouling, while the bulk fluid between the membrane disks moves very little.

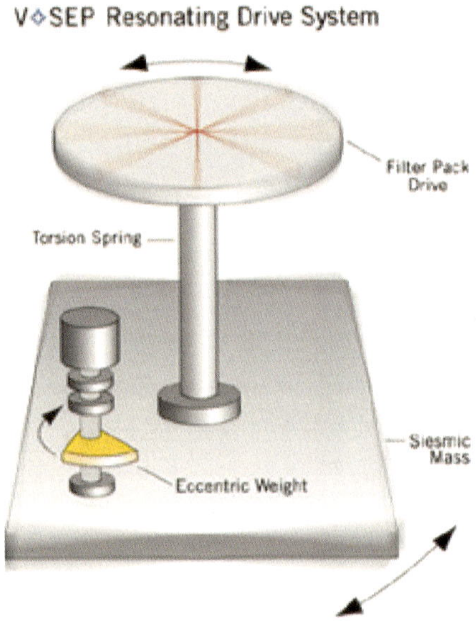

Because VSEP does not depend on feed flow induced shearing forces, the feed slurry can become extremely viscous and still be successfully dewatered. The concentrate is essentially extruded between the vibrating disc elements and exits the machine once it reaches the desired concentration level. Thus, VSEP Systems can be run in a single pass through the system, eliminating the need for costly working tanks, ancillary equipment and associated valving.

The disc pack holdup volume of a system with 1,400 ft2 (130 sq. meters) of membrane area, is less than 50 gallons (189 liters). As a result, product recovery in batch processes can be extremely high. Waste after draining the stack is less than 3 gallons (11 liters).

VSEP System Operation

At startup, the VSEP system is fed with a slurry and the concentrate valve is closed. Permeate is produced and suspended solids in the feed are collected inside the VSEP filter pack. After a programmed time interval, valve one is opened to release the accumulated concentrated solids. The valve is then closed to allow the concentration of additional feed material. This cycle repeats indefinitely.

Membrane selection is the single most important parameter that affects the quality of the separation. Other important parameters that affect system performance are pressure, temperature, vibration amplitude, and residence time. All of these elements are optimized during testing and entered into the programmable logic controller (PLC) which controls the system.

The operating pressure is created by the feed pump. VSEP machines can routinely operate at pressures as high as 1,000 psig (68.95 BAR). While higher pressures often produce increased permeate flow rates, they also use more energy. Therefore, an operating pressure is used that optimizes the balance between flow rates and energy consumption.

In most cases, the filtration rate can be further improved by increasing the operating temperature. The temperature limit on a standard VSEP system is 175° F (79°C), significantly higher than competitive membrane technology. Even higher temperature constructions are also available.

The vibration amplitude and corresponding shear rate can also be varied which directly affects filtration rates. Shearing is produced by the torsion oscillation of the filter stack. Typically the stack oscillates with an amplitude of 3/4 to 1 1/4 inches (1.9 to 3.2 cm) peak to peak displacement at the rim of the stack. The oscillation frequency is approximately 53 Hz and produces a shear intensity of about 150,000 inverse seconds.

Feed residence time is set by the frequency of the opening and closing of the exit valve (valve one). The solids level in the feed increases as the feed material remains in the machine. Occasionally, a cleaner is added to the membrane stack and continued oscillation helps clean the membrane in minutes. This process can be automated and only consumes approximately 50 gallons (189 liters) of cleaning solution thus reducing cleaner disposal problems inherent with other membrane systems.

References

- Concentration-polarization-311: corrosionpedia.com, Retrieved 16 July 2018
- Methods-to-reduce-membrane-fouling, learning-center: synderfiltration.com, Retrieved 11 May 2018
- What-is-ultrafiltration: crystalquest.com, Retrieved 15 March 2018
- What-is-nanofiltration, water-and-wastewater: filtsep.com, Retrieved 25 May 2018
- What-is-reverse-osmosis: puretecwater.com, Retrieved 15 April 2018

Permissions

All chapters in this book are published with permission under the Creative Commons Attribution Share Alike License or equivalent. Every chapter published in this book has been scrutinized by our experts. Their significance has been extensively debated. The topics covered herein carry significant information for a comprehensive understanding. They may even be implemented as practical applications or may be referred to as a beginning point for further studies.

We would like to thank the editorial team for lending their expertise to make the book truly unique. They have played a crucial role in the development of this book. Without their invaluable contributions this book wouldn't have been possible. They have made vital efforts to compile up to date information on the varied aspects of this subject to make this book a valuable addition to the collection of many professionals and students.

This book was conceptualized with the vision of imparting up-to-date and integrated information in this field. To ensure the same, a matchless editorial board was set up. Every individual on the board went through rigorous rounds of assessment to prove their worth. After which they invested a large part of their time researching and compiling the most relevant data for our readers.

The editorial board has been involved in producing this book since its inception. They have spent rigorous hours researching and exploring the diverse topics which have resulted in the successful publishing of this book. They have passed on their knowledge of decades through this book. To expedite this challenging task, the publisher supported the team at every step. A small team of assistant editors was also appointed to further simplify the editing procedure and attain best results for the readers.

Apart from the editorial board, the designing team has also invested a significant amount of their time in understanding the subject and creating the most relevant covers. They scrutinized every image to scout for the most suitable representation of the subject and create an appropriate cover for the book.

The publishing team has been an ardent support to the editorial, designing and production team. Their endless efforts to recruit the best for this project, has resulted in the accomplishment of this book. They are a veteran in the field of academics and their pool of knowledge is as vast as their experience in printing. Their expertise and guidance has proved useful at every step. Their uncompromising quality standards have made this book an exceptional effort. Their encouragement from time to time has been an inspiration for everyone.

The publisher and the editorial board hope that this book will prove to be a valuable piece of knowledge for students, practitioners and scholars across the globe.

Index